ORGANIZATION OF ENERGY-TRANSDUCING MEMBRANES

ORGANIZATION OF ENERGY-TRANSDUCING MEMBRANES

Edited by
MAKOTO NAKAO and LESTER PACKER

UNIVERSITY PARK PRESS
Baltimore · London · Tokyo

UNIVERSITY PARK PRESS
Baltimore · London · Tokyo

Library of Congress Cataloging in Publication Data
Main entry under title:

Organization of energy-transducing membranes.

Proceedings of a USA-Japan joint seminar held May 22–26, 1972, in Tokyo.
Includes bibliographical references.
1. Cell membranes—Congresses. 2. Energy metabolism—Congresses. 3. Adenosinetriphosphate—Congresses. 4. Adenosinetriphosphatase—Congresses. I. Nakao, Makoto, ed. II. Packer, Lester, ed. [DNLM: 1. Biophysics—Congresses. 2. Energy transfer—Congresses. 3. Membranes—Congresses. QT34 068 1972]
QH601.073 574.8′75 73-10341
ISBN 0-8391-0718-8

UTP 3043-67846-5149
Printed in Japan.

Originally published by
UNIVERSITY OF TOKYO PRESS

PREFACE

A U.S.-Japan joint seminar on the organization of energy-transducing membranes was held in Tokyo from May 22–26, 1972. This book is based essentially on the proceedings of the seminar.

One of the characteristics of life is effective transformation of energy and information. Membranes are places in the cell where enzymes may be firmly bound. This binding of enzymes like Na^+, K^+-ATPase and other proteins lends a high degree of efficiency and directionality to metabolism. Highly conspicuous among membrane-associated functions is the conversion of one type of energy to another; for example from photoenergy to chemical energy, from oxidation to phosphorylation, from chemical energy to osmotic energy, from chemical energy to mechanical energy, *etc.*

Currently, we can cite various parallels among energy transduction mechanisms, particularly in various ATPases including those in myosin, mitochondria, chloroplasts, plasma membranes, and sarcoplasmic membranes. The molecular sizes of active subunits, ATP hydrolysis mechanisms or phosphorylation and dephosphorylation mechanisms of various ATPases in these organizations are remarkably similar to one another. Particularly prominent is a strong and subtle interaction of allosteric sites with allosteric effectors as well as active sites, for example Na^+ and K^+ sites of Na^+, K^+-ATPase with Na^+ and K^+, Ca^{2+} and Mg^{2+} sites with Ca^{2+} and Mg^{2+} and the head of myosin with actin. In the case of oxidative phosphorylation and photophosphorylation, the condition may be similar. At any rate, unmasked ATPase interacts with other factors to display physiological functions. Allosteric site effects are more important than active site effects for considerations of energy and information transduction deriving from the physiological point of view. This machinery may be a general device to convert one type of free energy to another in life.

Although the scope of the seminar was broadly defined, interest quickly focused on the ATPase ion carrier protein and a discussion of the growing arsenal of methods used to elucidate the properties of this protein. Helpful insights were gained from reports of recent studies of the Ca^{2+} pump in sarcoplasmic reticulum, a description of the isolation of "ATP synthetase," and freeze-fracture and spin label investigations of mitochondrial membranes.

Contraction and translocation appear to differ from ion transport in chemical plan (especially the nature of the intermediate); nevertheless cross fertilization was provided during discussions of studies on ATP analogs and other probes of the contractile system.

A concensus was developed on mechanisms of energy transduction involving the reversible ATPase-ATP synthetase system which functions in different directions in different membranes. This pointed out the membrane asymmetry problem and revealed the need for a better knowledge of the structural organization of the ATPase complex.

Thus, the seminar aimed at communication and discussion between scientists of different fields using different experimental materials, concerning the different characteristics of the various membranes converting energy into driving various functions. The discussions were successful in revealing insights into common mechanisms and for this reason publication of this book has been undertaken.

Makoto Nakao
Lester Packer

CONTENTS

Ca^{2+} TRANSPORT AND Ca^{2+} ATPase

INTRODUCTION

Energy Transformations in Biology

D. C. Tosteson

Department of Physiology and Pharmacology, Duke University Medical Center, Durham, North Carolina, U.S.A.

The subject of this symposium is part of the general problem of energy transformations in biology. It is appropriate to begin with a brief review of such processes, with particular emphasis on those which are discussed in further detail in later papers.

This paper begins with a brief summary of the most important kinds of energy transformations known to occur in biology. These include interconversions between light, chemical bond energy and heat (photosynthesis and respiration), between chemical bond energy and transport of atoms and molecules (active transport), between chemical bond energy and changes in length and shape of cells (contraction), and between chemical potential differences and electric current (excitation and propagation). The paper concludes with a more detailed consideration of some of the characteristics of the Na^+-K^+ active transport process which occurs in the plasma membranes of most animal cells. This consideration is illustrated with examples of observations made on mammalian red cell membranes.

The fundamental energy transformation occurring in living systems is summarized by the equation

$$nCO_2 + nH_2O + \text{Light} \underset{\text{respiration}}{\overset{\text{photosynthesis}}{\rightleftharpoons}} (CH_2O)_n + nO_2\,. \qquad (1)$$

The reaction from left to right is called photosynthesis and leads to the production of carbohydrate from CO_2 and H_2O. It is driven by energy in the form of light derived from nuclear reactions occuring on the sun. The fundamental process involves the transfer of an electron from an oxygen to a carbon nucleus. The reaction in reverse direction is called fire or respiration and involves the transfer of an electron from a carbon nucleus to an oxygen nucleus to produce CO_2, H_2O, and heat from carbohydrate or other biological polymers. In this frame of reference it is evident that the fundamental reactions in living cells involve a transformation of energy from light to chemical bonds and from chemical bonds to heat. Except for water and wind power, which depend on the sun in other ways, these reactions have been the sole source of energy available to all forms of living organisms including man and man-made devices until the discovery of nuclear fission and fusion. However, both in living systems and in machines, function requires coupling of these fundamental reactions with other chemical reactions and other energy transformations. Most machines except electric motors are heat engines, that is, energy released in spontaneous chemical reactions is converted first to heat and thence to some other form of work. Biological systems take advantage of more subtle and specific transformations.

What are some of the specific kinds of energy transformations of special interest and relevance in biology? First, electron transfer reactions involved in the fundamental process can, as we know, be coupled to other classes of chemical reaction. An example of particular interest is the formation and hydrolysis of pyrophosphate bonds in compounds such as ATP. This reaction can be summarized by the equation

$$mADP + mP_i + (CH_2O)_n + nO_2 \rightleftharpoons mATP + nCO_2 + (n+1)H_2O\,. \qquad (2)$$

The reaction in the direction left to right is called oxidative phosphorylation. It can be recognized to involve a coupling of the fundamental process with reactions involving the addition or elimination of water. The mechanism of this coupling is not clearly understood but certainly involves enzymes in the membranes of mitochondria and is the subject of several papers in this conference (*e.g.*, those of L. Packer, B. Hagihara, and S. Muraoka). The reverse reaction, the hydrolysis of ATP, is coupled with many of the specific energy transformations through which living cells perform work.

A second specific energy transformation process in biology is the coupling of electron transfer reactions or reactions involving the addition or subtraction of water with uphill or active transport. One specific example of this kind of energy transformation can be written in the form potassium outside the cell (K_o^+) plus sodium inside the cell (Na_i^+) plus ATP leading to the products of potassium moved to the inside (K_i^+) of the cell plus sodium moved to the outside (Na_o^+) of the cell plus ADP, *i.e.*,

$$K_o^+ + Na_i^+ + ATP + H_2O \rightleftharpoons K_i^+ + Na_o^+ + ADP + P_i\,. \qquad (3)$$

Another example of this kind of energy transformation process is the movement of calcium ions in cytoplasm (Ca_{cyt}^{2+}) of muscle cells in the presence of ATP to calcium ions in the endoplasmic reticulum (Ca_{ER}^{2+}) plus ADP,

$$Ca_{cyt}^{2+} + ATP + H_2O \rightleftharpoons Ca_{ER}^{2+} + ADP + P_i\,. \qquad (4)$$

In both cases, a chemical reaction, the hydrolysis of ATP or some compound arising from ATP, leads to the translocation of ions from one side to the other of a membrane. In this case the membrane is the energy transducer. Both of these so-called active transport processes, Na^+ and K^+ by plasma membranes and Ca^{2+} by sarcoplasmic reticulum membranes, are treated extensively in this conference (*e.g.*, in the papers of R. Blostein, K. Nagano, M. Nakao, R. L. Post, H. Matsui, J. F. Hoffman, M. Fujita, G. Inesi, S. Harigaya, Y. Tonomura, S. Ebashi, O. Hatase, and S. Murakami).

A third kind of specific energy transformation is the coupling of a chemical reaction with movement to perform mechanical work, a process that is in close analogy to the kind of vectorial transport which I have just mentioned. In this case, it appears that there is the vectorial formation of chemical bonds between two macromolecules. This directionally oriented bond formation results in the movement of macromolecules in relation to one another. The papers of K. Yagi, F. Morita, M. F. Morales, and T. Sekine are related to this general subject.

Even though it is not considered in detail in this conference, I list one additional example of energy transformations in biology, namely the conversion of chemical potential differences for ions to electrical currents. This example is trivial from a physico-chemical point of view since it is easy to see how diffusion of charged molecules can produce electric current. However, from a biological point of view, it is of fundamental importance since the phenomena of excitation and propagation of impulses depends upon this kind of energy transformation.

So much then for an introduction to the general content of this conference. Let me now turn to a more detailed discussion of some of the characteristics of the active transport of sodium and potassium ions across red blood cell membranes. I present this subject in some detail because several of the later papers treat extensively the enzymology of this process. Since the goal of the biochemical experiments is to elucidate the molecular mechanism, not just of catalysis or hydrolysis of ATP, but of active transport of K^+ and Na^+ itself, it is appropriate to set out the main features of the system.

An important characteristic of this process is that it usually involves an exchange of ions. Ion movements in opposite directions across the membrane appear to be coupled. Under ordinary circumstances, this involves the coupling of external potassium moving inward with the outward movement of sodium (*1*, *2*).

$$K_o \rightleftharpoons Na_i$$

However, the exchange need not involve this particular direction of movement for these two ions. It can also involve any of the other possible combinations. For example, it can involve potassium exchange, $K_o \rightleftharpoons K_i$, or sodium exchange, $Na_o \rightleftharpoons Na_i$ (*3*), or even, as shown by Glynn and his colleagues, the reversal of the normal direction, sodium being pumped in and potassium being pumped out, $Na_o \rightleftharpoons K_i$ (*4*). An important point is that the transport system tends to be designed to accomplish cation exchange. This is not to say that it is impossible to obtain net cation movement through the system, that is, the ratio of the number of the sodium ions pumped outward to potassium ions pumped inward need not be one. In fact, Post and Jolly (*5*) showed a long time ago that in human red cells this ratio is about 1.5. It is a striking fact that this ratio appears to be unchanged in a given system over a wide range of conditions (*6*, *7*).

A second important characteristic of the active cation transport system is that it depends on both the potassium and the sodium concentrations on both sides of the membrane. For example, the rate of active transport of potassium into the cell is a function of both the external potassium concentration and the external sodium concentration (*7*). K_o^+ activates while Na_o^+ inhibits the process. The rate of operation of the pump is also a function, generally a different function, of both the concentration of potassium and sodium inside the cell. Algebraically,

$$^{i}M_{K}^{P} = F(K_o^+, K_i^+, Na_o^+, Na_i^+)\,,$$

where $^{i}M_{K}^{P}$ is the rate of active transport of K into the cell. On the inside, Na^+ activates while K^+ inhibits.

A third important general characteristic of the system is that activation and inhibition by external and internal ions appear to be independent processes (*7*). For example, the active transport for potassium can be described by the product of a function of the external ion composition $[F(\mathrm{O})]$ and a function of the internal ion composition $[G(\mathrm{I})]$,

$$^{\mathrm{i}}M_{\mathrm{K}}{}^{\mathrm{P}} = F(\mathrm{O}) \cdot G(\mathrm{I}) .$$

This is an example of the sidedness of the active transport system, a subject which will be treated in several aspects in this conference. This property seems to be quite important from the point of view of formulating models of the process. In particular, it is very difficult to find an example of a sequential model which satisfies this independence relation. In a sequential model, sodium interreacts with the transport system on the inside of the membrane, the loaded Na binding site is translocated to the outside and there converted to a form which releases Na and subsequently reacts with potassium and *vice versa*. It is this type of model which was first advanced for transport in red blood cells (*8*). Many similar models have been proposed. It is much easier to rationalize this independence principle with simultaneous models in which the two sides of the transport system are loaded independently and then translocation occurs by a single concerted mechanism which moves the sodium ions on the inside outward and the potassiums ion on the outside inward (*e.g.*, Refs. *9* and *10*).

A fourth important feature of the transport system is that more than one potassium or sodium ion are involved in the activation. In other words, using the nomenclature introduced above, the inward active transport of potassium is a function of the external concentration of potassium raised to some power, at least as high as two (*7, 11*). Likewise, the function for the internal ions must involve at least a square term. This feature means that activation by Na^+ and K^+ at neither the external nor the internal membrane surfaces can be described by a Michaelis-Menton type equation. A more complex mechanism is required.

A fifth property of active K^+-Na^+ transport is that ATP or some product of ATP metabolism is a necessary and sufficient source of energy for the process (*13*). Furthermore, there is a striking similarity between the kinetic properties of the cation active transport system in human red cells and those of an ATPase present in membranes of these cells (*16*). This correlation forms the basis for the line of research which led to the papers of M. Nakao, R. L. Post, J. F. Hoffman and others in this conference.

Two other properties can be listed as characteristics of the system. One is that genetically determined differences in the kinetic properties of the pump occur within individual species (*2*, *7*). For example, the K^+-Na^+ active transport systems in high K^+ (HK) and low K^+ (LK) sheep red blood cells differ not only in the number of active transport sites per cell (*12*), but also in the functions which describe the activation of the system by external and internal ions. This result suggests that the operation of the pump itself must be different in these different types of genetic variants. Finally, I list as a characteristic, that changes in shape of the membrane macromolecules which participate in active cation transport, that is conformation changes, seem to be involved in the process (*13*). One of the several kinds of evidence to support this conclusion is that in certain systems, antigen antibody reactions can be shown to modify the rate of active transport (*14*, *15*).

REFERENCES

1. I. M. Glynn, *J. Physiol.*, **134**, 278–310 (1956).
2. D. C. Tosteson and J. F. Hoffman, *J. Gen. Physiol.*, **44**, 169–194 (1960).
3. P. J. Garrahan and I. M. Glynn, *J. Physiol.*, **192**, 159–174 (1967).
4. P. J. Garrahan and I. M. Glynn, *J. Physiol.*, **192**, 237–256 (1967).
5. R. L. Post and P. C. Jolly, *Biochim. Biophys. Acta*, **25**, 118–128 (1957).
6. P. J. Garrahan and I. M. Glynn, *J. Physiol.*, **192**, 217–235 (1967).
7. P. G. Hoffman and D. C. Tosteson, *J. Gen. Physiol.*, **58**, 438–466 (1971).
8. T. I. Shaw, *J. Physiol.*, **129**, 464–475 (1955).
9. J. C. Skou, *Physiol. Rev.*, **45**, 596–617 (1965).
10. J. F. Hoffman, *in* "Biophysics of Physiological and Pharmacological Actions," Am. Assoc. Adv. Sci., Washington D.C., pp. 3–17 (1961).
11. J. R. Sachs and L. G. Welt, *J. Clin. Invest.*, **46**, 65–76 (1967).
12. P. B. Dunham and J. F. Hoffman, *Fed. Proc.*, **28**, 339 (1969).
13. J. F. Hoffman, *Circulation*, **26**, 1201–1213 (1962).
14. J. C. Ellory and E. M. Tucker, *Nature*, **222**, 477–478 (1969).
15. P. K. Lauf, B. A. Rasmusen, P. G. Hoffman, P. B. Dunham, P. Cook, M. L. Parmelee, and D. C. Tosteson, *J. Memb. Biol.*, **3**, 1–13 (1970).
16. R. L. Post, C. R. Merritt, C. R. Kinsolving, and C. D. Albright, *J. Biol. Chem.*, **235**, 1796–1802 (1960).

MOLECULAR ASPECTS OF Na^+, K^+-ATPase

Molecular Aspects of the Na^+, K^+-Pump in Red Blood Cells

Joseph F. Hoffman

Department of Physiology, Yale University School of Medicine, New Haven, Conneticut, U.S.A.

This paper will review recent work which we have been doing with regard to the molecular events associated with the active transport of Na^+ and K^+ by red blood cells. We will first survey features of the pump which can be assigned to one side of the membrane or to the other such as substrates, inhibitors, and immunologic reactions of the pump. Certain partial reactions of the pump will also be considered in an attempt to correlate translocation with transphosphorylation. Finally the separation of different parts of the pump labeled in different but specific ways will be described with an aim toward defining the organization of the membrane in the region of the pump.

Sidedness of the Pump

Since it is implicit in the concept of active transport that pumps contain both structural and functional asymmetry, study of the sidedness of the pump should provide insight into its molecular organization and mechanism. Thus the side assignments for Na^+, K^+, Mg^{2+}, and ATP are well known as are the location of the products of the reaction. This type of information is derived in part from studies using intact human red cells but mostly from work using reconstituted human red cell ghosts (*1–4*) in which the internal constituents can be varied at will while preserving the original orientation of the membrane. In this way the determinants of the pump activity with regard to their side dependence can be evaluated separately.

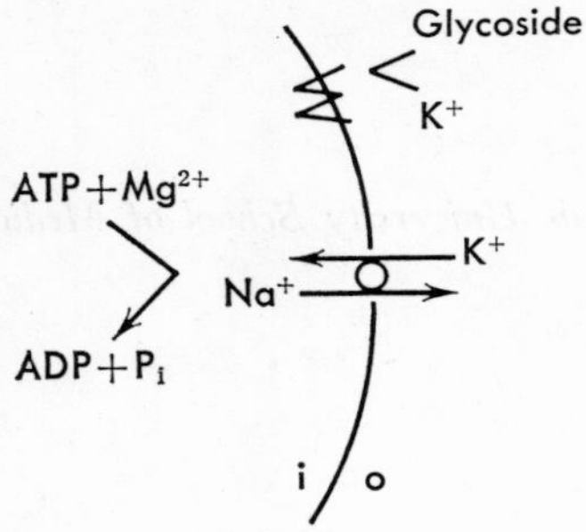

$$[\,ATP + E \xrightarrow{Na^+} E{\sim}P \xrightarrow{K^+} E + P_i\,]$$

FIG. 1. Outline of a model cell showing the aspects of sidedness of the basic features of the Na^+-K^+ pump. The membrane is indicated by the vertical line separating inside (i) from outside (o). The active transport system resides within the membrane pumping, at the expense of intracellular ATP, Na^+, and K^+ in the indicated directions. Cardiac glycosides bind to the outside of the membrane and inhibit the pump in a non-competitive manner. Glycoside binding is antagonized by external K^+. The bracketed reaction considers the pump as enzyme (E) and summarizes the transphosphorylation sequence thought to be associated with translocation of the ions. Taken from Ref. (*24*).

As summarized in Fig. 1, Mg^{2+} and ATP are required on the inside to drive Na^+ out and K^+ into the cell with the products, adenosine diphosphate (ADP)+inorganic orthophosphate (P_i), remaining inside during any single pump cycle. From studies dealing with the pump as Na^+, K^+-ATPase and the incorporation and release of ^{32}P from [γ-^{32}P]ATP (*5–8*) the role of substrates and products can be summarized by the bracketed equation given in Fig. 1. In this reaction Na^+ stimulates the phosphorylation of an enzyme-pump intermediate ($E{\sim}P$) while its dephosphorylation is stimulated by K^+. Translocation of the ions presumably occurs in concert with their effects on transphosphorylation perhaps as a result of sequential changes in the conformation of $E{\sim}P$ within the membrane. While there does not appear to be any active transport for the inward movement of Na^+ or the outward movement of K^+ there is competition between these two ions on both sides of the membrane (*9, 10*). Thus external Na^+ and internal K^+ alter the apparent affinities of the pump for K^+ and Na^+, respectively. In addition, detailed studies of the kinetics of activation and competition reveal that each pump cycle involves two or more Na^+ on the inside as well as two or more K^+ on the outside (*11–13*). The ratio of the number of Na^+ pumped in exchange for K^+ varies between 1 to 1 and 3 to 2 depending upon the manner of measurement and the conditions (*13–16*) as well as the type of red cell under study (*17*). Determination of the stoichiometry of the pump in terms of energy utilization indicates that 3 Na^+ are pumped per ATP hydrolyzed (*15*, *18*, *19*). In addition, the pump cycle can be reversed in the sense that ATP synthesis can take place at the expense of the downhill movements of Na^+ in and K^+ out (*20*, *21*). Thus it should be clear from the foregoing that the sensitivity and the subsequent action of Na^+ and K^+ on the inward as opposed to the outward facing aspects of the pump are different in addition to the fact that there is a tight coupling between those components responsible for translocation and those linked to the energetics of the system.

Other features of the sidedness of the pump can be accessed from the different ways in which the pump can be inhibited. Of particular interest are those inhibitors which act directly on the pump such as Ca^{2+} (*2*), cardiotonic steroids (*22*) and sulfhydryl reagents (*5*). These three agents illustrate the different types of access possible with different pump inhibitors in the sense that it appears that Ca^{2+} acts only from the inside (*2*), cardiotonic steroids only from the outside (*11*, *23*) while sulfhydryl reagents (*24*) act from either the inside and/or the outside of the membrane. It is presumed that the relevant context for these inside/outside relationships is with respect to the inward/outward facing aspects of the pump, *per se*.

It is of interest to briefly consider the mechanism of inhibition of the pump by internal Ca^{2+}. Since an outwardly oriented ATP-dependent Ca^{2+} pump has been identified (*25*) it is possible that Ca^{2+} converts the specificity of the pump from transporting Na^+ and K^+ to transporting Ca^{2+}. Recent evidence (*26*) makes this unlikely since the molecular weight of the Ca^{2+}-phosphoprotein, that is, the phosphorylated intermediate ($E \sim P$) found associated with the Ca^{2+}-ATPase is higher (150,000) than the molecular weight of the phosphoprotein associated with the Na^+-ATPase (103,000) as judged by acrylamide gel electrophoresis. On the other hand, the ^{32}P incorporated into the Na^+-phosphoprotein in the presence of Ca^{2+} is not decreased by the addition of K^+ indicating that the mode of Ca^{2+} inhibition of the Na^+-K^+ pump might be to inhibit the dephosphorylation step. If so, then this would be another indicator of the asymmetry of the pump in which internal Ca^{2+} alters the effect of external K^+ and points again to the connection between transphosphorylation and translocation of the transported species.

Ouabain Binding Site and Sidedness

Cardiotonic steroids, such as ouabain, have proved valuable in defining other asymmetries associated with the pump complex. These compounds are useful because they bind to the membrane with high affinity and with great specificity (*22*, *27*). As shown in Fig. 1, the sites to which cardiac glycosides bind are located on the outside and act allosterically to inhibit the pump mechanism (*11*). External K^+ competes with cardiac glycosides for binding (*11*, *27*, *28*) but the glycoside binding sites are spatially separated from the loci involved with the inward carriage of K^+. Even so, the glycoside binding site is a component part of the pump complex since glycosides bound to the pump before solubilization remain attached and inhibitory afterwards (*29*). The quantitative estimates of the number of glycoside molecules needed per cell to inhibit completely Na^+-K^+ pump activity can be made using [^{3}H]-ouabain (*28*, *30*, *31*). In human red cells the number of glycoside binding sites averages about 250 per cell. If it is assumed that there is a one-to-one correspondence between binding sites and pumps then the surface density of pumps approaches $2/\mu m^2$ and the turnover number approximates 6,000 ions pumped per site per minute.

In addition to the membrane locus of the glycoside binding site other pump asymmetries emerge from a determination of the factors which control glycoside (ouabain) binding. It should be emphasized that obtaining this type of information necessarily depends upon the use of a membrane preparation, such as reconstituted red cell ghosts, where the effects of alterations in the composition of the medium bathing one side of the membrane can be evaluated independently of the composition of the medium bathing the other side. Thus, while many of the different determinants of glycoside binding have been worked out in microsomal (*6*, *32–34*) or permeable (fragmented) red cell ghost (*28*) preparations, it is not possible to know the side-dependencies of these various determinants without studying their effects under conditions where the inside/outside relationship is established.

We have discussed previously (*24*) the side requirement for the stimulation of glycoside binding by nucleotide triphosphates (NTP), such as ATP or uridine triphosphate (UTP), in which it was shown (*28*, *35*) that NTP was required together with Mg^{2+}, on the inside to promote ouabain bin-

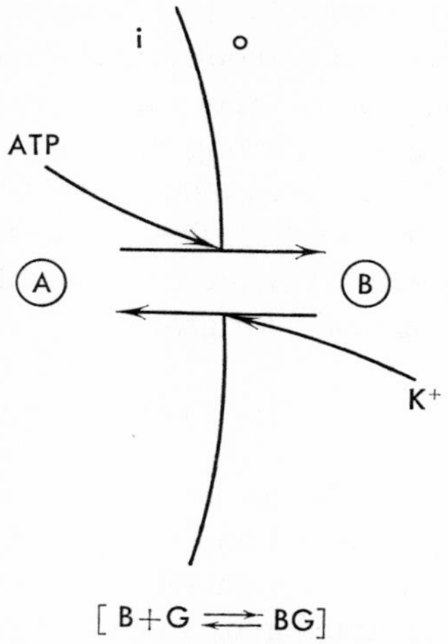

FIG. 2. A model of the glycoside binding site of the membrane. Different conformational states of the binding site are represented by A and B but glycosides (G) can only combine with the B form as indicated in the bracketed reaction. The B form has to be accessible from the outside in order for the binding reaction to occur. The set of the A ⇌ B equilibrium can be influenced for instance by intracellular ATP, which favors the B form, or by external K^+ which favors the A form. See text for further details. Taken from Ref. (*24*).

ding to the outside. This kind of side dependence strongly suggests that conformational changes of the pump complex underlie the binding reaction perhaps like the type implied in the model presented in Fig. 2. In this model the glycoside binding site can exist in two membrane forms, A or B, but since glycosides (G) can only combine with the B form to give BG, the availability of the B form defines the rate of glycoside binding. The pump, of course, is inactivated to the extent that BG is formed. (In red cells since the equilibrium constant is very large, BG once formed is almost irreversible, therefore it is only possible to study differences in rate of binding but not differences in equilibrium.) A shift to the right in the A ⇌ B equilibrium, which would result from the adsorption of NTP to its active site on the inside, would promote the change in conformation necessary for the appearance of the B form. Essentially all of the determinants of glycoside binding can be rationalized with this model in terms of their influence on the A ⇌ B equilibrium. Since glycoside binding to red cell ghosts is promoted equally well by pyrimidine as by purine NTP (*28*) but since only purine NTP (primarily ATP) can be broken down by the pump (*2*) it would appear that the binding of glycosides can occur without phosphorylation of the pump complex. This implies that phosphorylation of the pump complex is not a preferential requirement for glycoside binding or for changes in pump conformation associated with the B form even though glycosides can inhibit the dephosphorylation step once it is formed. Obviously, different configurational states of the pump complex can exist, presumably as different forms of E ~ P (*6*, *7*), compatible with this model for glycoside binding.

Partial Reactions and Sidedness

While the rate of glycoside binding in red cells can be markedly affected by either Na^+ or K^+, their effects are complicated and can only be sorted out in terms of the sidedness of their action on the membrane (*35*). This type of study was carried out using reconstituted ghosts in which the concentrations of Na^+ or K^+ on the two sides of the membrane can be varied separately using choline to keep the total isotonic concentration (Na^+ + K^+ + choline) constant (*26*). Let us consider first the effects of external Na^+ and K^+ on glycoside binding that is promoted by incorporated MgATP. It was found (*35*) as previously reported (*37*) that the rate of binding was increased as the concentration of Na^+ in the external medium increased (internal Na^+ and K^+ held constant). In this situation since the effectiveness of Na^+ is decreased as external K^+ is increased, it appears that external Na^+ acts indirectly on glycoside binding by decreasing the ability of K^+ to antagonize the binding of glycoside (*11*, *27*). According to the model (Fig. 2) this type of competition between Na^+ and K^+ increases the rate of glycoside by preventing K^+ from decreasing the availability of the B form. In contrast, internal Na^+ has a rather different effect on glycoside binding than external Na^+. Thus, under conditions where internal K^+ and external Na^+ and K^+ are held constant, the rate of glycoside binding is decreased as the concentration of internal Na^+ is increased. Again in terms of the model (Fig. 2) internal Na^+ would appear to interact with the conformation of the pump associated with the A form presumably to initiate phosphorylation as a requisite step for its translocation. On the other hand when internal K^+ is varied, at constant internal Na^+ and external Na^+ and K^+, the rate of binding is also decreased as the concentration of internal K^+ is increased. But, appearances to the contrary, this effect does not seem to be due to a competition between Na^+ and K^+ on the inside corresponding to their interaction on the outside, since it happens in the absence of internal Na^+ as well as external Na^+ and, further, this action of internal K^+ requires the presence of external K^+. These particular actions of internal Na^+ and K^+ are best understood when the different transport activities of the pump apparatus are considered. Thus, the effects of internal Na^+ and K^+ correlate with their effects in activating either $Na^+:K^+$ exchange or in activating the partial reactions of the pump which result in glycoside-sensitive $Na^+:Na^+$ exchange (*38*) or $K^+:K^+$ exchange (*39*). (Incidentally, glycoside once bound to the membrane inhibits all three types of fluxes to the same extent (*35*) indicating homogeneity of glycoside binding sites and transport activity.) Therefore, as either the Na^+-K^+ pump or one of its partial reactions is turned on by raising intracellular Na^+ or K^+, the rate of glycoside binding is decreased, thereby establishing a reciprocal relationship between the rate of glycoside binding and the turnover of the pump machinery. In terms of the model (Fig. 2) any increase in the activity of the pump apparatus would shift the $A \rightleftharpoons B$ equilibrium to the left and thereby decrease the availability of the B form.

Study of the partial reactions of the pump, by the way, offer another type of approach to the pump mechanism (*40*). For instance, the nucleotide dependence of $Na^+:Na^+$ exchange differs from $Na^+:K^+$ exchange in that $Na^+:Na^+$ exchange appears to be controlled by the level of ADP while $Na^+:K^+$ exchange depends upon the level of ATP (*41*). Since oligomycin stimulates [^{14}C]ADP: ATP exchange (*42*) but inhibits $Na^+:Na^+$ exchange (*43*), $Na^+:Na^+$ exchange must evidently be associated with a step that follows phosphorylation. Thus, in $Na^+:Na^+$ exchange, ADP would appear to reverse a reaction of which it is the product. This implies that the reversible translocation of Na^+ is correlated with the reversible transphosphorylation of the pump. Presumably the Na^+-ATPase and the formation of the Na^+-sensitive component of the phosphoenzyme (*8*), as referred to before, represent the biochemical correlates of this transport reaction.

Another indication of the complexity of interpreting results obtained say with a permeable or broken ghost preparation, without independent assessment of sidedness, can be seen when ATP-promoted glycoside binding is studied varying the concentrations of Mg^{2+} and Na^+ at constant K^+. Thus when the concetration of Mg^{2+} is high (equal to the ATP concentration) the addition of Na^+ results in a decrease in the rate of glycoside binding. On the other hand, when the concentration of Mg^{2+} is markedly reduced (relative to the same ATP concentration) then the addition of Na^+

results in an increase in the rate of glycoside binding. It is from studies using the reconstituted ghost system as described before that it can be shown that the effect of Na^+ addition at high Mg^{2+} is due to *internal* Na^+ while at low Mg^{2+}, it is due to *external* Na^+. Evidently, the affinity of the outside aspect of the pump for K^+ can be altered by internal Mg^{2+}.

As found previously for microsomal preparations (*44*) glycoside binding to ghosts can also be promoted, in the absence of Na^+, by P_i. And the sidedness of this type of effect together with its Mg^{2+} requirement, can be assessed using reconstituted ghosts (*35*). It was of course of interest to determine not only the sidedness of the inhibiting action of Na^+ but also the side from which P_i exerted its effects. This latter consideration presented a special problem since reconstituted ghosts like intact cells are rather permeable to P_i. This difficulty was overcome by pretreating reconstituted ghosts with 4-acetamido-4′-isothiocyano-stilbene-2,2′-disulfonic acid (SITS) an agent known (*45*) to decrease the membrane permeability to P_i without altering glycoside binding. Using this system it was established that P_i could only promote glycoside binding when it was present on the inside and not on the outside. And then, in a fashion similar to that already described, the separate effects of internal and external Na^+ were studied on P_i-promoted binding. As before it was found that increasing the external concentration of Na^+ increased the rate of glycoside binding by competition with external K^+ and that increasing the concentration of internal Na^+ decreased the rate of glycoside binding. While the action of internal Na^+ appears similar in the P_i-and ATP-promoted systems, the different conformational states that the pump complex can assume under these two circumstances, while exhibiting some overlap, must be different since P_i binds to the membrane as glycoside is bound.

Sidedness of Sulfhydryl Groups

Turning to other determinants of glycoside binding, it is possible to define at least two different classes of sulfhydryl groups associated with the pump from the differential action with regard to membrane sidedness of three different sulfhydryl reagents (*24*). The three reagents used were ethacrynic acid (EA), N-ethylmaleimide (NEM) and *p*-chloromercuriphenyl sulfonic acid (PCMBS). All three reagents inhibit the pump whether their locus of action is from the inside or outside or both. However, glycoside binding can only be prevented or inhibited by sulfhydryl reagents acting from the inside. Thus, EA which appears to interact with membranes only on the outside is without effect on glycoside binding. NEM, an agent which can penetrate the membrane and therefore act from both sides, prevents glycoside binding. Since the membrane is impermeable to PCMBS (*46*) its action can be restricted to the outside by using intact cells or reconstituted ghosts or it can be made to interact from both sides by using permeable ghosts. When its action is limited to the outside, PCMBS is without effect on the rate of glycoside binding but when PCMBS has access to the inside glycoside binding is prevented. Thus, there are at least two classes of sulfhydryl groups associated with the pump and these two classes are located on opposite sides of the membrane. Reaction with the outside groups alone is sufficient to cause inhibition of the pump but it is only by reaction with the class associated on the inside that glycoside binding is prevented. This type of result again emphasizes that alterations from the inside can be reflected through the membrane to change the outside reactivity of the pump complex.

Effect of Antibody

Another type of membrane sidedness has been characterized with regard to the action of an antibody, anti-L (AL), on low potassium-type sheep (*47*) and goat (*48*) red cells. AL acts on these types of cells to increase the active transport rates of Na^+ and K^+ by increasing the number of pumps (*49, 50*) as well as by changing the apparent affinities of the pump for Na^+ and K^+ (*49–52*). In the present context it is the sidedness of AL action that is of special interest in that AL bound to the outside changes the affinity of the inward—but not the outward—facing aspect of the pump for K^+ and Na^+ in both sheep and goat red cells. It is not clear at the present time how this effect of AL is mediated by the membrane or what rearrangement of the pump components are required for a change in its ionic affinities.

Separation of Different Parts of the Pump

In addition to the specification of pump asymmetries as defined in the foregoing discussion, it is also of interest to consider the functional organization of the pump within the membrane with regard to the utilization of ATP and the flow of energy through the system. There is now substantial evidence (*53*) pointing to the existence of a membrane pool of ATP which is preferentially used by the Na^+-component of the Na^+, K^+-ATPase and in the formation of its phosphoenzyme, distinct from those activities obtained in the presence of Mg^{2+} alone. The basis for this idea comes from experiments (*53*) in which permeable or broken ghost preparations were first preincubated with MgATP and then thoroughly washed prior to labeling the phosphoenzyme or assaying ATPase by incubation with [γ-^{32}P]ATP at 0°C in a standard way (*8*). The subsequent incorporation of ^{32}P into either the Na^+-component or the Mg^{2+}-component was not affected when the preincubation with nonradioactive ATP was carried out at 0°C. However, when the preincubation with ATP was carried out at 37°C there was no ^{32}P incorporated into the Na^+-component even though there was a normal amount of ^{32}P incorporated into the Mg^{2+}-component. When ATPase activities were determined after comparable preincubation conditions (assayed by $^{32}P_i$ liberation from [γ-^{32}P]ATP after prolonged incubation at 0°C), it was found that for ghosts preincubated with nonradioactive ATP at 0°C the Na^+-ATPase and Mg^{2+}-ATPase activities were normal. Again, however, for ghosts preincubated at 37°C, while the Mg^{2+}-ATPase activity was normal, there was no Na^+-ATPase activity demonstrable. These results are compatible with the view that preincubation filled a membrane compartment with ATP (nonradioactive), the compartment being accessible only at 37°C. ATP present in this pool is used in preference to bulk [γ-^{32}P]ATP for either ^{32}P incorporation or ATPase activity. Further evidence that this interpretation is correct comes from experiments in which the membrane pool is either filled or emptied in different ways. For this type of study the presence of ATP in the membrane compartment can be determined by whether or not ^{32}P can be incorporated into the Na^+-component: thus, the compartment is empty when labeling of the Na^+-component can take place and filled, when it cannot. In this way, it was found that the compartment could be emptied prior to assay by prolonged incubation at 37°C in the presence of Na^+. Alternatively if the incubation were carried out in the presence of K^+ as well as Na^+ then the emptying time for the compartment was greatly accelerated implying that the Na^+, K^+-ATPase as well as the Na^+-ATPase had access to the pool of ATP. But more interesting were experiments in which the lability of the membrane compartment could be influenced by the phosphoglycerate kinase (PGK) reaction:

Where TPD is the triose-phosphate dehydrogenase, NAD and NADH are the oxidized and reduced forms of nicotinamide-adenine dinucleotide, 1,3-DPG is 1,3-diphosphoglycerate and 3-PGA is 3-phosphoglycerate. Thus, when the compartment had already been filled with ATP it was possible to empty it quickly by running the reaction backward (at 37°C) by adding only 3-PGA and NADH, utilizing the PGK and TPD known to be present in the membrane (*54*) and assuming the presence of ATP in the pool. Alternatively, it was possible to fill the pool with ATP by running the reaction forward by adding P_i, triose-phosphate, NAD and ADP. (The amount of ADP added, or the amount of ATP subsequently synthesized, was insufficient by itself to affect the pool.) These results while providing rather direct support for the concept of a membrane compartment for ATP also indicate the location of this compartment within the membrane and its interrelationship between the ATP-generating reaction (PGK) and its utilization by the Na^+-K^+ pump as Na^+, K^+-ATPase. In addition, these results also imply that the Mg^{2+} and Na^+ components are different entities since their activities can be varied independently. It is interesting that the existence of such a membrane compartment and reaction ensemble was previously postulated by us (*55*) on other grounds.

That the Na^+-dependent component represents a different entity from the Mg^{2+}-component can be demonstrated (*56*) by the fact that the two components can be separated by sucrose density centrifugation, after labeling with ^{32}P from [γ-^{32}P]ATP and after solubilization with sodium dodecyl sulfate. The distribution of the Na^+-component in the gradient is the same as the distribution of the solubilized Na^+, K^+-ATPase as well as the distribution of the solubilized ouabain-binding component (*29*) of the membrane and is clearly distinct from (and more dense than) the distribution of the Mg^{2+}-component. Since it is possible to separate the Na^+-ATPase from the Mg^{2+}-ATPase physically as well as kinetically they must be taken as representing different enzyme systems.

Acknowledgments
The work reviewed in this paper was supported by NIH grants HE-09906 and AM-05644 and NSF grant GB-18924.

REFERENCES

1. G. Gardos, *Acta Physiol. Hung.*, **6**, 191 (1954).
2. J. F. Hoffman, *Circulation*, **26**, 1201 (1962).
3. I. M. Glynn, *J. Physiol.*, **160**, 18P (1962).
4. R. Whittam, *Nature*, **196**, 134 (1962).
5. J. C. Skou, *Physiol. Rev.*, **45**, 596 (1965).
6. R. W. Albers, G. J. Koval, and G. J. Siegel, *Mol. Pharmacol.*, **4**, 324 (1968).
7. R. L. Post, S. Kume, T. Tobin, B. Orcutt, and A. L. Sen, *J. Gen. Physiol.*, **54**, 306s (1969).
8. R. Blostein, *J. Biol. Chem.*, **243**, 1957 (1968).
9. R. L. Post, C. R. Merritt, C. R. Kinsolving, and C. D. Albright, *J. Biol. Chem.*, **235**, 1796 (1960).
10. J. F. Hoffman, *J. Gen. Physiol.*, **45**, 837 (1962).
11. J. F. Hoffman, *Am. J. Med.*, **41**, 666 (1966).
12. J. R. Sachs and L. G. Welt, *J. Clin. Invest.*, **46**, 65 (1967).
13. J. R. Sachs, *J. Gen. Physiol.*, **56**, 322 (1970).
14. R. L. Post and P. C. Jolly, *Biochim. Biophys. Acta*, **25**, 118 (1957).
15. P. J. Garrahan and I. M. Glynn, *J. Physiol.*, **192**, 217 (1967).
16. T. J. Gill and A. K. Solomon, *Nature*, **183**, 1127 (1959).
17. D. C. Tosteson and J. F. Hoffman, *J. Gen. Physiol.*, **44**, 169 (1960).
18. A. K. Sen and R. L. Post, *J. Biol. Chem.*, **239**, 345 (1964).
19. R. Whittam and M. E. Ager, *Biochem. J.*, **97**, 214 (1965).
20. P. J. Garrahan and I. M. Glynn, *J. Physiol.*, **192**, 237 (1967).
21. I. M. Glynn and V. L. Lew, *J. Physiol.*, **207**, 393 (1970).
22. H. J. Schatzmann, *Helv. Physiol. Pharmacol. Acta*, **11**, 346 (1953).
23. P. C. Caldwell and R. D. Keynes, *J. Physiol.*, **148**, 8P (1959).
24. J. F. Hoffman, *in* "Role of Membranes in Secretory Processes," ed. by L. Bolis, R. D. Keynes, and W. Wilbrandt, North-Holland, Amsterdam, p. 203 (1972).
25. H. J. Schatzmann and F. F. Vincenzi, *J. Physiol.*, **201**, 369 (1969).
26. P. A. Knauf, F. Proverbio, and J. F. Hoffman, Abstr., Int. Biophys. Congr., Moscow (1972).
27. I. M. Glynn, *J. Physiol.*, **136**, 148 (1957).
28. J. F. Hoffman, *J. Gen. Physiol.*, **54**, 343s (1969).

29. P. B. Dunham and J. F. Hoffman, *Proc. Natl. Acad. Sci. U.S.*, **66**, 936 (1970).
30. C. J. Ingram, Ph. D. Dissertation, Yale University (1971).
31. P. B. Dunham and J. F. Hoffman, *J. Gen. Physiol.*, **58**, 94 (1971).
32. A. Schwartz, H. Matsui, and A. H. Laughter, *Science*, **160**, 323 (1968).
33. J. C. Skou and C. Hilberg, *Biochim. Biophys. Acta*, **185**, 198 (1969).
34. O. Hansen, *Biochim. Biophys. Acta*, **233**, 122 (1971).
35. H. Bodemann and J. F. Hoffman, unpublished.
36. H. Bodemann and H. Passow, *J. Memb. Biol.*, **8**, 1 (1972).
37. H. J. Schatzmann, *Biochim. Biophys. Acta*, **94**, 89 (1965).
38. P. J. Garrahan and I. M. Glynn, *J. Physiol.*, **192**, 159 (1967).
39. I. M. Glynn, V. L. Lew, and U. Lüthi, *J. Physiol.*, **207**, 371 (1970).
40. I. M. Glynn, J. F. Hoffman, and V. L. Lew, *Phil. Trans. Roy. Soc. Lond. B.*, **262**, 91 (1971).
41. I. M. Glynn and J. F. Hoffman, *J. Physiol.*, **218**, 239 (1971).
42. R. Blostein, *J. Biol. Chem.*, **245**, 270 (1970).
43. P. J. Garrahan and I. M. Glynn, *J. Physiol.*, **192**, 189 (1967).
44. G. E. Lindenmayer, A. H. Laughter, and A. Schwartz, *Arch. Biochem. Biophys.*, **127**, 187 (1968).
45. P. A. Knauf and A. Rothstein, *J. Gen. Physiol.*, **58**, 190 (1971).
46. P. J. Garrahan and A. F. Rega, *J. Physiol.*, **193**, 459 (1967).
47. J. C. Ellory and E. M. Tucker, *Nature*, **222**, 477 (1969).
48. J. C. Ellory and E. M. Tucker, *Biochim. Biophys. Acta*, **219**. 160 (1970).
49. P. K. Lauf, B. A. Rasmusen, P. G. Hoffman, P. B. Dunham, P. B. Cook, M. L. Parmelee, and D. C. Tosteson, *J. Memb. Biol.*, **3**, 1 (1970).
50. J. C. Ellory, J. R. Sachs, P. B. Dunham, and J. F. Hoffman, *in* "Biomembranes," ed. by F. Kreuzer and J. F. G. Slegers, Plenum Press, New York, Vol. 3, p. 237 (1972).
51. J. C. Ellory, I. M. Glynn, V. L. Lew, and E. M. Tucker, *J. Physiol.*, **217**, 61P (1971).
52. R. Blostein, P. K. Lauf, and D. C. Tosteson, *Biochim. Biophys. Acta*, **249**, 623 (1971).
53. F. Proverbio and J. F. Hoffman, *Fed. Proc.*, **31**, 215 (1972).
54. S. L. Schrier and L. S. Doak, *J. Clin. Invest.*, **42**, 756 (1963).
55. J. C. Parker and J. F. Hoffman, *J. Gen. Physiol.*, **50**, 893 (1967).
56. F. Proverbio and J. F. Hoffman, unpublished.

The Na^+, K^+-ATPase Molecule

Makoto Nakao, Toshiko Nakao, Hidehiko Ohta, Fumiko Nagai, Koichi Kawai, Yoko Fujihira, and Kei Nagano

Department of Biochemistry, Tokyo Medical and Dental University School of Medicine, Tokyo, Japan

NaI enzyme was treated with Lubrol WX and the supernatant was treated with various modified celluloses; then the preparation was applied to a cellulose or AE-Sepharose column. The elution pattern showed 3 separate peaks. Each peak showed practically only 1 band in sodium dodecylsulfate (SDS) disk gel electrophoresis. Specific activities of the 3 peaks were sometimes higher than 4,000 μmoles P_i/mg/hr. Polar modification of membrane enzymes was attempted by inhibitors of various molecular sizes. Finally, we synthesized a dextran derivative of molecular weight 250,000 to which *p*-chloromercuribenzoate (PCMB) was bound by the bridge of the aminoethyl group. Although it was a potent inhibitor of ATPases in ghosts, it inhibited only Na^+, K^+-ATPase but not Na^+, K^+-insensitive ATPase when applied to intact red cells. It halted the ouabain-sensitive transport of ^{86}Rb to red cells.

Two different approaches have been tried in our laboratory for several years. Both are continuing, but already some of the results may be quite suggestive for clarification of active transport mechanisms.

Purification of Na^+,K^+-ATPase

According to modern practice enzymes should generally be purified before further investigations. Many laboratories (*1–5*) have attempted to purify Na^+,K^+-ATPase, but the results are not yet satisfactory. The highest specific activity obtained so far is less than 2,000, and is in most cases less than 1,000 μmoles P_i liberation/mg protein/hr. The purity of the preparation is less than 50% according to some authors (*1*, *3*). The reasons why purification is extremely difficult are as follows. First of all, it is difficult to obtain pure membranes. Although the purest membrane preparations are obtained from red cells of animals, the ATPase activity is very low, 2 or 3 orders lower than that of microsome fractions. Secondly, the solubilization of the enzyme from the membrane is very difficult since the enzyme is not simply bound to the membrane, but is an essential constituent of the membrane. Once the enzyme is solubilized with detergents, the activity usually becomes very labile. Several laboratories (*6*, *7*) adopted Lubrol WX for solubilization of the enzyme, but the elution pattern of the enzyme from an

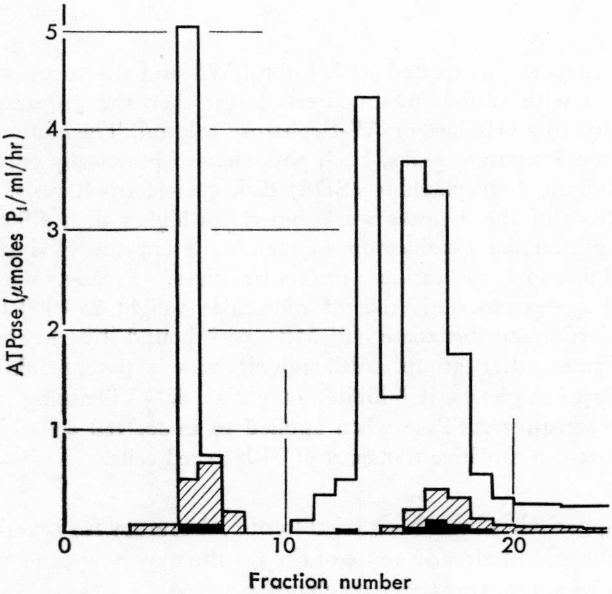

FIG. 1. The Na^+,K^+-ATPase fraction obtained from deoxycholate (DOC)-treated microsomes of pig brain was treated with 2 M NaI and then with Lubrol WX. The supernatant was applied to an aminoethyl cellulose column and eluted with ATP solution containing 14 mM veronal buffer at pH 6.6 with a concentration gradient of NaCl from 0 to 0.1 M. □ enzyme activity; ▨ protein based on Lowry's determination; ■ protein based on staining with Amido Black (T. Nakao *et al.*, *Anal. Biochem.*, in press.)

agarose column (*7*) was not symmetrical, presumably owing to nonspecific interaction between the enzyme and the agarose gel. It is difficult to tell whether a sample which is not spun down at 100,000 $\times g$ is really a solubilized enzyme or merely a complex micelle of the detergent with the enzyme, especially in the case of a protein containing a large amount of lipid.

During purification of the enzyme, very high specific activity was attained in our laboratory and the active enzyme fraction was separated to at least 3 different peaks in column chromatography as follows. Pig brain microsomes which were obtained by a procedure with deoxycholate (DOC) essentially according to Skou (*8*) were treated with 2 M NaI. The latter treatment, which was developed by us in 1963 (*9, 10*), was very effective in reducing ouabain-insensitive ATPase activity, *i.e.*, so-called Mg ATPase activity. NaI enzyme which had a specific activity from 100 to 200 μmoles P_i liberation/hr/mg protein was treated with Lubrol under carefully selected conditions. The supernatant after centrifugation at 160,000 $\times g$ was treated by negative adsorption with modified cellulose and finally the solution was developed on a column of aminoethyl cellulose. Elution was carried out with an NaCl gradient plus ATP. As in Fig. 1, 3 activity peaks were observed. They are tentatively designated as α, β, and γ. The activities of all 3 were labile and the activity of β was extremely labile, sometimes missing.

TABLE I
ATPase Fractions

Fraction	α	β	γ
Na^+,K^+-ATPase specific activity	1,000–5,000 (3,400)	0–5,000	700–7,000 (2,400)
Percent of total	10–40	?	10–40
Ouabain inhibition	Complete	Complete	Complete
pNPPase	Active	Active	Active
Lability	Very labile	Extremely labile	Very labile
S	7.1S	—	
S.G.	1.09	—	1.12
G200-M.W.	$4–5\times10^5$	—	$>\alpha$?
Lipid / protein	5–6 : 1	1 : 1	2–3 : 1
Lubrol / protein	Large	—	4–5 : 1
M.W. of peptide (SDS gel electrophoresis)	$10–12\times10^4$	$10–12\times10^4$	$10–12\times10^4$
Rechromatography	$\alpha\rightarrow\gamma$		$\gamma\not\rightarrow\alpha$

Except for these points, the separation of these peaks was reproducible. The most conspicuous features are the specific activities.

Table I shows some characteristics of these peaks. The figures for the specific activities were dispersed over a rather wide range, but the numbers in parenthesis, indicating the mean of more than 20 experiments, are relatively large. The first peak α gave a high specific activity but unfortunately, free Lubrol was eluted very close to or overlapping this peak. Around the third peak γ, many protein peaks appeared. Therefore, the specific activity of the γ peak varied from one case to another. Owing to the lability of the enzyme activity, the quantitative recovery of each peak was not constant, but very high specific activities were sometimes observed in each peak. The activities had to be determined within a few hours, otherwise they decreased so rapidly that they were less than several % of their original values within several hours, even after the addition of 50% glycerol or other compounds while the specific activities were extremely high. On the other hand, specimens having a low specific activity, for example 1,000, remained relatively stable for several days. The yield of activity from the column is usually from 40 to 80%, while that of protein is less than 10%. After rechromatography of the α peak on the same column, γ peaks were observed as well as α peaks but after rechromatography of the γ peak, nothing was observed at the α position and the γ peak was the only one present.

All of the products hydrolyzed *p*-nitrophenophosphate in the presence of K^+ and were completely inhibited by 10^{-5} M ouabain. Sodium dodecyl sulfate (SDS) disk electrophoresis according to Weber and Osborn (*11*) showed a main peak at about 11 or 12×10^4 molecular weight using dansylated albumin as an internal standard and there were several small bands weakly stained with Coomasie brilliant blue.

Some Properties of 3 Active Fractions

All 3 products showed the same or very similar bands. Although the study of ^{32}P-phosphorylated intermediates is now in progress, a preliminary result showed only 1 peak which coincided with the main band besides that of free $AT^{32}P$. In this experiment the yield was only 8%, because electrophoresis was carried out with a phosphate buffer at pH 7.0, where the binding of phosphate to Na^+, K^+-ATPase protein was very labile in contrast to the situation at acidic pH (*12, 13*). Several years ago, Nagano and others at the Jichi Medical School assumed the molecular size of phosphorylated peptide as 200,000–300,000 using SDS-saturated G-200 or Sepharose columns. The molecular weight involved an amount of SDS bound to the protein. If the amount of bound SDS is 1.4 times the weight of the protein, an average for ordinary proteins, the real molecular weight of peptide is calculated to be about 80,000 to 120,000, which is very close to the present result. Amino acid analysis of fractions with a relatively high specific activity of 1,000 to 3,000 for the 3 peaks showed very similar amino acid patterns. Probably, the α, β, and γ protein moieties consist of the same or very similar peptides.

However, their lipid contents were different, as shown in Table I. The α peak contained an anomalous amount of lipids and Lubrol. This might be due to the overlapping of Lubrol micelles. The molecular weight of native protein was difficult to estimate, partly because it aggregated easily and interacted with the Sepharose column. This caused unsymmetrical elution patterns, as found in other studies (*7*), and also they contained various amounts of lipids. Therefore, G-200 which is not adequate for the estimation of large sizes molecular was used. The enzyme activity showed a single peak which had an approximately symmetric pattern. The peak was observed somewhat behind the void volume, and the molecular weight was estimated to be roughly 500,000. By the technique of "sink and flow," which we developed (*14*) some years ago to assess the partial specific volume of impure enzymes, we obtained a specific gravity of 1.09 for the γ peak and 1.12 for the α peak, and by application of the sucrose gradient technique of Martin and Ames (*15*), we obtained an S value of 9S (not corrected for specific gravity) for the γ peak using catalase and albumin as internal standards. This is not much different from the molecular weight assumed from the G-200 column. Taking the specific gravity of phospholipids in the enzyme as 1.03 and that of protein as 1.34, the light specific gravities of α and γ were well accounted for.

Thus this enzyme is concluded to be a single peptide having a molecular weight of a little more than 100,000 with a large amount of lipids. It may not contain a so-called subunit structure, in disagreement with our previous opinion which was expressed several years ago (*6*), and other authors who suggested various subunit structures, though small subunits having a molecular weight of less than 20,000 cannot be ruled out. So far, various figures for the molecular weight have been proposed, such as 1,000,000, 700,000, 500,000 (*14, 16*), 300,000 (*17*) and 8,000 (*5*). However, all of them were obtained using impure preparations, with specific activities less than 1,000, in contrast to the present experiments.

Generally speaking, the molecular weight of lipoproteins is not necessarily well defined. Lipids bound to the protein equilibrate with free lipids in the medium. This fact obscures the structure of the molecule. One cannot tell the exact molecular weight of such molecules. According to preliminary electron micrographic observations, no long rods of any size were observed. The difference between α, β, and γ materials might be due to a difference of phospholipid content, but it is not clear why 3 sharp peaks were observed instead of a single sharp peak or a single wide peak consisting of a number of different peaks of different lipid composition. Lipid composition and aggregation and disaggregation of the molecules should also be investigated.

Observation of the Enzyme in Situ

Though purification of the enzyme and reconstitution of active membranes may be one of the final steps in the biochemical approach to physiological phenomena, the behavior of the entity *in situ* must be studied even for this purpose. Besides kinetic studies, the observation of the enzyme molecule in the plasma membrane *in situ* is also necessary. We have been trying to modify Na^+, K^+-ATPase from one side of the membrane. This may be called polar modification of the enzyme (*18*), although the stoichiometry of the modified groups is not known yet.

First of all, we tried to synthesize large SH-inhibitors. Hg fluorescein, though larger than *p*-chloromercuribenzoate (PCMB) and *p*-chloromercuribenzoic sulfate (PCMBS), could penetrate the cell membrane easily as could the smaller molecules. Red cells were incubated with 0.1 mM Hg fluorescein acetate at 37°C for 30 min. After washing the medium with albumin and hemolysis by freezing and thawing, a fair amount of fluorescein

TABLE II
Permeability of Hg Compounds

Compound	M. W.	Active transport	Na^+,K^+-ATPase	Ouabain-insensitive ATPase	Detection in hemolysate supernatant
PCMB	377	Inhibition	Inhibition	Inhibition	Yes
Hg fluorescein	533	,,	,,	,,	,,
Hg bromosulphalein	1,000	—	,,	,,	,,
Hg anilinodextran	70,000	—	,,	,,	,,
Hg benzoyl amino-ethyl dextran	250,000	Inhibition	,,	No	—

was found in a trichloroacetic acid (TCA) extract of the hemolysate supernatant. A more dissociable and larger molecule, bromosulphalein, which has 4 bromine and 2 sulphonic groups was mercurated, but the compound, previously labeled molecular inhibitors were synthesized. Hg aniline bound to a dextran with ^{203}Hg, could easily penetrate the cell membrane.

Therefore, macromolecular inhibitors were synthesized. Hg aniline was bound to a dextran having an average molecular weight of 70,000 with BrCN. The compound inhibited Na^+,K^+-ATPase as well as ouabain-insensitive ATPase in ghosts. This huge compound was applied to intact red cell suspensions, and the remaining reagent was washed 3 times. The red cells were hemolyzed with a large amount of monoiodacetate to minimize translocation of the compound from one SH to another during hemolysis. ATPase activities in the cell membranes obtained were estimated. Both ATPases, ouabain-sensitive and -insensitive were inhibited. Therefore, these compounds might penetrate the cell membrane relatively easily. If ^{14}C-aniline and ^{203}Hg were used to synthesize the compound, a doubly labeled compound could be obtained. Table III shows the distribution of the compound between hemolysate and membranes. It is evident that the compound was found in hemolysate. Some authors (*19–21*) reported that a compound having a molecular weight of less than 1,000 cannot cross the red cell membrane because of similar results. The reason which they gave was as follows. Since the protein in the red cell membrane comprises only 1–2% of the total protein in the hemolysate supernatant, the specific activity of ^{203}Hg or ^{14}C/mg protein in the cell membranes was over 100 times that in the hemolysate supernatant. However, two-thirds of the total activity found

TABLE III
Distribution of an Hg Aniline Derivative of Dextran Having a Molecular Weight of 70,000

Fraction / addition	1.8×10^{-5} ^{14}C-Dextran Hg aniline (glucose: C: Hg=1 : 3.3 : 3)	1.3×10^{-5} ^{203}Hg-Dextran Hg aniline (glucose : Hg = 1 : 11 : 3)
Red cells	15×10^{9} moles/ml cells	19
Ghost	8.5	6.5
Hemolysate	7.3	13

Total 4 ml, containing 100 mM NaCl, 2 mM $MgCl_2$, 25 mM Tris-HCl (pH 7.4), 5 mM glucose, 50 mM sucrose, and red cells (Hematocrit 5%) was incubated at 37°C for 45 min and washed.

in the supernatant must have actually penetrated the cell membranes, or been redistributed between the supernatant and the residue on hemolysis. The latter possibility cannot be excluded completely, but no evidence supporting this hypothesis has been obtained yet either. The total activity, but not the specific activity, must be taken into consideration in this case. The difference between the specific activities cannot be regarded as evidence for nonpenetration.

The following experiment, however, gave a different result. The short-circuit current (SCC) of frog bladder (*22*) was completely blocked by 10^{-5} M ouabain or digitoxigenin in contrast to toad bladder, which is quite insensitive to the cardiac glycoside. The same concentration of $HgCl_2$ was applied to the mucosal side of the membrane as a control. In this case the SCC was completely blocked. When an Hg compound having a molecular weight of 70,000 was applied to the mucosal side (*23*), the compound did not change the SCC at all. This fact might mean that the compound cannot penetrate the bladder epithelium, although there remained the possibility that the compound could enter the cell but was consumed by high concentrations of cytoplasmic proteins.

Finally, we synthesized a much larger compound. A type of dextran having a molecular weight of 250,000 was reacted with aminoethyl sulfate in strong alkali to produce aminoethyl dextran. This was used for various modifications of dextran as a starting material. PCMB was attached to the compound with the aid of diethylcarbodiimide (DCC) in dimethylsulphate or N(3-diethylaminoethyl), N-ethylcarbodiimide in water. It was dialyzed against alcohol/water, acetone/water and water exhaustively. The compound obtained usually contained approximately 20 Hg per molecule. The absence of free Hg or PCMB was checked with a G-50 column. The compound inhibited Na^+,K^+-ATPase from pig brain strongly and both ouabain-sensitive and -insensitive ATPase in red cell membranes were also strongly inhibited when it was applied to red cell ghosts which were previously prepared. On the other hand, if the compound was added to intact cell suspensions preincubated for 30 min and the cell membranes were obtained after washing and hemolysis, Na^+,K^+-ATPase activity was inhibited, but not that of ouabain-insensitive ATPase, as shown in Table IV. Moreover, ^{86}Rb uptake was observed using intact cells after preincubation with or without the compound. The compound strongly inhibited the uptake as indicated in Fig. 2. Hemolysis during the experiment was not significant. Generally speaking, Hg compounds are so-called reversible inhibitors. In

TABLE IV
Inhibition of Membrane ATPases by Hg Benzoylamidoethyl Dextran Having a Molecular Weight of 250,000

Applied to	Inhibitor	ATPase activity	
		Ouabain-insensitive	Ouabain-sensitive
Ghosts	—	0.61	0.87
	+	0.28	0.47
	—	1.07	1.35
	+	0.50	1.02
Intact cells	—	0.54	0.47
	+	0.61	0.14
	—	0.77	0.33
	+	1.00	0.08
	—	1.15	0.75
	+	0.91	0.27

ATPase activity was measured after washing the intact cells, hemolysis, and washing the cell membranes.

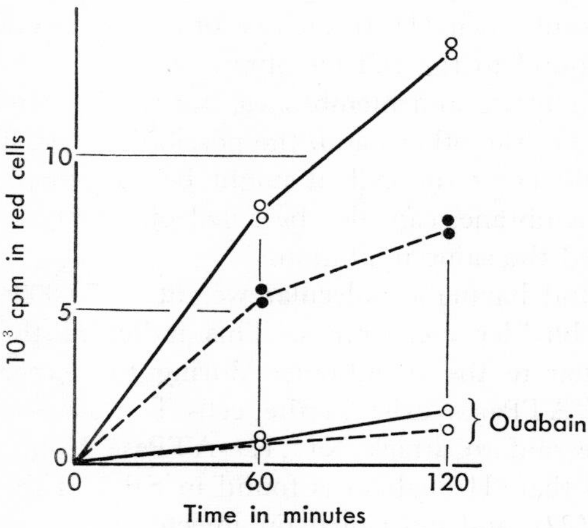

FIG. 2. Effect of macromolecular Hg compound on the cation uptake of human red cells. K+-depleted red cells (0.3 ml) suspended in a medium containing 1 mM ^{86}RbCl (0.3 μCi) were preincubated with (----) or without (——) 55 μM macromolecular Hg compound (PMDT-250) having a molecular weight of 250,000 at 0°C for 15 min. Then the red cells were collected, washed, and lyzed by adding a small amount 0.1% SDS at an alkaline pH.

the case of these inhibitors, the inhibition is eliminated by the addition of a large amount of cysteine.

An irreversible inhibitor was also synthesized using aminoethyl dextran having a molecular weight of 250,000 as the starting material. Monoiodoacetate was bound to this molecule by water-soluble DCC or N(3-diethylaminoethyl), N-ethylcarbodiimide in water. This large monoiodoacetate derivative gave the same results in experiments using intact cells as the large Hg compound. Prior to hemolysis, a large amount of cysteine was added to the washing medium to minimize a number of SH groups remaining, but the results were the same.

These observations are summarized as follows:

1) In contrast to previously hemolyzed ghosts, only Na^+, K^+-ATPase was inhibited when large inhibitor molecules were applied to the cell suspension. Ouabain-insensitive ATPase, which was also one of the membrane ATPases, was not affected at all even at high concentrations. This indicated that Na^+, K^+-ATPase has a reactive group outside the cells, however, in the case of ouabain-insensitive ATPase, no reactive group exists outside the cells.

2) Both reversible and irreversible inhibitors having a large molecular weight showed the same results as in (1). In the case of reversible inhibitors, the Hg compound was bound to the cell membrane and might be redistributed between the hemolysate and membranes, but the irreversible inhibitor is not transferred. On the other hand, the possibility that, if the irreversible compound could enter the cell, it might be consumed during penetration across the membrane, can also be ruled out because the reversible compound showed the same inhibition.

3) Even the Hg compound having a molecular weight of 70,000 did not affect the SCC in frog bladder membranes. This indicated that these compounds did not penetrate the membranes during the experiments. That the active sites of ATPases exist in the cells is well established and K^+ and ouabain are said to attack Na^+, K^+-ATPase from outside. However, it was reported that ^3H-ouabain is found in SR and the supernatant of heart muscle (*24*), and most ouabain injected intravenously is excreted in the bile (*25*). These results suggest that ouabain can penetrate the cell membrane and K^+, of course, enters the cell by passive diffusion. Therefore, the present experiments provide evidence that Na^+, K^+-ATPase faces not only the inside of the cells but also the outside.

According to experiments discussed in the previous section, the protein of Na^+, K^+-ATPase has a molecular weight of 110,000 or 120,000 without any other subunit chains of similar size. It may be concluded that this relatively small peptide faces both sides. Recently, Ohta and others in our laboratory found that trypsin attacks Na^+, K^+-ATPase only from outside, but this is not the case with insensitive ATPase, while both activities decreased markedly when trypsin was applied to ghosts. These results also supported the idea that a fairly large surface area of Na^+, K^+-ATPase is exposed on the outer surface of the membranes. Polar modification of the cell membrane enzyme from one side of the membrane is expected to be a useful tool for the elucidation of orientation, posture, shape and behavior of the enzymes.

REFERENCES

1. L. J. ϕrgensen, J. C. Skou, and K. P. Somonsen, *Biochim. Biophys. Acta*, **233**, 381 (1970).
2. D. W. Towle and J. H. Copenhaver, Jr., *Biochim. Biophys. Acta*, **203**, 124 (1970).
3. S. Uesugi, N. C. Dulak, J. F. Dioxon, T. D. Henum, J. L. Dahl, J. F. Pendus, and L. E. Hokin, *J. Biol. Chem.*, **246**, 531 (1971).
4. J. Kyte, *J. Biol. Chem.*, **246**, 4157 (1971).
5. A. Atkinson, A. G. Gatenly, and G. Lowe, *Nature*, **233**, 145 (1971).
6. P. D. Swanson, H. F. Bradford, and H. McIlwain, *Biochem. J.*, **92**, 325 (1964).
7. F. Medzihradsky, M. H. Kline, and L. E. Hokin, *Arch. Biochem. Biophys.*, **121**, 311 (1967).
8. J. C. Skou, *Biochim. Biophys. Acta*, **58**, 314 (1962).
9. T. Nakao, N. Nagano, K. Adachi, and M. Nakao, *Biochem. Biophys. Res. Commun.*, **13**, 444 (1963).
10. T. Nakao, Y. Tashima, K. Nagano, and M. Nakao, *Biochem. Biophys. Res. Commun.*, **19**, 755 (1965).
11. K. Weber and M. Ostorn, *J. Biol. Chem.*, **244**, 4406 (1969).
12. K. Nagano, T. Kanazawa, N. Mizuno, Y. Tashima, T. Nakao, and M. Nakao, *Biochem. Biophys. Res. Commun.*, **19**, 759 (1965).
13. K. Nagano, N. Mizuno, M. Fujita, Y. Tashima, T. Nakao, and M. Nakao, *Biochim. Biophys. Acta*, **143**, 239 (1967).
14. N. Mizuno, K. Nagano, T. Nakao, Y. Tashima, M. Fujita, and M. Nakao, *Biochim. Biophys. Acta*, **168**, 311 (1968).
15. R. G. Martin and B. N. Ames, *J. Biol. Chem.*, **236**, 1372 (1961).

16. M. Nakao, K. Nagano, T. Nakao, N. Mizuno, Y. Tashima, M. Fujita, H. Maeda, and H. Matsudaira, *Biochem. Biophys. Res. Commun.*, **29**, 588 (1967).
17. G. R. Kepner and R. L. Macy, *Biochem. Biophys. Res. Commun.*, **30**, 582 (1968).
18. H. Ohta, J. Matsumoto, K. Nagano, M. Fujita, and M. Nakao, *Biochem. Biophys. Res. Commun.*, **42**, 1127 (1971).
19. M. S. Bretscher, *J. Mol. Biol.*, **58**, 775 (1971).
20. M. S. Bretscher, *Nature*, **321**, 229 (1971).
21. W. W. Bender, H. Garan, and H. C. Berg, *J. Mol. Biol.*, **58**, 783 (1971).
22. Y. Asano, Y. Tashima, H. Matsui, K. Nagano, and M. Nakao, *Biochim. Biophys. Acta*, **219**, 169 (1970).
23. Y. Asano and M. Nakao, unpublished.
24. S. Dutta and B. H. Barks, *J. Pharmacol. Exp. Therap.*, **164**, 10 (1968).
25. E. Cox and S. E. Wright, *J. Pharmacol. Exp. Therap.*, **126**, 117 (1959).

Active Site of Phosphorylation of Na^+, K^+-ATPase

Robert L. Post and Betty Orcutt

Department of Physiology, Vanderbilt University Medical School, Nashville, Tennessee, U.S.A.

The active site of Na^+ and K^+ transport adenosine triphosphatase is identifiable at present only by a labile radioactive phosphate group attached to it. Quantities are insufficient for purification and amino acid analysis. Nevertheless, the electrophoretic mobility and chemical reactivity of overlapping [^{32}P]-phosphopeptides released by proteolytic digestion from the denatured phosphoenzyme have allowed tentative identification of an active site phosphotripeptide and the relative location of three marker groups in the neighborhood. The phosphotripeptide was released by limit digestion with pronase or by digestion of peptic peptides with trypsin and leucine aminopeptidase. It was sensitive to acetylation before digestion and to periodate oxidation after digestion. The tentative structure of the tripeptide is -(serine or threonine)-phosphoaspartate-lysine-. The aspartate residue was identified by comparison of the natural phosphotripeptide with two analogous synthetic phosphotripeptides with respect to sensitivity to digestion by carboxypeptidase B, isoelectric point, and pH hydrolysis profile. In each case the synthetic phosphoaspartyl tripeptide differed from the corresponding synthetic phosphoglutamyl tripeptide and resembled the natural phosphotripeptide from the enzyme.

The Na^+ and K^+ transport system converts the chemical energy of hydrolysis of the terminal phosphate bond of ATP into the energy of concentration gradients of Na^+ and K^+ in electrochemical disequilibrium across the plasma membrane of animal cells. It pumps Na^+ outward and K^+ inward.

At moderate to high rates of transport the stoichiometry is 3 Na^+ per 2 K^+ per ~P. Intracellular Mg^{2+} is required, and extracellular cardioactive steroids and erythrophleum alkaloids are inhibitory. Other monovalent inorganic cations substitute for K^+ but not Na^+; other nucleotides substitute only poorly for ATP (except dATP). In preparations of broken membranes the activity appears as an Na^+, K^+-dependent ATPase which is inhibited by ouabain, the most commonly used cardioactive steroid. Na^+, K^+-ATPase is very tightly bound to the membrane; it requires phospholipid for activity.

Na^+, K^+-ATPase acts in at least 2 steps. In the presence of Na^+ and Mg^{2+} it accepts the terminal phosphate group of ATP to form a phosphoenzyme. In a second step, addition of K^+ accelerates hydrolysis of the phosphoenzyme. When solubilized, this phosphoprotein has a molecular weight of about 100,000, and the phosphate is attached to a carboxyl group in a mixed anhydride linkage. This bond is that of an acyl phosphate, which would be a high-energy bond in a small molecule free in solution. Within the enzyme this bond at a single active site appears to have 2 energy levels, which may depend on the conformation of the surrounding active center. In 1 conformation the phosphate group reacts reversibly with ADP to form ATP; in the other conformation it reacts reversibly with water to form P_i. In each conformation the phosphorylation kinetics are modified by ligands such as inorganic cations, ouabain, or oligomycin. When either reactive form of the [^{32}P]-phosphoenzyme is denatured, digested with proteolytic enzymes and subjected to electrophoresis and radioautography, then identical fingerprints have been found (see Refs. *1–5*).

Previous Work on the Active Site

The enzyme-phosphate bond at the active site has been characterized as an acyl phosphate on the basis of its pH hydrolysis profile, sensitivity to dephosphorylation by hydroxylamine, acyl phosphatase or molybdate, and by its phosphorylation of alcohols in acid solution (*6–8*). Kahlenberg *et al.* (*9*) favored specifically the γ-carboxyl group of a glutamyl residue on the basis of an *n*-propylhydroxamate derivative which they obtained preferentially, but in low yield, from the phosphoenzyme with the dephosphoenzyme taken as a control. A proteolytic digest of the phosphoenzyme was treated with *n*-propylhydroxylamine to form the hydroxamate derivative.

Proteolytic digestion of the denatured [^{32}P]-phosphoenzyme releases various [^{32}P]-phosphopeptides overlapping the active site. All of these contain a dicarboxylic amino acid and a basic amino acid other than histidine. Some of them also contain a cysteine (*10, 11*).

Current Work on the Active Site

1. Technical difficulties

At present the phosphoenzyme is available only in impure preparations in pico- to nanomole quantities and has been identifiable only by a radioactive phosphate group attached only by a labile bond. At O°C and optimal pH (about 3), the loss of the active site phosphate from the denatured enzyme is about 1% per hr.

2. Fingerprints of phosphopeptides

In order to characterize the active site we have applied paper electrophoresis to [^{32}P]-phosphopeptides overlapping the active site and have studied their radioactive fingerprints. The peptides were released from the denatured [^{32}P]-phosphoenzyme by digestion with pronase or pepsin (*4, 10*). Large peptides appeared early and smaller ones appeared later during digestion. In naming the peptides we have assigned larger numbers to the smaller peptides produced by each enzyme. Some peptides were sensitive to treatment with acetic anhydride, periodate, or trypsin. From these sensitivities we tentatively deduced part of the structure of a limit phosphotripeptide. On the basis of the partial structure we synthesized congeners of the natural tripeptide. Comparison of the natural and synthetic peptides with respect to sensitivity to carboxypeptidase B has probably identified the active site itself as the β-carboxyl group of an aspartic acid residue.

3. A lysine residue

All phosphopeptides overlapping the active site move during electrophoresis toward the cathode at pH 2 and consequently are positively charged and contain a basic amino acid (*10, 11*). The basic amino acid was identified as a lysine by its sensitivity to acetylation before digestion. Histidine was excluded earlier (*10*). Arginine is not acetylated. Acetylation of the ϵ-amino group of lysine neutralizes its positive charge. Acetylation prior to digestion reduced the electrophoretic mobility of all small peptides to zero at pH 2. At pH 4.2 peptides released by digestion with pronase were separated and compared with their singly and doubly acetylated derivatives produced by acetylation after digestion. Acetylation prior to digestion pro-

TABLE I
Acetylation of the Phosphoenzyme Near the Active Site

Phosphopeptides	Electrophoretic mobility (cm)			
	Control	Acetylated		
		Prior to digestion	After digestion	
			Singly	Doubly
Pr1	0	−5.2	—	—
Pr2	−1.3	−8.6	−10	−17
Pr3	−3.4	−12.5	−15	−25
Pr4	−4.1	−13.7	−17	−26
P_i	−37	—	—	—

The [^{32}P]-phosphoenzyme was denatured with acid, and part of it was acetylated with acetic anhydride at pH 7 in a solution of 4 M guanidine·HCl dissolved in saturated sodium acetate. Acetylated and untreated denatured protein were washed and digested with 1/5 or 5-fold their weight of pronase in 0.2 M K_2HPO_4 at 0°C for 1.75 hr. The resulting soluble [^{32}P]-phosphopeptides were electrophoresed on paper at pH 4.2 and 2°C for 2 hr at 80 V/cm. The phosphopeptides were located by autoradiography and designated as in Fig. 20 of Ref. (*4*). The low dose of pronase produced mostly Pr2 and Pr3; Pr1 was faint; the high dose produced mostly Pr3 and Pr4. The numbers below are distances in cm from the origin toward the cathode. For a given peptide the greater the distance the less the positive charge and the fewer the number of free amino groups. Acetylation prior to digestion produced only singly acetylated peptides. When acetylation was performed after digestion, both singly and doubly acetylated phosphopeptides were found.

duced only singly acetylated derivatives so that the α-amino groups of these peptides were blocked prior to digestion (Table I). (Further evidence for a block of peptides, Pr2, Pr3, and Pr4, is the position of the cysteine, which these peptides lack (*4*, *10*), farther out on the amino-terminal side of the active site, as shown later.)

The lysine residue was positioned on the carboxy-terminal side of the active site phosphate by the response of peptic peptides to digestion with trypsin. Electrophoresis at pH 2 showed that tryptic digestion produced from the peptic peptides, P5 and P6, a new phosphopeptide, P-T, which still moved toward the cathode and contained the lysine (Fig. 1). Trypsin cuts peptides on the carboxy-terminal side of lysine or arginine residues. Consequently the active site phosphate is on the amino-terminal side of the lysine.

4. *A serine or threonine residue*

Digestion to the limit with pronase yields Pr4, which has a molecular weight of about 400 according to molecular sieving on Bio-Gel P2 (*11*). This value

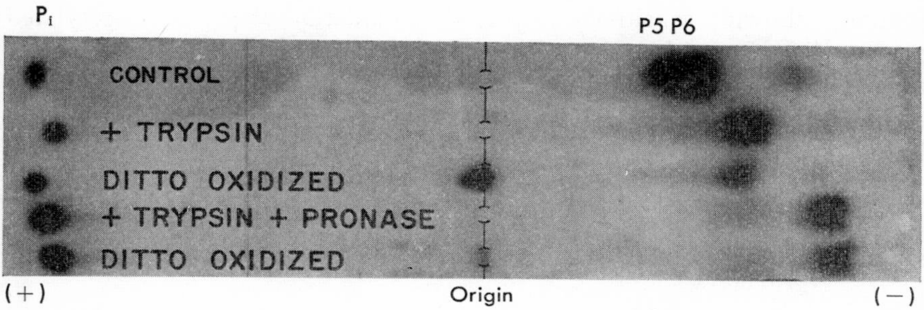

FIG. 1. Electropherogram of [^{32}P]-phosphopeptides overlapping the active site of Na^+,K^+-ATPase and obtained by sequential digestion with pepsin, trypsin, and pronase. First, 0.25 ml of a peptic digest (*10*) was incubated with 5 μl of 1 M 2-mercaptoethanol, 12.5 μl of 10% (w/v) NH_4HCO_3, and 1 mg of trypsin for 4 hr at 0°C. Then 0.6 mg of pronase was added and incubation was continued for 1.5 hr at 0°C. Performic oxidation (*10*) of the tryptic digest was incomplete. Samples of each digest were electrophoresed at pH 2 for 160 V hr $(cm)^{-1}$. The length of the paper shown is 53.4 cm. P5 and P6 are the smallest peptic peptides; P-T (not designated), produced from them by digestion with trypsin moved a little faster; Pr4 (not designated), produced from P-T by digestion with pronase, moved fastest of all.

TABLE II

Treatment with Periodate of [^{32}P]-Phosphopeptides Produced by Digestion with Pronase of Phosphorylated, Denatured Na^+,K^+-ATPase

[^{32}P]-phosphopeptide	Electrophoretic mobility (cm)	
	Control	Oxidized
Pr3	21.5	−0.3
Pr4	23.5	−0.3
P$_i$	−22.5	−22.5

Ten μl of a pronase digest of the phosphoenzyme (*10*) was incubated with 1 μl of glacial acetic acid and 2 μl of 1 M sodium metaperiodate in the dark for 2 hr at 23°C. The samples were electrophoresed on paper at pH 2 and 2°C for 160 V hr $(cm)^{-1}$ and autoradiographed. For control spots the sodium periodate solution was replaced with water. The mobility is taken with respect to caffeine, which was used as a neutral marker to indicate endosmosis. Positive mobility is toward the cathode.

suggests a tripeptide. Periodate oxidation of Pr4 removed 1 positive charge (Table II). Periodate oxidizes vicinal hydroxyl groups and vicinal hydroxyl and amino groups and so in peptides it is specific for amino-terminal serine or threonine or for hydroxylysine. Hydroxylysine was excluded by an experiment showing sensitivity to periodate oxidation of peptides acetylated prior to digestion with pronase. The amino-terminal position of the serine or threonine clearly positioned the active site acidic amino acid between the serine or threonine and the lysine. The phosphotripeptide, Pr4, appeared to have the structure:

$$\begin{pmatrix}\text{Thr}\\ \text{Ser}\end{pmatrix}\text{-X(P)-Lys}\,,$$

where X is Asp or Glu.

5. A phosphotripeptide

We compared the limit phosphopeptide, Pr4, with synthetic analogs, [^{32}P]-Pro-Glu(P)-Lys and [^{32}P]-Pro-Asp(P)-Lys. (The use of proline, in place of serine or threonine, was not optimal but was the only possibility at the time.) Electrophoresis at pH 2 showed the same mobility for all 3 phosphopeptides. pH 2 is sufficiently far from the p*K*'s of these peptides (*11*) that mobility should depend only on differences in molecular size and structure rather than on differences in electric charge. We concluded that Pr4 is a phosphotripeptide. We compared the isoelectric points of the 3 phosphopeptides by interpolation to zero mobility between pH 3.5 and pH 4.0. Those of Pr4 and Pro-Asp(P)-Lys were the same, namely 3.6; that of Pro-Glu(P)-Lys was different, namely 3.8. The p*K*'s of the terminal carboxyl group and the secondary ionization of the acyl phosphate are near these values (*11*). We considered that the difference in the acidic amino acid was probably responsible for the difference in the isoelectric points and that the active site phosphate was likely to be on an aspartyl residue. Furthermore, between pH 4 and pH 7 the rate of hydrolysis at 37°C was 2-fold greater for the aspartyl synthetic peptide than for the glutamyl synthetic peptide; Pr4 was almost the same as the aspartyl peptide between pH 0 and pH 12. Electrophoresis at pH 8 showed that the mobilities of the aspartyl and glutamyl synthetic phosphopeptides were respectively 14.4 and 12.0 cm toward the anode compared with 24.9 cm for that of Pr4. (P_i moved 46.8 cm.) This pH is in the range of the p*K*'s of α-amino groups of peptides (*11*). Clearly the amino-terminal proline was incorrect.

6. Aspartyl phosphate at the active site

A carboxy-terminal lysine should be split from a peptide by carboxypeptidase B. Loss of the positive charge of the lysine should be easily detected by paper electrophoresis. However, Pr4 was highly, but not completely, resistant to carboxypeptidase B. We tested our preparation of carboxypeptidase B and our reaction mixture on the 2 synthetic phosphotripeptides. Carboxypeptidase B easily attacked Pro-Glu(P)-Lys but had no detectable effect on Pro-Asp(P)-Lys, nor on Pr4 (Fig. 2). After dephosphorylation of Pro-Asp(P)-Lys, carboxypeptidase B had no difficulty in

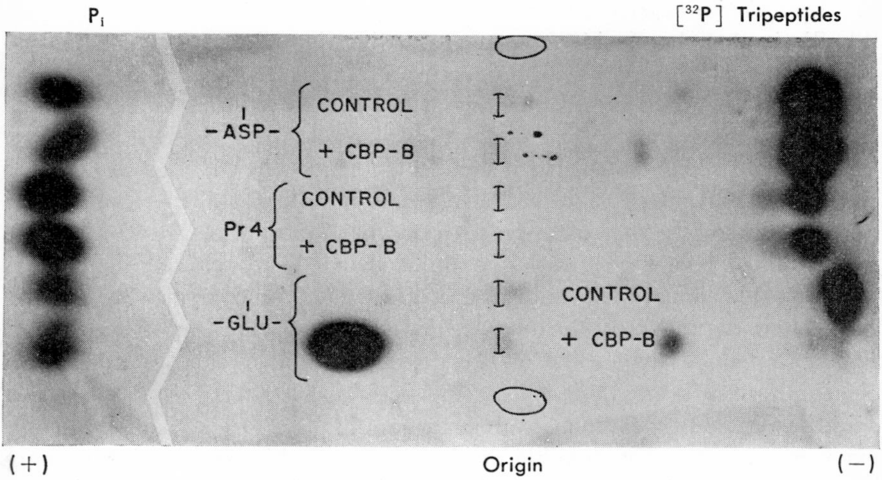

Fig. 2. Comparison of the sensitivity to carboxypeptidase B of the limit [^{32}P] phosphopeptide, Pr4, with those of synthetic [^{32}P]-phosphotripeptide congeners. [^{32}P]-Pro-Asp(P)-Lys and [^{32}P]-Pro-Glu (P)-Lys were synthesized by modifications of the method of Nishimura *et al.* (*12*) from the corresponding tripeptides having their amino groups blocked with carbobenzoxy groups and with their terminal carboxyl groups blocked as benzyl esters. The blocked tripeptides were purchased from Miles Laboratories and International Chemical and Nuclear Corporation, respectively. The yield of phosphopeptide was 8 to 32% of the initial amount of blocked tripeptide. Ten μl of a solution of each phosphopeptide was incubated with 2 μl of 1.3 M NH_4NCO_3 and 5 μl of 0.1 M NaCl with or without 5 μg of carboxypeptidase B. Incubation for 15 min at 23°C was stopped with 0.5 μl of formic acid. Samples were electrophoresed at pH 2.9 as in Fig. 1. -ASP- and -GLU- indicate the synthetic aspartyl and glutamyl phosphotripeptides, respectively. +CBP-B indicates treatment with carboxypeptidase B. The open circles at the origin show the position of caffeine as a neutral marker of endosmosis.

releasing lysine from the dephosphotripeptide. We concluded that phosphorylation of the penultimate aspartyl residue inhibited the action of carboxypeptidase B but phosphorylation of a corresponding glutamyl residue did not. We considered that this enzymatic reaction was a sensitive discriminator between these 2 amino acids. Finally we concluded that the amino acid sequence around the active site of Na^+, K^+-ATPase is probably:

$$\text{-}\begin{pmatrix}\text{Thr}\\ \text{Ser}\end{pmatrix}\text{-Asp(P)-Lys-.}$$

7. *Dansylation experiments*

The limit phosphopeptide from pronase digestion, Pr4, was made fluorescent by dansylation of its amino groups (*13*). The dansylated peptide was purified by repeated electrophoresis at pH 2, before and after loss of the labile phosphate group. Loss of the phosphate doubled the electrophoretic mobility and separated this peptide from all others. At pH 4.38 the isolated dansylated natural peptide had an electrophoretic mobility close to, but not identical with, that of any of 4 possible doubly dansylated synthetic dipeptides of lysine with an acidic amino acid. We concluded that it was itself doubly dansylated and not a dipeptide. Acid hydrolysis of the dansylated natural peptide followed by paper electrophoresis at pH 4.38 demonstrated ε-dansyllysine, as expected, and dansyl-free acid. In this electrophoresis system, dansyl serine is easily obscured by dansyl-free acid, but dansyl threonine, if present, should have been visible. Further experiments were not possible at the time. On this basis serine is more likely than threonine to be the amino-terminal amino acid in Pr4.

8. *Isolation of phosphopeptides at neutral pH*

It is always desirable to test the denaturation procedure which isolates a peptide at an active site. In order to find out if acid denaturation of the enzyme produced the acyl phosphate bond, the enzyme was denatured at neutral pH with a volume of cold 160 mM Tris dodecyl sulfate equal to that of the reaction mixture followed 10 sec later by 13 volumes of cold acetone: Water 2: 1 (v/v). The denatured membranes were washed twice with more cold acetone: water and then digested with pronase. Radioactive peptides were obtained after electrophoresis which had the same mobility as Pr3 and Pr4 at both pH 2 and pH 6. These phosphopeptides were as sensitive to hydroxylamine as were Pr3 and Pr4. It is unlikely that an artifact is intro-

TABLE III
Electrophoretic Mobilities of Peptic and Peptic-tryptic Phosphopeptides

Phosphopeptide	Electrophoretic mobility (cm)	
	pH 2	pH 6
P6	+11.9	−6.2
P5	+10.2	−5.7
P4	+17.1	+1.5
P3	+13.1	−2.3
P2	+20.6	+5.9
P-T	+13.8	−7.3
P3-T	+8.7	−9.2
P_i	−24.1	−37.0

Four mg of washed, denatured phosphoenzyme was digested with 1 mg of pepsin in 0.01 M HCl for 30 min at 23°C. After centrifugation 0.35 ml of the supernatant was added to 0.17 ml of 40 mM N-ethylmaleimide in 0.1 M histidine. The incubation lasted 15 min at 23°C and protected the sulfhydryl group from side reactions at neutral pH. A sample was saved for electrophoresis of the peptic peptides. To the remainder were added 23 μl of 10% (w/v) NH_4HCO_3 and 0.5 mg of trypsin. Digestion lasted about 3 hr at 0°C. For electrophoresis at pH 2 the buffer was 2% (v/v) formic acid. For electrophoresis at pH 6 the buffer was 30 mM succinic acid and 56 mM Tris. The symbols P2-P6 refer to peptic peptides (*10*). P-T refers to the peptic-tryptic peptide found in other experiments when only P4-P6 were present before tryptic digestion. P3-T refers to the peptic-tryptic peptide found in other experiments only when P3, with or without P2, was present before tryptic digestion. Mobility toward the cathode is positive; the distances are from the origin.

duced by exposure of the native phosphoenzyme to acid. This experiment extends a similar experiment by Nagano *et al.* (*6*).

9. *Three marker groups near the active site*

There is a cysteine in all the peptic peptides and also in the largest peptide produced by pronase, Prl (*4, 10*). The cysteine is easily detected by sensitivity of the phosphopeptide to oxidation with performic acid. The peptic-tryptic peptide, P-T, was sensitive to performic acid (Fig. 1) and so contains the cysteine on the amino-terminal side of the active site. If electrophoretic mobility is proportional to the 0.6th power of molecular weight (*14*) and if Pr4 has a molecular weight of 428, then P-T has a molecular weight of 691 approximately, and so should be a hexapeptide, give or take a residue. We think the cysteine may be amino-terminal in P-T.

There is an additional basic residue in peptic peptide, P4, and in the

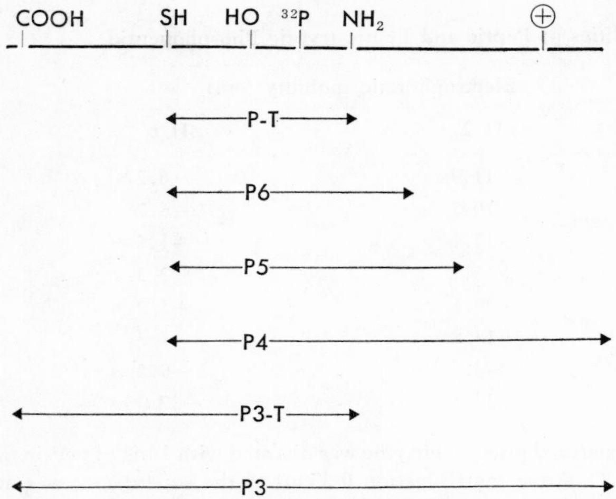

Fig. 3. Relative positions of marker groups on peptic and peptic-tryptic peptides overlapping the active site of phosphorylation of Na^+, K^+-ATPase. ⊕ indicates a basic amino acid. The double-headed arrows indicate the positions of peptic peptides, P3 to P6, and peptic-tryptic peptides, P3-T and P-T. The distance between the points of each arrow is approximately proportional to the molecular weight as estimated from the electrophoretic mobility at pH 2 (*14*).

larger peptic peptide, P3. (P2 is even more basic.) After performic oxidation of the cysteine, the charge of this basic residue provides at pH 2 electrophoretic mobility toward the cathode equal to one-half of that found in the untreated peptide (*10*). Lacking this basic residue, oxidized P5 and P6 do not move. Without performic oxidation, the mobility of P4 is 1.7-fold greater than that of P5 (Table III). One would expect double the mobility from double the charge, but P4 is slowed a little by its greater size relative to P5. Since P4, P5, and P6 are all cut down to P-T by trypsin (Table III), the neutral moieties which distinguish P6 from P-T and P5 from P6 are positioned successively toward the carboxy-terminal end away from the active site. The moiety with the extra basic residue is still farther out (Fig. 3).

There is an additional acidic residue in P3 which is not found in P4. At pH 6, P4 moves toward the cathode but P3 moves toward the anode and so is more negatively charged. The p*K* of this acidic residue is between pH 6 and pH 2 since it is uncharged at pH 2. At pH 2, P3 moves toward the cathode more rapidly than P6 (Table III) in spite of its greater size. The extra positive charge contributed by P4 is therefore still predominant and not neutralized by negative charge on the acidic residue. P3 is cut down to P3-T by trypsin. P3-T also contains this additional acidic residue. In comparison to P-T it runs more rapidly toward the anode at pH 6 (Table III), and so has more negative charge; on the basis of size alone it would run more slowly. A p*K* between pH 2 and pH 6 suggests a carboxyl group (Fig. 3).

10. Active site of the Ca^{2+}-ATPase of sarcoplasmic reticulum

Recently in this laboratory F. Bastide has performed all these experiments on the [^{32}P]-phosphoenzyme of the Ca^{2+}-ATPase. She found no difference in the pronase limit [^{32}P]-phosphopeptide, Pr4, from the 2 sources. Peptides larger than the peptic peptide, P6, showed clear differences.

Acknowledgments

The authors thank Frances Gonsoulin and Jean Sorrells for performing some of the experiments. The authors thank Richard Luben, Jerelyn Pitterman, Mary Jean Morris, Marianne Talbert, Jackie Waynick, and James Brineaux for preparing the membranes which contained the Na^+, K^+-ATPase.

This work was supported by a grant, No. 5R01 HE-01974, from the National Heart and Lung Institute, N.I.H., U.S.P.H.S.

REFERENCES

1. R. Whittam and K. P. Wheeler, *Ann. Rev. Physiol.*, **32**, 21–60 (1970).
2. J. C. Skou, *in* "Current Topics in Bioenergetics," ed. by D. R. Sanadi, Academic Press, New York, Vol. 4, pp. 357–398 (1971).
3. G. J. Siegel and R. W. Albers, *in* "Handbook of Neurochemistry," ed. by A. Lajtha, Plenum Press, New York, Vol. 4, pp. 13–44 (1970).
4. R. L. Post, S. Kume, T. Tobin, B. Orcutt, and A. K. Sen, *J. Gen. Physiol.*, **54**, 306s–326s (1969).
5. R. L. Post, S. Kume, and F. N. Rogers, *in* "Mechanisms in Bioenergetics," ed. by G. F. Azzone, Academic Press, New York (1973).
6. K. Nagano, T. Kanazawa, N. Mizuno, Y. Tashima, T. Nakao, and M. Nakao, *Biochem. Biophys. Res. Commun.*, **19**, 759–764 (1965).
7. L. E. Hokin, P. S. Sastry, P. R. Galsworthy, and A. Yoda, *Proc. Natl. Acad. Sci. U.S.*, **54**, 177–184 (1965).
8. H. Bader, A. K. Sen, and R. L. Post, *Biochim. Biophys. Acta*, **118**, 106–115 (1966).

9. A. Kahlenberg, P. R. Galsworthy, and L. E. Hokin, *Arch. Biochem. Biophys.*, **126**, 331–342 (1968).
10. H. Bader, R. L. Post, and D. H. Jean, *Biochim. Biophys. Acta*, **143**, 229–238 (1967).
11. D. H. Jean and H. Bader, *Biochem. Biophys. Res. Commun.*, **27**, 650–654 (1967).
12. J. S. Nishimura, E. A. Dodd, and A. Meister, *J. Biol. Chem.*, **239**, 2553–2558 (1964).
13. W. R. Gray, *in* "Methods in Enzymology", ed. by C. H. W. Hirs, Academic Press, New York, Vol. 11, pp. 139–151 (1967).
14. W. R. Gray, *in* "Methods in Enzymology", ed. by C. H. W. Hirs, Academic Press, New York, Vol. 11, pp. 469–475 (1967).

ATP as a Modulator of Na^+, K^+-ATPase

Kei Nagano,* Yoko Fujihira, Yukichi Hara, and Makoto Nakao

*Department of Biochemistry, Tokyo Medical and Dental University School of Medicine, Tokyo, Japan, and Laboratory of Biology, Jichi Medical School, Tochigi, Japan**

Modulation of K^+-*p*-nitrophenylphosphatase (pNPPase) and Na^+, K^+-ATPase activities and their sensitivity to ouabain by ATP have been investigated using partially purified, but highly specific, Na^+, K^+-ATPase preparations. ATP increased K_m for K^+ in the case of Na^+,K^+-inosine triphosphatase (ITPase) and decreased K_i for ouabain in the cases of both Na^+,K^+-ITPase and K^+-pNPPase. Cytidine triphosphate (CTP) also appeared to have these modulating effects to some extent, but ITP did not. It was suggested that this property of ATP may be of special importance in relation to active transport of Na^+ and K^+ across the cell membrane.

p-Nitrophenylphosphate (pNPP), acetyl phosphate, and a limited number of nucleoside triphosphates substitute for ATP as substrates of Na^+, K^+-ATPase (*1*, *2*). The rate of hydrolysis of these substrates is much slower than that of ATP, but the activities are definitely dependent on the existence of Na^+ and K^+ and are inhibited by ouabain, so the identity of the molecular mechanism of these activities is generally accepted.

Various effectors have been reported to modulate these cation-dependent activities, and the "allosteric" nature of these modifications has been suggested. Cations, *e.g.*, Na^+, K^+ or Mg^{2+} in ATP hydrolysis (*3*), or Na^+ in ouabain binding (*4*, *5*), behave as modulators of the enzyme. In K^+-dependent pNPPase and acetylphosphatase activities, complex modulation by cations and/or ATP or cytidine triphosphate (CTP), added separately or in combination, has been the subject of several investigations (*6–9*). ATP may also have a modulating role in the case of Na^+, K^+-ATPase activity (*3, 10, 11*), but analysis of this effect is complicated by the fact that the substrate and the modulator are the same molecule.

CTP is hydrolyzed at a considerable rate, but it shares the modulating property with ATP, as suggested by its effect on pNPPase and acetylphosphatase activities (*6, 7, 9*) and on ouabain inhibition of these activities (*8*). We found, in accord with several reports (*12, 13*), that inosine triphosphate (ITP) was hydrolyzed by this enzyme (*10*). This substrate appeared to be devoid of modulating effect. Hence, studies of the effect of ATP on Na^+, K^+-ITPase (and *vice versa*) and the effect of ATP and ITP on K^+-pNPPase were expected to be informative. Some of the results obtained during this line of investigation will be described.

K_m *Values of* Na^+*,* K^+*-ATPase, -ITPase, and* K^+*-pNPPase for* Na^+ *and* K^+

The enzyme preparations we used were NaI-treated pig brain microsomal fractions (*14*). One of the advantages of NaI treatment was that the concomitant Mg^{2+}-ATPase (Na^+, K^+-independent ATPase) activity in the microsomal fraction was essentially removed. Thus, accurate kinetic analysis was possible in cases such as Na^+, K^+-ITPase, where the activity was rather low (Table I).

TABLE I
Specificity Range of the Na^+, K^+-ATPase Preparation

Activity (substrate)	Specific activity in the presence of: (μmoles/mg protein/hr)			
	NaCl (140 mM) +KCl (14 mM)	NaCl (140 mM)	—	Ouabain (0.1 mM)
ATPase (ATP: 3 mM)	98.0	2.1	0.4	0.3
ITPase (ITP: 3 mM)	9.5	3.4	0.9	0.8
CTPase (CTP: 3 mM)	15.8	3.1	0.8	0.9
GTPase (GTP: 3 mM)	2.9	1.7	—	0.4
pNPPase (pNPP: 18 mM)	(13.5)[a]	—	—	(1.0)[b]

NaI-treated pig brain microsomes (*14*) were used as enzyme preparation. Standard assay mixture contained 50 mM Tris-HCl buffer (pH 7.8), 5 mM $MgCl_2$, 0.5 mM Tris-glycoletherdiaminetetraacetic acid (GEDTA), and Na^+, K^+, substrates, and ouabain as indicated. Reactions were performed with enzyme concentrations appropriate for hydrolyzing 0.1–0.5 μmoles of the substrate during 30-min incubation period at 37°C.
[a] KCl: 18 mM. [b] Ouabain: 1.0 mM.

Sufficient concentrations of substrates and cations were used unless otherwise indicated.

K_m of the enzyme for ITP (0.19 mM) was of the same order as for ATP

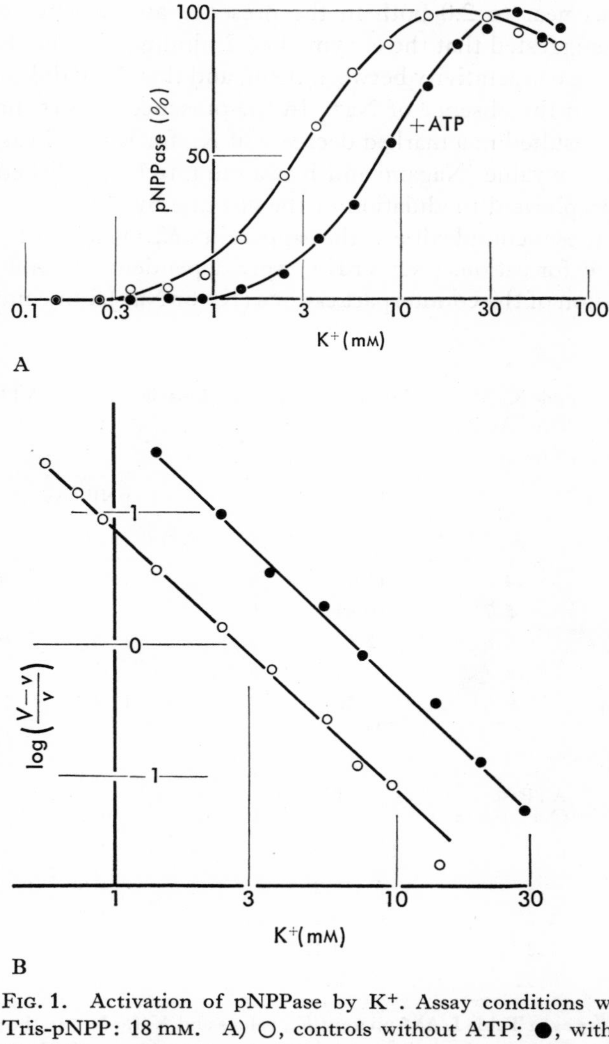

FIG. 1. Activation of pNPPase by K^+. Assay conditions were as in Table I. Tris-pNPP: 18 mM. A) ○, controls without ATP; ●, with 1 mM Tris-ATP. The highest values in each series were expressed as 100%. B) Hill's plot of the data shown in (A). n=1.98 for ○; n=2.03 for ●.

(0.062 mM). K_m for pNPP was rather high (2.0 mM), and remained the same at different K^+ concentrations (K^+=14 mM or 5.6 mM).

K_m for K^+ in K^+-pNPPase was also high, and shifted even higher when ATP was added to the system without Na^+ (Fig. 1,A). However, the *n* value of Hill's plot was near to 2.0 both in the presence and absence of ATP (Fig. 1,B). This suggested that the enzyme had 2 binding sites for K^+, that there existed strong cooperativity between them, and that ATP did not affect the cooperativity in the absence of Na^+. In the presence of Na^+, however, addition of ATP resulted in a marked decrease of K_m for K^+ (*6, 7*) as well as modification of the *n* value (Nagano and Kawamura, to be published). This illustrates the complicated modulation of the enzyme by ATP.

Apart from these complexities, the apparent K_m values of various activities (Table I) for cations (Na^+ or K^+) were dependent, to some extent, on the concentration of the counterpart cations (K^+ or Na^+), but comparison

TABLE II
Comparison of K_m and K_i Values for Monovalent Cations in Na^+,K^+-ATPase and -ITPase, and K^+-pNPPase Activities

A) K_m or K_i for Na^+ (mM)

Ion present	ATPase		ITPase		pNPPase	
	(mM)	K_m (Na^+)	(mM)	K_m (Na^+)	(mM)	K_m (Na^+)
Cs^+	(40)	14	(10)	6.1	(40)	20
NH_4^+	(26)	5.7	(6.4)	6.8		
K^+	(14)	8.8	(2.6)	8.2	(14)	16
Rb^+	(16)	12	(0.64)	8.2		
Tl^+	(1.6)	9.0	(0.16)	7.0	(1.6)	24

B) K_m for cation (mM)

Ion present	ATPase (Na^+ 140 mM)	ITPase (Na^+ 100 mM)	pNPPase (without Na^+)
Cs^+	10	1.3	30
NH_4^+	9.4	1.2	4.0
K^+	1.3	0.14	2.7
Rb^+	1.2	0.12	1.4
Tl^+	0.32	0.044	0.30

Assay conditions were as in Table I. A) Apparent K_m values and K_i for Na^+ in the presence of K^+ or K^+-equivalent ions at the concentration shown in parentheses. B) Apparent K_m values for K^+ in the presence of Na^+ at the concentration shown in parentheses.

of K_m values around the optimal conditions may be meaningful (Table II). The apparent K_m values of Na^+, K^+-ATPase and -ITPase for Na^+ were approximately the same. K^+-pNPPase was inhibited by Na^+, and the apparent K_i was similar to K_m values for Na^+ activation of nucleotide hydrolysis.

In contrast with this coincidence, the apparent K_m of Na^+, K^+-ITPase for K^+ was about 10 times smaller that of Na^+, K^+-ATPase. Rb^+, Cs^+, Tl^+ or NH_4^+ could substitute for K^+, and there existed similar differences between K_m values of ATPase and ITPase for these cations. This parallelism among different cations provides circumstantial evidence for the identity of the molecular mechanism in Na^+, K^+-ATPase and -ITPase activities.

Kinetics of Activation by Na^+ and K^+

Curves of activation of ATP and ITP hydrolysis by Na^+ in the presence of K^+, and those by K^+ in the presence of Na^+ are shown in Figs. 2 and 3. When we replot the data according to Hill's equation, the n value is generally not an integer but is a fractional value between 1 and 2. Moreover, Hill's plot did not always give straight lines but sometimes gave curved ones, when a wide range of cation concentration was covered (Figs. 2,A and 3,B; however, *cf.* Ref. *7*). Moreover, the concentrations of Na^+ and K^+ and their ratio affected the n value and the degree of curvature.

These facts may suggest some complicated "allosteric" interactions between cationic sites, but the interpretation is not conclusive since the activation and mutual inhibition by Na^+ and K^+ may, alternatively, be expressed by the following general form (*cf.* Refs. *3* and *15*):

$$v = \frac{V_{\max}}{[1+K_1/a\cdot(1+b/K_3)]^p[1+K_2/b\cdot(1+a/K_4)]^q}$$

This equation contains 2 dissociation constants, 2 inhibition constants, and 2 powers p and q, but no "allosteric" interactions are implied. If we assume the number of ions transported in one cycle of the reaction sequence ($Na^+=3$; $K^+=2$; Refs. *16* and *17*) and the number of ions necessary for enzyme activation are the same, p and q should be 3 and 2, respectively. This assumption, however, seems to be neither a necessary prerequisite nor endorsed by direct determinations yet.

Results with Na^+ and K^+ activation and inhibition are not in full accord

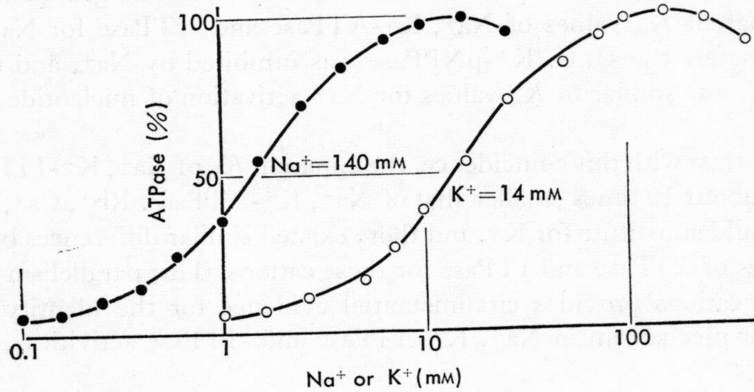

A

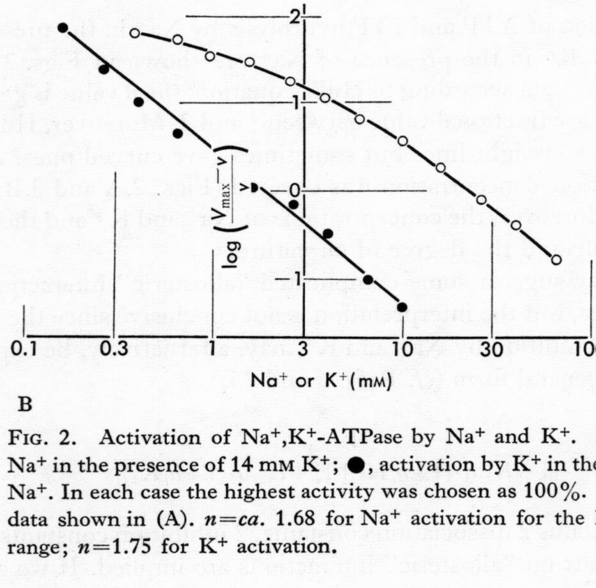

B

FIG. 2. Activation of Na^+,K^+-ATPase by Na^+ and K^+. A) ○, activation by Na^+ in the presence of 14 mM K^+; ●, activation by K^+ in the presence of 140 mM Na^+. In each case the highest activity was chosen as 100%. B) Hill's plot of the data shown in (A). n=*ca.* 1.68 for Na^+ activation for the higher concentration range; n=1.75 for K^+ activation.

among different laboratories. Some results reported on Na^+, K^+-ATPase activity (*15*) fit Hill's diagram with $n=2$ for Na^+ and $n=1$ for K^+, suggesting 2 sites for Na^+ and 1 for K^+, respectively, and a strong cooperative interaction between the Na^+ sites. Results obtained by us (Figs. 1 and 4) and some other investigators (*18*) indicate $n=2$ for K^+ activation and Na^+ inhibition

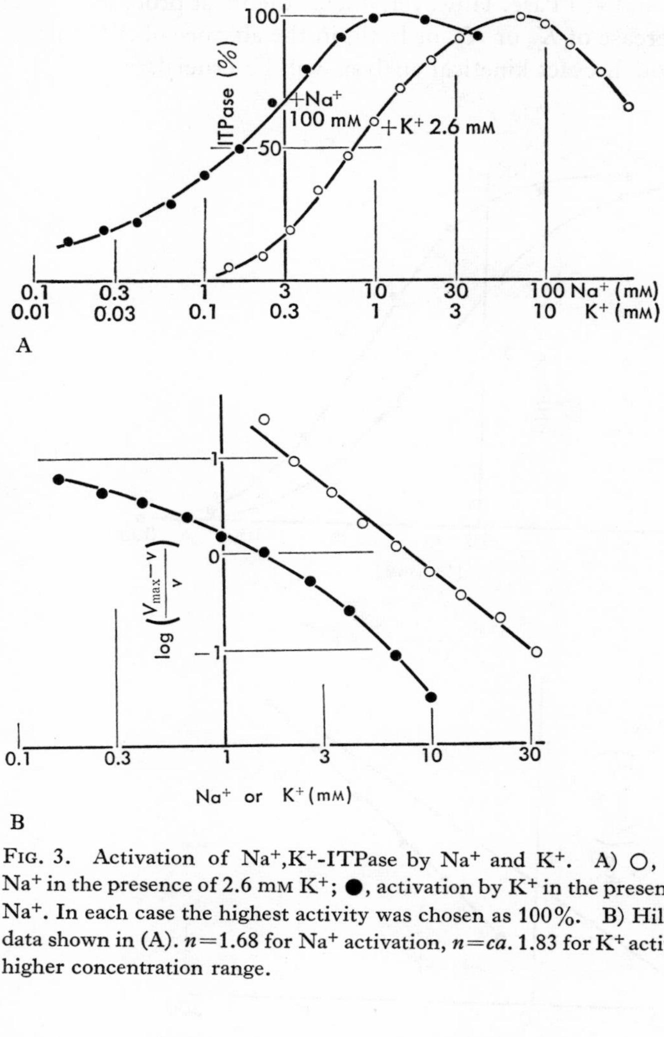

FIG. 3. Activation of Na^+,K^+-ITPase by Na^+ and K^+. A) ○, activation by Na^+ in the presence of 2.6 mM K^+; ●, activation by K^+ in the presence of 100 mM Na^+. In each case the highest activity was chosen as 100%. B) Hill's plot of the data shown in (A). n=1.68 for Na^+ activation, n=*ca.* 1.83 for K^+ activation for the higher concentration range.

of pNPPase activity, although in the latter case n was less than 2 when K^+ concentration was not high enough.

Thus we have different sets of data and several different possibilities for explaining the effects of cations and ATP. Consequently, this makes our interpretation inconclusive regarding the different K_m values for K^+ of Na^+,

K^+-ATPase and -ITPase. However, one of the most probable mechanisms may be a decrease of K_2 or K_3, or both, in the absence of ATP, if we adopt Eq. (1), although exact kinetical analysis will be crucial.

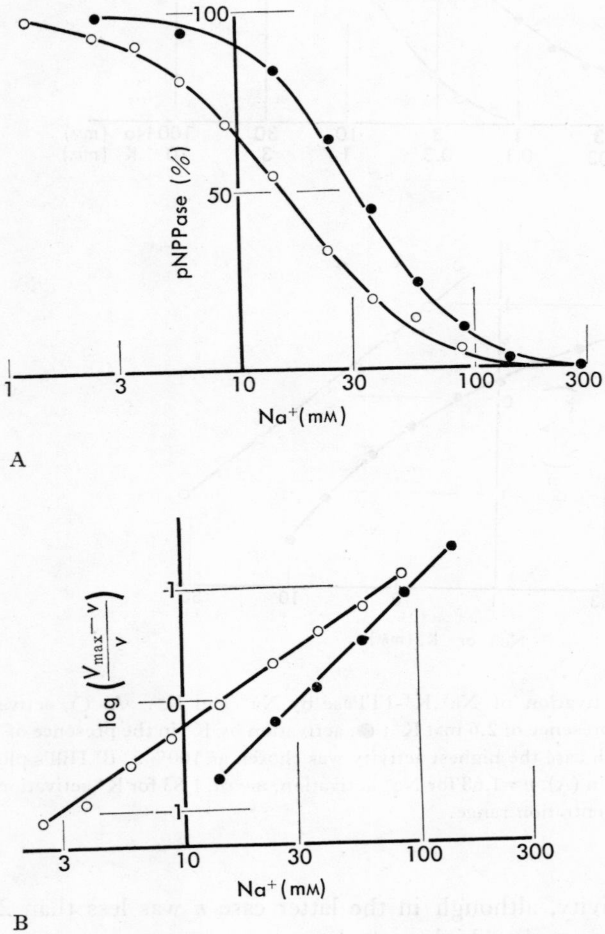

FIG. 4. Inhibition of K^+-pNPPase by Na^+. A) Inhibition curves by Na^+. Assay conditions were as in Table I. K^+=18 mM (●), or 5.6 mM (○). pNPP= 18 mM. B) Hill's plot of the data shown in (A). n=1.43 for ○, 2.04 for ●.

Insensitization for K^+ by ATP

From the tentative interpretation above, it was expected that K_m for K^+ in the Na^+, K^+-ITPase reaction would increase when ATP was added to the system. Thus, hydrolysis of $IT^{32}P$ was determined in the presence and

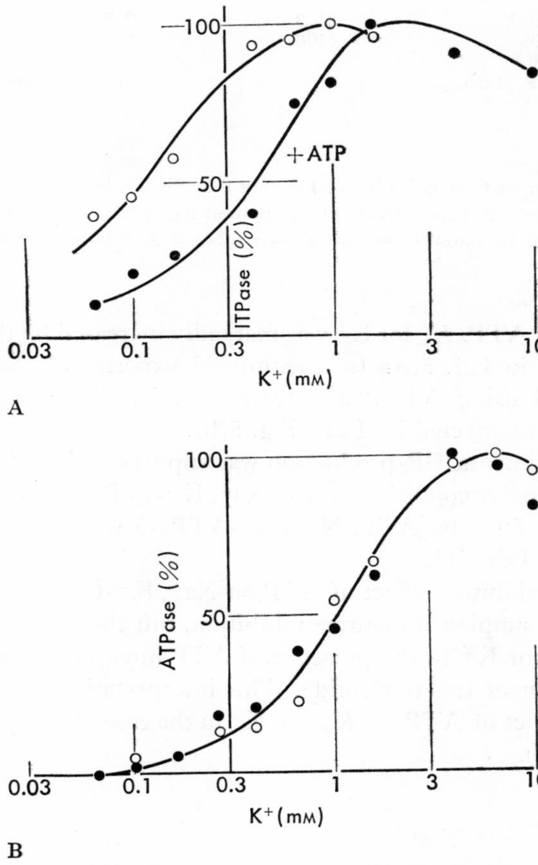

FIG. 5. Effect of ATP and ITP on K_m values for K^+ of Na^+,K^+-ITPase and -ATPase, respectively. A) Na^+,K^+-ITPase. Activation by K^+ without (○) and with (●) 1 mM Tris-ATP. K_m values were 0.11 mM and 0.43 mM, respectively. B) Na^+,K^+-ATPase. Activation by K^+ (○). K_m was 1.1 mM and was not affected by addition of 3 mM ITP (●). Substrates were 3 mM ITP or 1 mM ATP. Other assay conditions were the same as in Table I.

TABLE III
Reciprocal Inhibition of Na^+,K^+-ATPase or -ITPase by ITP or ATP, Respectively

Na+,K+-dependent activity (μmoles/mg protein/hr)			
ITPase (IT^{32}P : 3 mM)		ATPase (AT^{32}P : 1 mM)	
−ATP	+ATP (1 mM)	−ITP	+ITP (1 mM)
8.7 (100)	3.9 ; 3.3 (45 ; 38)	82.0 (100)	42.6 ; 45.9 (52 ; 56)
	+ATP (2 mM)		+ITP (2 mM)
	1.5 (17)		59.0 (79)

Assay conditions were similar to those in Table I except that ^{32}P-labeled ATP or ITP was used as substrate. Released $^{32}P_i$ was extracted and counted using Martin-Doty's method with slight modifications. In parentheses are shown relative activities referred to the uninhibited control.

absence of nonlabeled ATP. K_m for K^+ was markedly increased by the addition of ATP as seen in Fig. 5,A. In a reciprocal experiment, Na^+, K^+-ATPase was assayed using AT^{32}P as substrate and ITP as the added nucleotide. K_m was not affected by ITP (Fig. 5,B).

In the above experiments, $^{32}P_i$ production was suppressed by addition of the cold nucleotide. Under conditions where Na^+, K^+-ITPase was inhibited by as much as about 50% by ATP, Na^+, K^+-ATPase was also inhibited about 50% by ITP (Table III).

In summary, the inhibitory effect of ATP on Na^+, K^+-ITPase (and *vice versa*) may be due to simple competitive inhibition, but the increase of K_m of Na^+, K^+-ITPase for K^+ in the presence of ATP must be attributed to the modulatory action of this nucleotide. This interpretation is in accord with the observed effect of ATP on K_m for K^+ in the case of K^+-pNPPase (Fig. 4; Refs. *6* and *8*).

Inhibition of K^+-pNPPase by ATP

Adenosine triphosphate is a strong inhibitor of K^+-pNPPase (*6–8*). Results and interpretation are not in agreement as to whether this inhibition may be a simple competitive one or not. To obtain further information, the effects of ATP and ITP were compared (Fig. 6). K_i for ATP (without Na^+) was 0.3 mM, while K_i for ITP was more than 3 mM, in contrast to the only

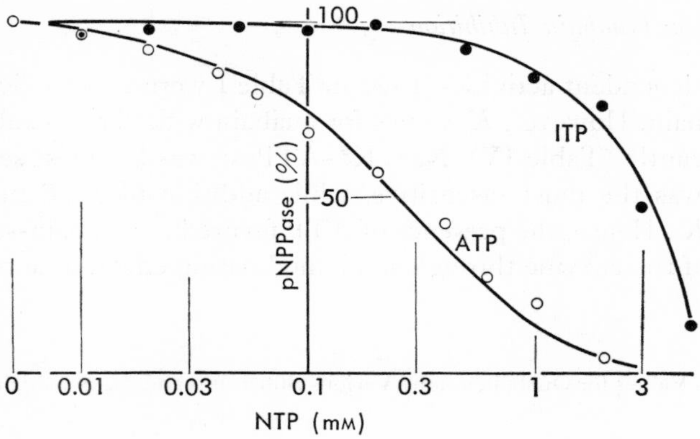

A

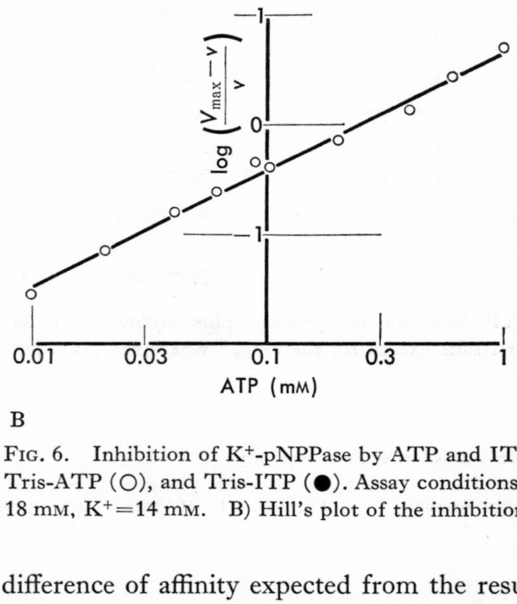

B

FIG. 6. Inhibition of K^+-pNPPase by ATP and ITP. A) Inhibition curves for Tris-ATP (○), and Tris-ITP (●). Assay conditions were as in Table I. pNPP= 18 mM, K^+=14 mM. B) Hill's plot of the inhibition curve for ATP. n=1.07.

3-fold difference of affinity expected from the results in Table III. Thus some sensitization mechanism may be involved here, too. The inhibition by ATP itself, however, followed first-order inhibition kinetics (Fig. 6,B). The mechanism of the sharp drop of K^+-pNPPase activity when ITP was over 3 mM requires further elucidation.

Effect of ATP on Ouabain Inhibition

The Na^+, K^+-dependent activities listed in Table I were all specifically inhibited by ouabain. However, K_i values for ouabain with various substrates differed significantly (Table IV). Na^+, K^+-ATPase was the most sensitive. K^+-pNPPase was the most insensitive, while addition of ATP markedly decreased the K_i. Hence, the presence of ATP favored an ouabain-sensitive conformation of the enzyme through some modulating effect. The relation-

TABLE IV
Comparison of K_i Values for Ouabain among Various Substrates and the Sensitizing Effect of ATP

Activity	K_i for ouabain (M)	Cationic and other conditions (mM)
Na^+,K^+-ATPase (ATP : 3 mM)	3.1×10^{-7}	Na^+ 140 ; K^+ 14
Na^+,K^+-CTPase (CTP : 3 mM)	1.0×10^{-6}	Na^+ 140 ; K^+ 14
Na^+,K^+-ITPase (ITP : 3 mM)	1.9×10^{-6}	Na^+ 100 ; K^+ 14
,,	1.9×10^{-7}	Na^+ 100 ; K^+ 14, ATP 1
,,	1.9×10^{-7}	Na^+ 100 ; K^+ 14, ATP 0.1[a]
K^+-pNPPase (pNPP : 18 mM)	8.0×10^{-6}	K^+ 18
,,	9.0×10^{-6}	K^+ 18, ITP 3.3
,,	2.5×10^{-6}	Na^+ 0.6, K^+ 18, ATP 0.3[b]

[a] Recycling system for ATP (phosphoenolpyruvate plus pyruvate kinase) added. Nucleotides were used as Tris-salts except for the case, [b] where the disodium salt was used.

TABLE V
Effect of Nucleotide Addition on K^+-pNPPase Activity

Nucleotide addition	Without ouabain	+Ouabain (1×10^{-5} M)	Inhibition (%)
—	(100)	85	15
UTP[a]	97	84	13
GTP[b]	99	85	14
CTP	89	70	21
ATP	80	64	20

Assay conditions were similar to those in Table I. pNPP: 18 mM, K^+: 18 mM, Na^+: 5 mM. Nucleoside triphosphates (Tris-salt): 9×10^{-5} M. [a] Uridine triphosphate. [b] Guanosine triphosphate.

ship between the effect of ATP on K_m for K^+ and that on K_i for ouabain remains to be clarified.

Addition of ITP to the K^+-pNPPase system, even at 10 times higher concentration than ATP, produced no sensitization (Table IV).

CTP also sensitized the enzyme for ouabain at a low concentration, while GTP did not (Table V).

Inhibition of Na^+,K^+-ATPase and K^+-pNPPase by Phenolphthalein-diphosphate

Phenolphthalein diphosphate was not detectably hydrolyzed by Na^+,K^+-ATPase, but it was an inhibitor of K^+-pNPPase. Inhibition was observed even at the saturating concentration of the substrate, with K_i of 1.2 mM (Fig. 7). The inhibition obeyed first-order inhibition kinetics. Na^+,K^+-ATPase activity was not inhibited up to 3 mM of the inhibitor. Above 5 mM of diphosphate, it strongly interfered with the P_i assay.

The difference in susceptibility to phenolphthalein diphosphate inhibition may be interpreted as one of the examples of modification of the enzyme conformation induced by ATP.

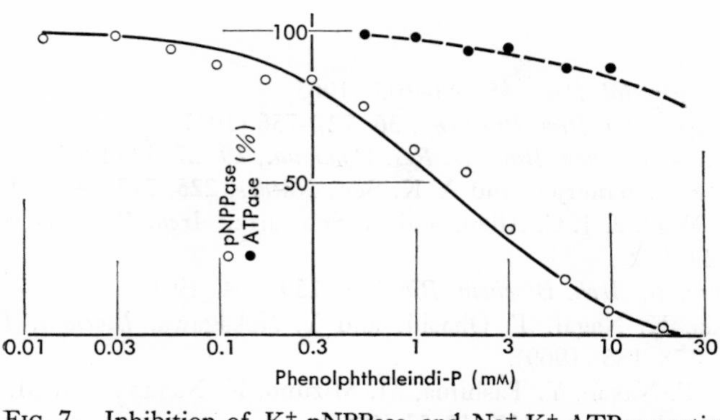

FIG. 7. Inhibition of K^+-pNPPase and Na^+,K^+-ATPase activities by phenolphthaleindiphosphate. Assay conditions were as in Table I. ○, K^+-pNPPase. pNPP=18 mM and K^+=18 mM. ●, Na^+,K^+-ATPase. ATP=3 mM, Na^+=140 mM, and K^+=14 mM. Above 5 mM of diphosphate, values were uncertain because of its interference with P_i assay.

Concluding Remarks

The modulation of K^+-pNPPase activity and its sensitivity to ouabain have been confirmed and discussed by several authors. From the results described above, ATP is suggested to play such a role in the cation-dependent hydrolysis of another substrate, ITP. It is reasonable to infer that ATP has a similar role also in the natural reaction of the enzyme, *i.e.*, Na^+, K^+-ATPase activity.

The modulatory activity of ATP was based on modification of K_m values for cations and K_i for ouabain of the enzyme. This property of ATP may be of special interest, since such modulation of sensitivity toward Na^+ and K^+, occurring cyclically along with some conformational change of the enzyme, will result in net transport of these cations across the cell membrane where the enzyme is located.

Acknowledgments
This investigation was supported by research grants for the study of membrane transport and energy transduction in biological systems #A-738015 and #A-738033 from the Ministry of Education, Japan.

REFERENCES

1. J. C. Skou, *Physiol. Rev.*, **45**, 596–617 (1965).
2. R. W. Albers, *Ann. Rev. Biochem.*, **36**, 727–756 (1967).
3. R. F. Squires, *Biochem Biophys. Res. Commun.*, **19**, 27–32 (1965).
4. T. Tobin, S. P. Banerjee, and A. K. Sen, *Nature*, **225**, 745–746 (1970).
5. W. R. van Winkle, J. C. Allen, and A. Schwartz, *Arch. Biochem. Biophys.*, **151**, 85–92 (1972).
6. J. D. Robinson, *Arch. Biochem. Biophys.*, **139**, 164 (1970).
7. H. Yoshida, K. Nagai, T. Ohashi, and Y. Nakagawa, *Biochim. Biophys. Acta*, **171**, 178–185 (1969).
8. M. Fujita, T. Nakao, Y. Tashima, M. Mizuno, K. Nagano, and M. Nakao, *Biochim. Biophys. Acta*, **117**, 42–53 (1966).
9. B. Formby and J. Clausen, *Z. Physiol. Chem.*, **349**, 909–919 (1968).
10. M. Nakao and K. Nagano, *in* "Molecular Mechanisms of Enzyme Action," ed. by Y. Ogura *et al.*, University of Tokyo Press, Tokyo, pp. 297–314 (1972).
11. T. Kanazawa, M. Saito, and Y. Tonomura, *J. Biochem.*, **61**, 555 (1967).
12. A. Askari and D. Koyal, *Biochem. Biophys. Res. Commun.*, **32**, 227–232 (1968).

13. W. Schöner, R. Bensch, and R. Kramer, *Eur. J. Biochem.*, **7**, 102–110 (1968).
14. T. Nakao, Y. Tashima, K. Nagano, and M. Nakao, *Biochem Biophys. Res. Commun.*, **19**, 755–759 (1965).
15. K. Ahmed, J. D. Judah, and P. G. Scholefield, *Biochim. Biophys. Acta*, **120**, 351–360 (1966).
16. R. L. Post, A. K. Sen, and A. S. Rosenthal, *J. Biol. Chem.*, **240**, 351–360 (1964).
17. I. M. Glynn, V. L. Lew, and U. Luthi, *J. Physiol.*, **207**, 371–391 (1970).
18. G. Toda, H. Koide, and Y. Yoshitoshi, *J. Biochem.*, **69**, 73–82 (1971).
19. P. J. Garrahan, M. I. Pouchan, and A. F. Rega, *J. Membrane Biol.*, **3**, 26–42 (1970).
20. A. L. Green and C. B. Taylor, *Biochem. Biophys. Res. Commun.*, **14**, 118–123 (1964).

Inhibition of Na^+, K^+-ATPase by Fusidic Acid

Hideo Matsui, Atsuko Nakagawa,* and Makoto Nakao*

*Department of Biochemistry, Kyorin University School of Medicine, Tokyo, Japan, and Department of Biochemistry, Tokyo Medical and Dental University School of Medicine, Tokyo, Japan**

Fusidic acid, which was known as a guanosine triphosphatase (GTPase) inhibitor blocking the translocation of peptidyl-tRNA on ribosomes, inhibited Na^+,K^+-ATPase as well as K^+-dependent *p*-nitrophenylphosphatase activity, and both K_i's were about 5×10^{-4} M. Unlike ouabain, K_i for fusidic acid was not influenced by K^+ concentration levels. In spite of the fact that the inhibition by ouabain and fusidic acid was additive, fusidic acid had no influence on ouabain binding supported by ATP, Mg^{2+}, and Na^+, and this result indicates that the binding site of the enzyme for fusidic acid differs from the site for ouabain. However, Mg^{2+}, P_i-dependent ouabain binding was decreased by fusidic acid. Na^+-dependent phosphorylation of the enzyme by γ-^{32}P-ATP was not affected by the addition of 1 mM fusidic acid in the presence or absence of K^+. On the other hand, fusidic acid inhibited ouabain-dependent phosphorylation of the enzyme by $^{32}P_i$ when ouabain concentration was at about K_i level. This inhibition, however, was overcome by elevation of the ouabain concentration to 0.1 mM. These results suggest that the inhibition point of fusidic acid may lie in the latter part of the ATPase reaction, possibly the P_i-association step or a conformational change of free enzyme.

It is well established that Na^+, K^+-ATPase takes an essential part in active cation transport across the cell membrane. The main problem with Na^+, K^+-ATPase as a constituent of energy-transducing membranes is to show how the energy from ATP is converted into the translocation of Na^+ and K^+

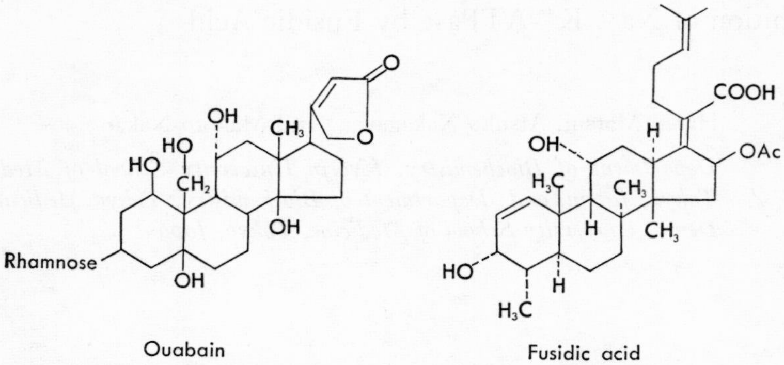

FIG. 1. Comparative structure of ouabain and fusidic acid.

by the enzyme. One of the approaches to this question has been to study the reaction mechanism of Na^+, K^+-ATPase, particularly by the use of various inhibitors of the enzyme such as cardiac glycosides (*1*), SH-inhibitors (*2*) and oligomycin (*3*). In addition to these, the authors found that the enzyme was also inhibited by fusidic acid, which was previously known as a guanosine triphosphatase (GTPase) inhibitor blocking the translocation of peptidyl-tRNA on ribosomes (*4*). It is therefore of interest to study the mode of inhibition of the enzyme by fusidic acid.

Fusidic acid, the structure of which is shown in Fig. 1, is classified as a steroid antibiotic, and ouabain also has a steroid structure as an essential constituent. Therefore, the first question is whether fusidic acid inhibition is of the ouabain type or not. The Na^+,K^+-ATPase used was obtained

TABLE I
Inhibition of Na^+,K^+-ATPase by Fusidic Acid

Additions	Activity (%)
Complete	100
2×10^{-4} M fusidic acid	83.6
6×10^{-4} M fusidic acid	48.5
10^{-3} M fusidic acid	25.3

Preparation of Na^+,K^+-ATPase from pig brain and the assay method for enzyme activity were described previously (*5*). The reaction mixture contained 2 mM ATP, 5 mM $MgCl_2$, 100 mM NaCl, 10 mM KCl, 1 mM ethylenediaminetetraacetic acid (EDTA), 50 mM Tris-HCl buffer (pH 7.4) and about 10 μg of enzyme protein in the presence or absence of fusidic acid.

from pig brain, by treatment with deoxycholate (DOC) and NaI (*5*), and more than 99% of the total activity was ouabain-sensitive. As shown in Table I, the enzyme activity was clearly inhibited by 10^{-4} M fusidic acid. K_i for fusidic acid was about 5×10^{-4} M, and this K_i level was almost the same order for the inhibition of GTPase (*4*).

Ouabain inhibition apparently competes with K^+, and this phenomenon is specific for ouabain-type inhibition (*6*). In the experiments shown in

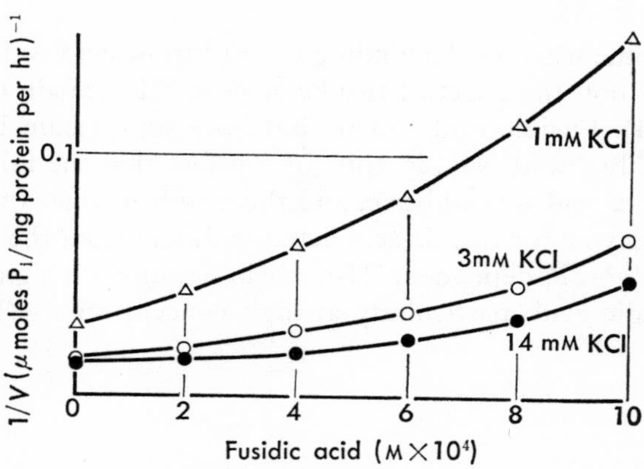

FIG. 2. Inhibition of Na^+,K^+-ATPase by fusidic acid at various concentration of K^+.

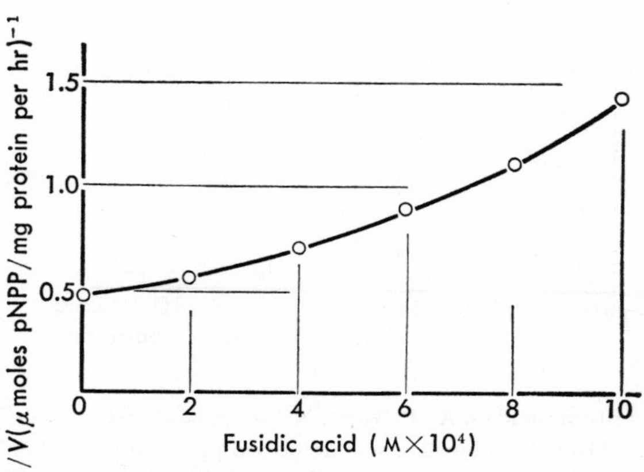

FIG. 3. Inhibition of K^+-dependent phosphatase by fusidic acid. The reaction mixture contained 6 mM *p*-nitrophenylphosphate instead of ATP and Na^+.

Fig. 2, K_i for fusidic acid was determined at various K^+ concentrations. Unlike ouabain, K_i for fusidic acid was not influenced by K^+ concentration levels. However, a Dixon plot of the inhibition (Fig. 2) did not fit a straight line, suggesting a rather complicated inhibition mechanism. Fusidic acid also inhibited *p*-nitrophenylphosphatase activity, with the same level of K_i as for the ATPase inhibition (Fig. 3). A Dixon plot of *p*-nitrophenylphosphatase inhibition also exhibited the same form as for ATPase. These results suggest that the mode of inhibition by fusidic acid is quite different from that by ouabain.

In order to determine whether the binding site of fusidic acid is the same as that of ouabain or not, the effect of fusidic acid on ^{3}H-ouabain binding was observed (Fig. 4). Fusidic acid had no influence on ouabain binding supported by ATP, Mg^{2+} and Na^+, in spite of the fact that the inhibition by ouabain and fusidic acid was additive, and this result indicates that the binding site of the enzyme for fusidic acid actually differs from the site for ouabain. However, Mg^{2+}, P_i-dependent ^{3}H-ouabain binding was significantly decreased by fusidic acid, particularly at high concentrations (Fig. 5),

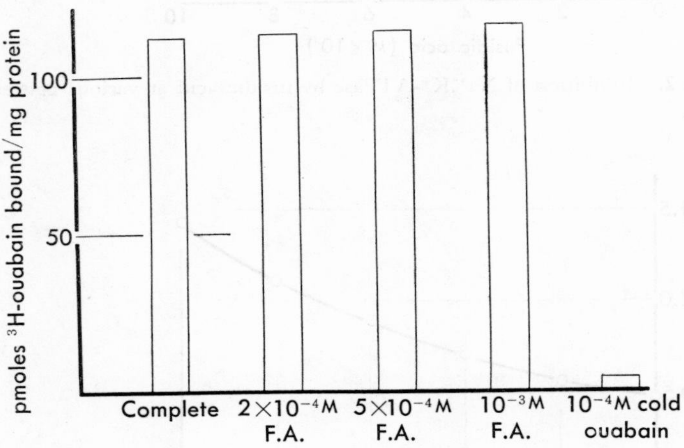

FIG. 4. Effect of fusidic acid on ATP, Mg^{2+}, Na^+-dependent ^{3}H-ouabain binding. Assay method for ^{3}H-ouabain binding was described previously (*1*). The reaction mixture, in a total volume of 1 ml, contained 4 mM ATP, 5 mM $MgCl_2$, 100 mM NaCl, 1 mM EDTA, 50 mM Tris-HCl buffer (pH 7.4), 4×10^{-7}M ^{3}H-ouabain, and about 1 mg enzyme protein with or without fusidic acid. F.A.: fusidic acid.

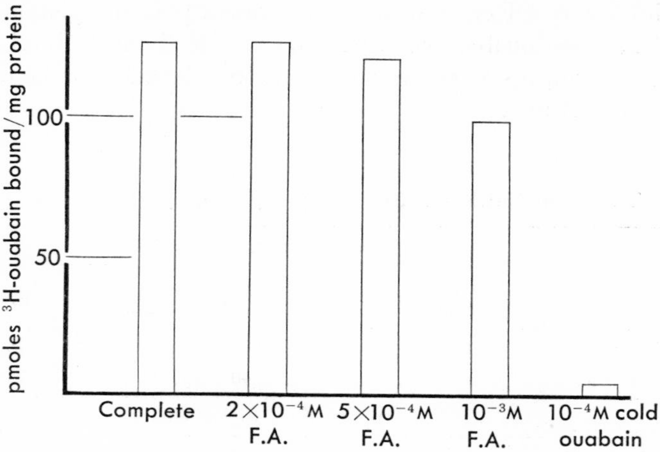

FIG. 5. Effect of fusidic acid on Mg^{2+}, P_i-dependent ^3H-ouabain binding. ATP and Na^+ were replaced by 4 mM P_i in the standard reaction mixture. F.A.: Fusidic acid.

TABLE II
Effect of Fusidic Acid on Na^+,K^+-ATPase Phosphorylation by γ-^{32}P-ATP

Condition of phosphorylation	pmoles ^{32}P/mg protein	
	No fusidic acid	10^{-3} M fusidic acid
$AT^{32}P+Mg^{2+}+K^+$	132	60
$AT^{32}P+Mg^{2+}+Na^+$	255 (123)	173 (113)
$AT^{32}P+Mg^{2+}+Na^++K^+$	169 (37)	106 (46)

The assay method for enzymed phosphorylation was described previously (*8*). Numbers in parentheses indicate net phosphorylation levels which were calculated by subtraction of nonspecific phosphorylation level in the presence of ATP, Mg^{2+}, and K^+.

suggesting some difference in the binding processes of ouabain between ATP, Mg^{2+}, Na^+-dependent and Mg^{2+}, P_i-dependent bindings.

To determine the inhibition point of fusidic acid to the enzyme, the effect of fusidic acid on the partial reaction of ATP hydrolysis was tested. Na^+-dependent phosphorylation of the enzyme by γ-^{32}P-ATP was not affected by the addition of 1 mM fusidic acid (Table II). Fusidic acid was also ineffective on the depressed phosphorylation level following the addition of K^+ to the standard system containing Na^+ alone. On the other hand, fusidic acid showed an inhibitory effect on ouabain-dependent phosphorylation of the enzyme by $^{32}P_i$ when ouabain concentration was at about K_i level (Table III). This inhibition, however, was overcome by elevation of the ouabain concentration to 0.1 mM.

TABLE III
Effect of Fusidic Acid on Na^+,K^+-ATPase Phosphorylation by $^{32}P_i$

Additions	pmoles ^{32}P/mg protein		
	(A) −Ouab.	(B) 4×10^{-7} M Ouab.	(B)–(A)
No fusidic acid	24	72	48
5×10^{-4} M fusidic acid	15	40	25
		10^{-4} M Ouab.	
No fusidic acid	24	142	118
5×10^{-4} M fusidic acid	15	133	118

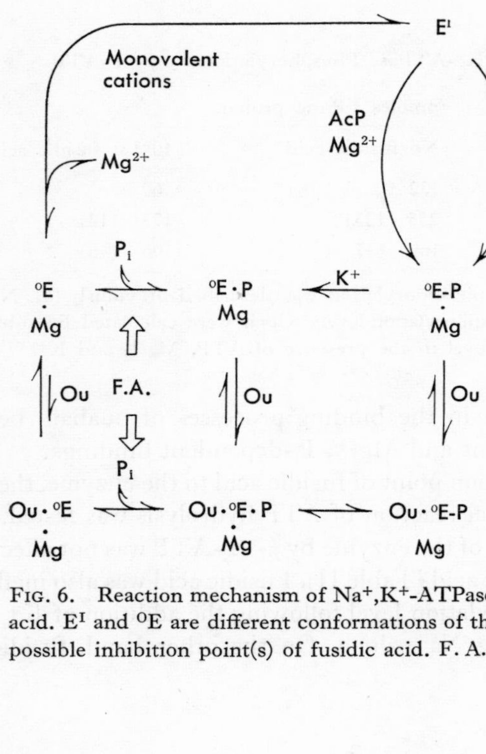

FIG. 6. Reaction mechanism of Na^+,K^+-ATPase and inhibition point of fusidic acid. E^I and ^OE are different conformations of the enzyme. Open arrows suggest possible inhibition point(s) of fusidic acid. F. A.: Fusidic acid. Ou: Ouabain.

Before discussion of the reaction point of fusidic acid, the reaction mechanism of Na^+, K^+-ATPase is summarized on the basis of the interaction with ouabain in Fig. 6. The normal route of Na^+, K^+-dependent ATP hydrolysis is as follows. Free enzyme (E) is phosphorylated by ATP in the presence of Mg^{2+} and Na^+, producing the phosphorylated intermediate (E-P). Then, E-P decomposes to E and P_i, and this step is accelerated by K^+. Free enzyme normally takes the I-conformation to which ouabain cannot bind (*1*, *7*). The phosphorylated intermediate takes the O-conformation to which ouabain can bind. The inhibition point of fusidic acid could be deduced as follows. Fusidic acid neither interferes with the phosphorylation by ATP, nor with dephosphorylation by K^+ (Table II), suggesting that the inhibition point may not lie in the early part of the reaction sequence. Since fusidic acid inhibits Mg^{2+}, P_i-dependent ouabain binding (Fig. 5) and ouabain-dependent P_i incorporation (Table III), the inhibition point may lie in the latter part of the reaction, possibly the P_i-association step or a conformational change of free enzyme.

REFERENCES

1. H. Matsui and A. Schwartz, *Biochim. Biophys. Acta*, **151**, 665 (1968).
2. S. Fahn, J. G. Koval, and R. W. Albers, *J. Biol. Chem.*, **243**, 1993 (1968).
3. R. J. Blostein, *J. Biol. Chem.*, **243**, 1957 (1968).
4. N. Tanaka, T. Kinoshita, and H. Masukawa, *Biochem. Biophys. Res. Commun.*, **30**, 278 (1968).
5. N. Mizuno, K. Nagano, T. Nakao, Y. Tashima, M. Fujita, and M. Nakao, *Biochim. Biophys. Acta*, **168**, 311 (1968).
6. H. Matsui and A. Schwartz, *Biochem. Biophys. Res. Commun.*, **25**, 147 (1966).
7. A. K. Sen, T. Tobin, and R. L. Post, *J. Biol. Chem.*, **244**, 6596 (1969).
8. K. Nagano, N. Mizuno, M. Fujita, Y. Tashima, T. Nakao, and M. Nakao, *Biochim. Biophys. Acta*, **143**, 239 (1967).

Na^+-Dependent ATP Hydrolysis in Mammalian Erythrocyte Membranes

Rhoda Blostein

Division of Hematology, Royal Victoria Hospital, and Department of Experimental Medicine, McGill University, Montreal, Canada

The properties of human erythrocyte membrane Na^+-stimulated ATPase and associated partial reactions are generally similar to those described for a variety of cell membranes. In experiments with membranes of sheep red cells possessing low K^+ (LK) or high K^+ (HK) levels it was observed that this genetic modification manifests itself not only in lower Na^+-ATPase activity in LK compared to HK membranes, but also in a kinetic distinction. This distinction was first suggested by experiments that showed variation in the HK: LK activity ratio for Na^+-ATPase and associated partial reactions. Further evidence is based on the following: 1) With low ATP concentration the responses of LK and HK Na^+-ATPase to K^+ are markedly different. HK membranes are more sensitive to activation and less sensitive to inhibition by K^+ than are LK membranes. 2) Ouabain inhibition of Na^+-ATPase is greater for HK than for LK membranes. 3) The response to oligomycin (OM) is dependent on the concentrations of ATP and Na^+ and is different for LK and HK membranes, activation by OM of LK Na^+-ATPase being observed with 50 mM Na^+ and ATP$\leq$0.2 μM, that of HK Na^+-ATPase with ATP$\leq$0.02 μM and / or lower [Na^+]. Inhibition occurs with higher concentrations of ATP. OM increases the phosphorylated intermediate markedly in LK membranes ($\sim$3-fold) as compared with HK membranes ($\sim$1.5-fold). There is interaction of OM and K^+ with Na^+-ATPase in LK and HK cells, *i.e.*, OM counteracts K^+ inhibition and K^+ counteracts OM inhibition. These observations are consistent with a reaction sequence involving OM-sensitive conformational changes ($E_iP \longleftrightarrow E_oP$ and probably $E_i \longleftrightarrow E_o$). In the presence of OM, hydrolysis of E_iP predominates and at very low catalytic rates appears as a stimulation of Na^+-ATPase. The differences in behavior of HK and LK Na^+ATPase are compatible with differences in the affinities for Na^+ and K^+ of the two systems which, in turn, affect the affinity for ATP and *vice versa*. It is further plausible that a Na^+,K^+-transport unit consists of interacting subunits whose affinities for Na^+, K^+, and ATP are related to the fraction of active subunits per transport unit.

The purity of erythrocyte membranes and the occurrence of a genetic modification of the ion transport system in red cells of certain species have made these preparations particularly suitable for studies of the enzymic basis of ion transport.

Human Erythrocyte Membranes

Previous studies in this laboratory have shown that Na^+-activated erythrocyte membrane ATPase involves phosphorylation and then dephosphorylation of some membrane-bound component (*1*, *2*), in agreement with studies with other preparations (*3–6*). However, studies with red cell membranes have been carried out at very low ATP concentrations to avoid nonspecific membrane phosphorylation; under these conditions Na^+ alone activates the Mg^{2+}-dependent ATP hydrolysis. The properties of the Na^+-dependent hydrolytic, phosphorylation, and [^{14}C] ADP-ATP exchange reactions measured at low substrate concentration have been described

TABLE I
Properties of Human Erythrocyte Membrane Na^+-ATPase and Associated Partial

Additions		Na^+-dependent component measured	
		^{32}P-" intermediate "	
	ATP conc.:	Low	High
		a	b
1. Na^+ (control)		Present	Undetected
2. plus K^+		Decreased	
3. plus ouabain		Decreased	
4. plus oligomycin		Present, $\geq$1a	

previously for human erythrocyte membranes (*1*, *2*) and are summarized in Table I.

HK and LK Sheep Erythrocyte Membranes

More recently, our attention has been directed to a comparative study of the Na^+-ATPase in high potassium (HK) and low-potassium (LK) sheep red cell membranes. A summary of some of the properties of these HK and LK red cells as described by several investigators (*7–13*) is shown in Table II. The ouabain-sensitive active transport of Na^+ and K^+ operates 4 times faster in HK than in LK red cells, and the ratio of the rate constant for K^+ leakage is about twice as high in LK as that in HK cells. The differences in ion transport are associated with differences in both the antigenic properties and the Na^+, K^+-activated, ouabain-sensitive ATPase of the two types of cells. One antigen, M, is found in homozygous HK sheep red cells; another, L, in homozygous LK cells. As shown, the activity of the Na^+,K^+-ATPase is 4–6 times greater in HK than in LK cells, and the number of ^{3}H-ouabain binding sites is about 7 times greater in HK than LK cells. When the intracellular K^+ is low or reduced in both HK and LK cells, the HK pump operates 5–9 times as fast as that of LK cells. This variation in pump ratios is probably related to a kinetic difference in the pump of the two cell types (*13*).

It seemed to us that if the difference(s) in HK and LK red cells was due to a difference in either the number of pumping sites or to some kinetic difference, or both, further examination of the partial reactions involved in the Na^+, K^+-ATPase would help distinguish between these possibilities and should not only help elucidate the nature of this genetic modification but

Reactions

Na^+-ATPase		[^{14}C] ADP-ATP exchange	
Low	High	Low	High
c	d	e	
Present	Undetected	Present	Not measured
≤1c	Present, ≫1c	Present	
Decreased	Decreased	Decreased	
Decreased	Decreased	Increased	

TABLE II
Comparative Properties of HK and LK Red Cells

Properties tested	HK	LK
Antigen (*7–9*)	MM	LM
Na^+,K^+-transport (*10*)	4	: 1
K^+ leakage (*10*)	1	: 2
Na^+,K^+-ATPase (*11*)	4–6	: 1
^3H-ouabain binding (sites/cell) (*12*)	7(42)	: 1(6)
K^+-pump rate at low K_{in} (*13*)	5–9	: 1
[K^+] for half-maximal stimulation of the pump (*13*)	3 (0.6 mM)	: 1(0.2 mM)

also contribute to the understanding of the enzymic basis of active Na^+ and K^+ transport.

Using procedures described previously (*14*) and a low ATP concentration (0.2 μM) to maximize specific ^{32}P labeling, the phosphorylated intermediate of Na^+-ATPase was measured in HK and LK membranes in both cell types as shown previously and summarized in Table III. It was found that Na^+ greatly stimulates the steady-state level of bound ^{32}P (^{32}P-"intermediate") in HK whereas little stimulation was seen in LK cells. The HK: LK ratio of the Na^+-stimulated increment of ^{32}P-intermediate is approximately 7: 1. Hydrolytic activity and [^{14}C] ADP-ATP exchange activity were also measured at low ATP concentrations, and the results are summarized in Table III. The much higher Na^+-ATPase activity of HK as compared with LK membranes is in accord with the previous data of Tosteson (*11*), although the

TABLE III
Comparative Activities of HK and LK Membranes

Parameter measured	LK	HK	HK : LK activity ratio
Na^+-ATPase (pmoles/min/mg)	7±3	69±19	10
Na^+,K^+-ATPase (pmoles/min/mg)	9±3	111±35	13
Na^+-stimulated (^{14}C) ADP-ATP exchange (pmoles/min/mg)	115±37	303±118	2.7
^{32}P-" intermediate " (pmoles/mg)	0.10±0.05	0.70±0.17	7

Data are summarized from previous studies (*14*).

ratio of the two activities here is considerably greater and increased further in the presence of K^+. The Na^+-stimulated [^{14}C] ADP-ATP exchange activities shown were determined at 0°C and in the presence of oligomycin (OM) to minimize ATP hydrolysis. The HK : LK exchange ratio obtained is considerably lower than that observed for the hydrolytic activities. This difference between the two rates suggested that the partial reactions in the two systems were kinetically different rather than only quantitatively different.

A striking difference between HK and LK membranes was found in the response to K^+ at very low ATP concentration (Fig. 1). Whereas Na^+-ATPase in HK membranes is stimulated by low levels of K^+ (1–5 mM), Na^+-ATPase in LK membranes is markedly inhibited at concentrations of added K^+ greater than 1 mM. In HK membranes, inhibition by K^+ occurs at much higher K^+ concentrations. We have recently examined the kinetics of Na^+-activation of the two types of cells and again observed a difference in their response to increasing Na^+ concentration.*

Another difference between the two types of membranes was found in

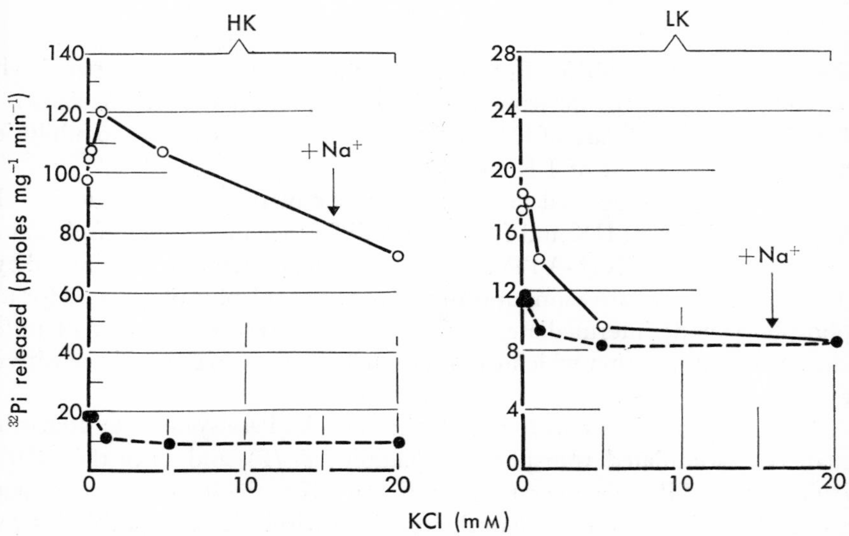

FIG. 1. Effect of KCl on Mg^{2+}- and ($Mg^{2+}+Na^+$)-ATPase. γ-^{32}P-ATP, 0.2 μM, NaCl, 50 mM, and KCl, at the indicated concentrations, were added as shown. By permission from *J. Biol. Chem.*, **2346**, 3518 (1971).

* R. Blostein and E. S. Whittington, unpublished observations.

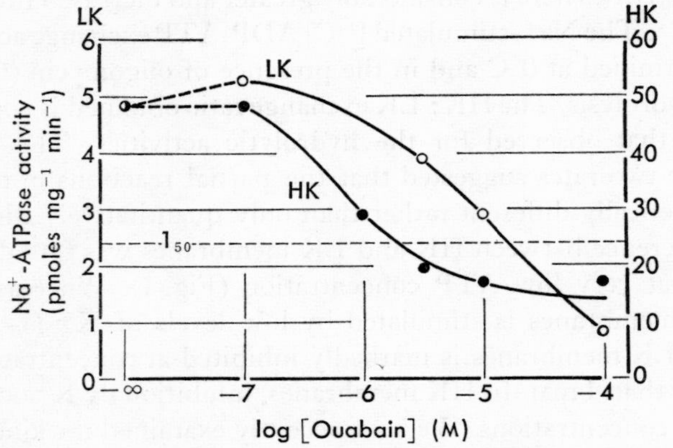

FIG. 2. Effect of ouabain on LK and HK Na^+-ATPase. Assays were carried out with 0.2 μM γ-^{32}P-ATP, 12 μM MgCl, and (a) 50 mM NaCl in the presence of ouabain as indicated or (b) 50 mM KCl plus 10^{-4} M ouabain. Na^+-ATPase represents the difference (a–b). The results given (average of duplicates) are representative of 6 experiments.

their response to ouabain. Both types of membranes were incubated with 0.2 μM γ-^{32}P-ATP for the same length of time, 3 min at 37°C. As shown in Fig. 2, the Na^+-ATPase of HK membranes was more sensitive to inhibition by ouabain than that of LK membranes.

These results suggested to us that a kinetic distinction characterizes the genetically different HK and LK cells. A difference in the magnitude of one rate constant in the Na^+-ATPase reaction sequence could be manifested by a difference in the relative amounts of two forms of the phosphorylated (or unphosphorylated) intermediate, and possibly the relative number of pump sites in a conformation which can bind ouabain and have specific sensitivity to Na^+ or K^+ ions.

Since oligomycin seems to inhibit the Na^+-ATPase system at a step other than Na^+-stimulated transphosphorylation (*15–17*), and since the relative [^{14}C] ADP-ATP exchange rates for HK and LK with oligomycin present are more similar in magnitude than the relative hydrolytic rates (Table III), we have examined the Na^+-dependent hydrolysis and phosphorylation in the presence of oligomycin in an effort to further delineate the kinetic distinction between HK and LK cells.

As shown in Table IV, at 0.2 μM ATP oligomycin markedly increased the

TABLE IV
Effects of OM on Na^+-ATPase and the ^{32}P-intermediate of HK and LK Membranes

Parameter measured	ATP conc. (μM)	Activity			
		LK		HK	
		Control	+ OM	Control	+ OM
Na^+-ATPase (% of control)	0.02	100	440	100	150
	0.20	100	220	100	60
	2.0	100	100	100	30
	5.0	100	30	100	—
^{32}P-intermediate (% of control)	0.02	100	396±67	100	173±16
	0.20	100	353±137	100	127±17
Catalytic center activity (min^{-1})	0.20	16±6	17±4	50±20	25±6

Na^+-ATPase activity and the ^{32}P-intermediate were determined as described previously (*14*) except that the assays for Na^+-ATPase were carried out with HK membranes diluted 1:1 and incubated 0.25, 0.5, and 4 min at γ-^{32}P-ATP concentrations of 0.02, 0.20, and 2.0 μM, respectively, and LK membranes diluted 1:1 and incubated 1, 4, 10, and 15 min at ATP concentrations of 0.02, 0.20, 2.0, and 5.0 μM, respectively. Catalytic center activities were determined from the amount of ^{32}P-intermediate and Na^+-ATPase measured in the same reaction vessel and for short (10-sec) incubation periods. Results shown are either a representative experiment or the mean± S.D., replicate ($\gtrsim$4) experiments.

level of phosphorylated intermediate in LK (about 4-fold), but not in HK membranes. At lower ATP concentrations (0.02 μM)oligomycin increased labeling in HK cells as well, but to a lesser extent ($\sim$1.7-fold). The next question was whether the increase in phosphorylated intermediate was associated with an increased or decreased turnover of the phosphorylated intermediate. Therefore, we determined the rate of Na^+-ATPase under similar conditions and over a range of ATP concentrations, as indicated in Table IV. The Na^+-ATPase activity observed without oligomycin is the control and is represented as 100%. Note that at a single concentration of ATP, *e.g.*, 0.2 μM, inhibition of HK and stimulation of LK Na^+-ATPase were observed. The pattern for both systems is similar, *i.e.*, stimulation is observed at the lowest ATP concentrations followed by a tendency toward inhibition at higher ATP concentrations. The Na^+-ATPase activity obtained divided by the Na^+-stimulated increment in phosphorylated intermediate is the turnover or catalytic center activity and is shown for experiments carried out at 0.2 μM ATP.

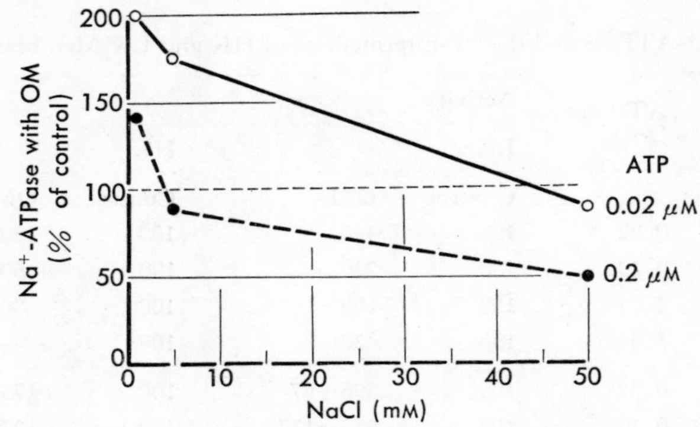

FIG. 3. Effect of Na^+ on oligomycin sensitivity of HK Na^+-ATPase. Na^+-ATPase was determined as described in Fig. 2. The results given (average of duplicates) are representative of 5 experiments.

In the absence of oligomycin, the catalytic center activity of HK Na^+-ATPase is greater than that of LK. In the presence of oligomycin, the turnover of LK is not changed, *i.e.*, both hydrolysis and phosphorylation are about equally stimulated. However, the catalytic center activity of HK is decreased. As a result, the turnover of LK and HK in the presence of oligomycin are similar.

Table IV shows that the oligomycin response is dependent on ATP concentration. When the Na^+ concentration was varied, it was found that the action of oligomycin is dependent also on Na^+ concentration as shown for HK membranes in Fig. 3. As in Table IV, the inhibitory effect of oligomycin is greater at 0.2 than at 0.02 μM ATP; in addition, less inhibition and even stimulation occur as the Na^+ concentration is decreased. As a result, a decrease in both Na^+ and the ATP concentrations magnifies the stimulatory action of oligomycin. The effects with varying Na^+, however, may reflect the influence of oligomycin on the kinetics of cation activation, *i.e.*, reducing the concentration for half-maximal activation for Na^+, as described by Robinson (*18*).

It has been shown previously by others and also in our laboratory that oligomycin inhibits K^+-stimulated dephosphorylation of the ^{32}P-intermediate. We examined the effect of oligomycin at 0.2 μM ATP, 50 mM Na^+, and 5 mM K^+ added as indicated (Fig. 4). As shown previously (*14*), K^+ stimulated HK Na^+-ATPase and markedly inhibited LK Na^+-ATPase. In these experiments K^+ inhibition of LK is counteracted by oligomycin, and

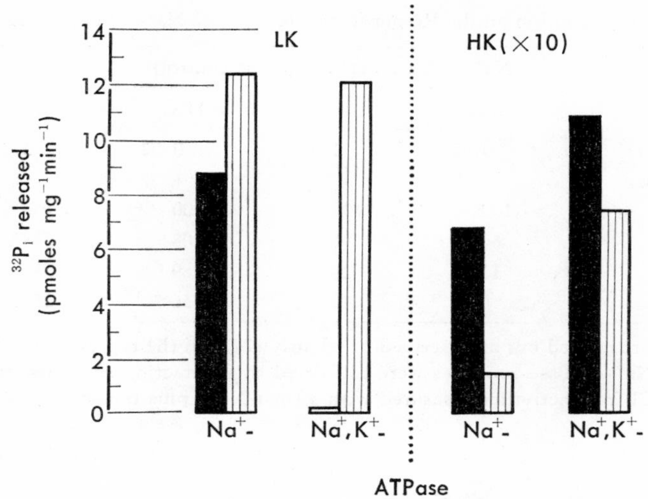

FIG. 4. Effects of oligomycin and K^+ on Na^+-ATPase of HK and LK membranes. Experiments were carried out as described previously (*14*). The results given (average of duplicates) are representative of 5 experiments. ■ control; ▥ OM.

oligomycin inhibition of HK Na^+-ATPase is counteracted by K^+, *i.e.*, the percentage inhibition is less with K^+ present. These results suggest that oligomycin interferes with the transformation of an intermediate or site to a K^+-sensitive state.

Since the oligomycin response is a function of the ATP concentration, a function of the Na^+ concentration, and is counteracted by K^+ and *vice versa* and since the interaction of the ATPase system with ATP (ATP binding) is decreased by K^+, an effect counteracted by Na^+ (*19*), the question is whether the difference between HK and LK is simply one of affinity for ATP which in turn affects the K_m for Na^+ and K^+ and vice versa.

Preliminary experiments with varying ATP, in a low concentration range, have not revealed a significant difference in the K_m for ATP for LK compared to HK Na^+-ATPase. One difficulty has been the variability in values for preparations obtained from different animals. However, we have examined the effects of ATP on the K^+ response. Although the response was dependent on ATP concentration, particularly with HK membranes, the K^+-response pattern for LK remained different from that of HK membranes

TABLE V
Effect of ATP Concentration on the Response of HK and LK Na^+-ATPase to K^+

Additions	Na^+-ATPase activity (% of control)			
	LK		HK	
ATP conc. (μM):	0.02	2.0	0.02	2.0
KCl added (mM)				
None (control)	100	100	100	100
1	60	96	108	338
5	13	26	89	343
20	0	10	56	250

Experiments were carried out as described previously (*14*). In the representative experiments shown, Na^+-ATPase activities were calculated by subtracting activities measured with 50 mM KCl from activities measured with 50 mM NaCl plus the amounts of added KCl indicated.

even when the ATP concentration was varied by two orders of magnitude as shown in Table V.

Conclusions

We have considered these results in terms of a plausible model of the Na^+, K^+-ATPase system, compatible with observations in a number of laboratories (Fig. 5). In complete inhibition of Na^+, K^+-ATPase by oligomycin has been observed by Robinson (*18*) as the ATP concentration is decreased, and he has suggested two routes of hydrolysis of two forms of phosphorylated intermediate, as shown in Fig. 5. His model does not necessarily include an equilibrium between E_i and E_o. With oligomycin, hydrolysis *via* pathway B is inhibited while that via A still functions and can appear as a stimulation particularly if interaction of ATP with E_i is no longer decreased by the tendency of E_i to equilibrate with E_o. Interaction of K^+ with the system would similarly be counteracted by oligomycin (Fig. 4, LK membranes), and oligomycin inhibition counteracted by K^+ (Fig. 4, HK membranes). Some support for the scheme is indicated when the action of oligomycin is considered in terms of the absolute Na^+-ATPase activity in the absence of oligomycin; that is, when the Na^+-ATPase rate is greater than about 20 pmoles mg^{-1} min^{-1}, the effect of oligomycin is inhibitory. At lower rates, oligomycin stimulates Na^+-ATPase.

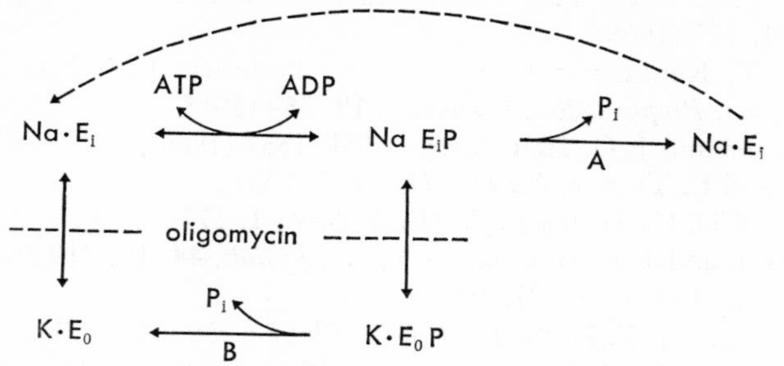

FIG. 5. Diagram illustrating sequence of partial reactions involved in Na^+-ATPase. E_i and E_o represent 2 forms of unphosphorylated intermediate, E_iP and E_oP, 2 forms of phosphorylated intermediate. "A" refers to ATPase activity *via* hydrolysis of E_iP and "B" refers to the ATPase activity *via* the sequence $E_iP \rightarrow E_oP \rightarrow E_o + P_i$.

The allosteric nature of the system has been indicated in a number of studies in various laboratories and is suggested by the change in the properties of the system as the ATP concentration is increased. It is plausible that a difference in affinity for ATP, perhaps secondary to different affinities for Na^+ and K^+ or *vice versa* is the basis for the kinetic difference between LK and HK Na^+-ATPase. It is further plausible that a Na^+,K^+-transport unit consists of interacting subunits whose affinities for ATP, Na^+, and K^+ are a function of the fraction of active subunits per transport unit. If the quantitative differences in HK and LK transport ATPase have a common kinetic basis, then this type of allosteric system would be attractive.

Acknowledgments
The contributions of Mr. Emerson S. Whittington are gratefully acknowledged. This work was supported by the Medical Research Council of Canada.

REFERENCES

1. R. Blostein, *J. Biol. Chem.*, **243**, 1957 (1968).
2. R. Blostein, *J. Biol. Chem.*, **245**, 270 (1970).
3. R. L. Post, A. K. Sen, and A. S. Rosenthal, *J. Biol. Chem.*, **240**, 1437 (1965).
4. R. W. Albers, S. Fahn, and G. J. Koval., *Proc. Natl. Acad. Sci. U.S.*, **50**, 474 (1963).

5. L. E. Hokin, P. S. Sastry, P. R. Galsworthy, and A. Yoda, *Proc. Natl. Acad. Sci. U.S.*, **54**, 177 (1965).
6. K. Nagano, T. Kanazawa, N. Mizuno, Y. Tashima, T. Nakao, and M. Nakao, *Biochem. Biophys. Res. Commun.*, **19**, 759 (1965).
7. B. A. Rasmusen and J. G. Hall, *Science*, **151**, 1551 (1966).
8. J. C. Ellory and E. Tucker, *Nature*, **222**, 477 (1969).
9. P. K. Lauf and D. C. Tosteson, *J. Memb. Biol.*, **1**, 177 (1969).
10. D. C. Tosteson and J. F. Hoffman, *J. Gen. Physiol.*, **44**, 169 (1960).
11. D. C. Tosteson, *Fed. Proc.*, **22**, 19 (1963).
12. P. B. Dunham and J. F. Hoffman, *J. Gen. Physiol.*, **58**, 94 (1971).
13. P. G. Hoffman and D. C. Tosteson, *J. Gen. Physiol.*, **58**, 438 (1971).
14. E. S. Whittington and R. Blostein, *J. Biol. Chem.*, **246**, 3518 (1971).
15. G. J. Siegel and R. W. Albers, *J. Biol. Chem.*, **242**, 4972 (1967).
16. W. L. Stahl, *Neurochemistry*, **15**, 1511 (1968).
17. C. E. Inturrisi and E. Titus, *Mol. Pharmacol.*, **4**, 591 (1968).
18. J. D. Robinson, *Mol. Pharmacol.*, **7**, 238 (1971).
19. C. Hegyvary and R. L. Post, *J. Biol. Chem.*, **246**, 5234 (1971).

Localization of Ouabain-sensitive ATPase in Intestinal Mucosa

Michiya Fujita and Makoto Nakao

Department of Biochemistry, Tokyo Medical and Dental University School of Medicine, Tokyo, Japan

What seemed to be a microvillous membrane fraction and a basolateral membrane fraction were isolated from jejunal epithelial cell homogenate. The former was rich in alkaline phosphatase, sucrase, and leucylglycine hydrolase, whereas the latter was rich in ouabain-sensitive ATPase activity. These findings seem to conform to the hypothesis that the microvillous membrane acts as the absorptive-digestive surface, while the lateral and, possibly, basal membranes as the site of active extrusion of Na^+.

One of the fundamental cytological differences between epithelial and non-epithelial cells is the fact that the former form a continuous boundary layer. In intestinal mucosa the apical or luminal surface of the cell faces a different environment from that of the basal and lateral surfaces (Fig. 1). The latter membranes may be regarded as a common surface or a functional unit (*1*). As a corollary, the surface of an intestinal mucosal cell must be regarded as geometrically asymmetric. This is a situation not met in nonepithelial cells such as erythrocytes. This spatial asymmetry could have a profound implication in the cellular transport of solutes.

It seems evident from physiological studies that there is an active transport of Na^+ from mucosal to serosal medium, irrespective of the animal species (*2*). The measurements of transmural potential differences and short-circuit currents, their increases following luminal addition of amino acids or hexoses, and their inhibition by serosal addition of ouabain, all favor the assumption that a coupled transport system for Na^+ and nutrients

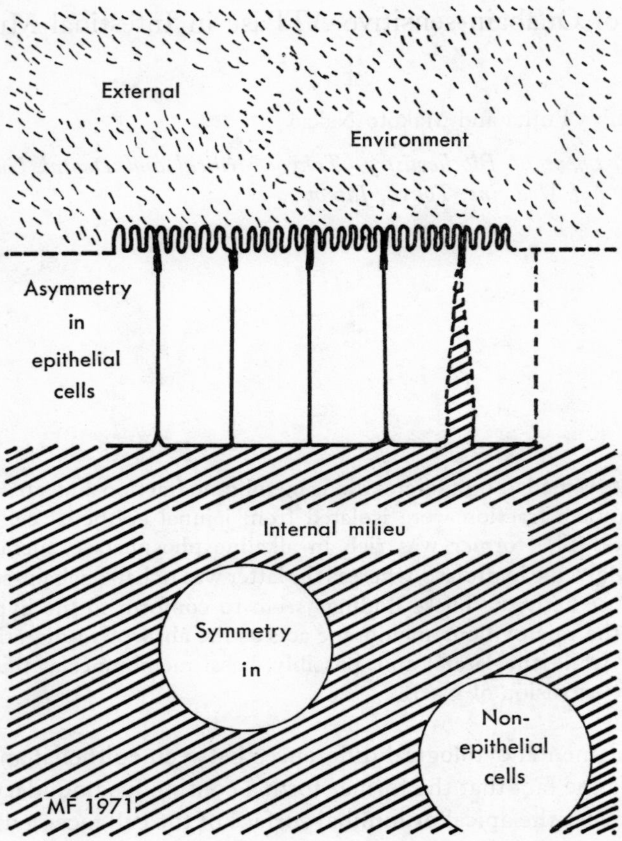

FIG. 1. Geometric asymmetry in intestinal mucosal epithelium.

is situated at the luminal border, whereas the Na^+ pump is situated at the serosal border of the cell (*3*). The asymmetric distribution of Na^+ pump activity of this type was first proposed by Koefoed-Johnsen and Ussing (*4*) to explain active Na^+ uptake by the frog skin.

Spatial asymmetry of Na^+ pump activity, however, does not necessarily require its absence in the brush border of the cell, provided more pump activity occurs in the serosal part of the cell membrane than in the luminal part.

In order to investigate the presumed asymmetry of Na^+ pump activity, we have employed a biochemical approach. The Na^+, K^+-activated ATPase or ouabain-sensitive ATPase activity may be taken as a biochemical measure of Na^{+} pump activity. We (*5*) have isolated the brush borders from the intestinal mucosa and carefully washed them by repeated homogenization and centrifugation to strip them of contaminating lateral membranes. Ouabain-sensitive ATPase activity was assayed with the brush border and the other subcellular fractions. Sucrase activity was used as an index of the brush border.

The mucosa was very gently homogenized with a Dounce-type homogenizer to minimize fragmentation of the brush borders. This homogenization technique is based on that of Boyd *et al.* (*6*). Centrifugation at a sufficiently low speed preferentially removed intact brush borders. More than 50% of nuclei remained in the supernatant after a single centrifugation. Earlier workers (*7–9*) have used faster centrifugation which resulted in a brush-border fraction heavily contaminated with lateral membranes. The gentle homogenization was repeated, effecting a careful washing of the brush borders. All supernatant fractions were saved and assayed for ouabain-sensitive ATPase and sucrase activities (Fig. 2).

Fraction 1 is the starting homogenate. Fractions 2 to 7 are supernatants recovered through the washing procedure. Fraction 8 is the final pellet or brush-border fraction. The bars above the horizontal line represent the ouabain-sensitive ATPase activity and those below it sucrase activity. Most

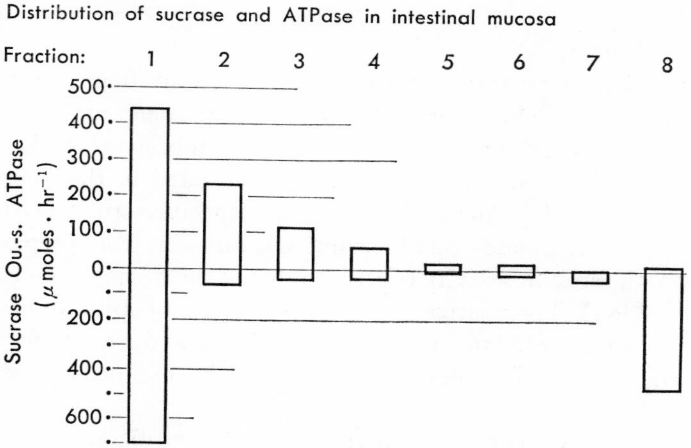

FIG. 2. Washing procedure and distribution of ouabain-sensitive ATPase and sucrase activities.

of the ATPase was lost into the first 3 supernatants, while sucrase remained with the pellet. This finding strongly suggested the association of sucrase activity with brush borders in accordance with the results of earlier workers (*10*).

On the other hand, it suggested a deficiency of ouabain-sensitive ATPase activity in the brush border. It is least conceivable that the activity was solubilized from the brush-border membrane, while sucrase remained with it. It must be located in the other part of the cell membrane. In order to confirm a possible localization of the ouabain-sensitive ATPase activity in the lateral (and probably basal) part of the cell membrane, *i.e.*, basolateral plasma membrane, we have further fractionated the combined supernatants after the isolation of brush borders.

A crude mitochondrial fraction was richest in ouabain-sensitive ATPase activity. It was therefore used as a starting material to separate a plasma membrane fraction with a high Na^+, K^+-ATPase activity. A crude mitochondrial fraction and a brush-border fraction, after further homogenization, were centrifuged in a sucrose gradient. The highest Na^+, K^+-ATPase activity was found at the interfacial band between the 30 and 40% (w/v) sucrose layers of the tube containing crude mitochondria. This fraction was considered to correspond to basal and lateral plasma membranes (*11*). In contrast, there was very little alkaline phosphatase activity in this fraction. It was richest in the 50% sucrose layer of fragmented brush borders. This fraction showed very little Na^+, K^+-ATPase activity. The results are summarized in Table I.

Electron microscopy of the 2 fractions confirmed the identity of the Na^+, K^+-ATPase-rich membrane with basolateral membrane and the alkaline phosphatase-rich membrane with microvillous membrane (Fig. 3). Sucrase and L-leucylglycine hydrolase were also concentrated in the microvillous membrane fraction (*11*). Although the absolute specific activity (activity per unit protein) of membrane-bound enzyme was different from preparation to preparation, the specific activity relative to that of the homogenate was fairly constant (Table I). The relative specific activity ratio of apical to basolateral plasma membrane was 22 to 1 for Na^+, K^+-ATPase and 1 to 180 for alkaline phosphatase. These differences suggest an extensive regional differentiation of cell membrane.

In agreement with enzymic and functional differentiation, the size distribution of protein components of apical and basolateral membranes contrasted strongly. The electrophoretic pattern was surprisingly simple for

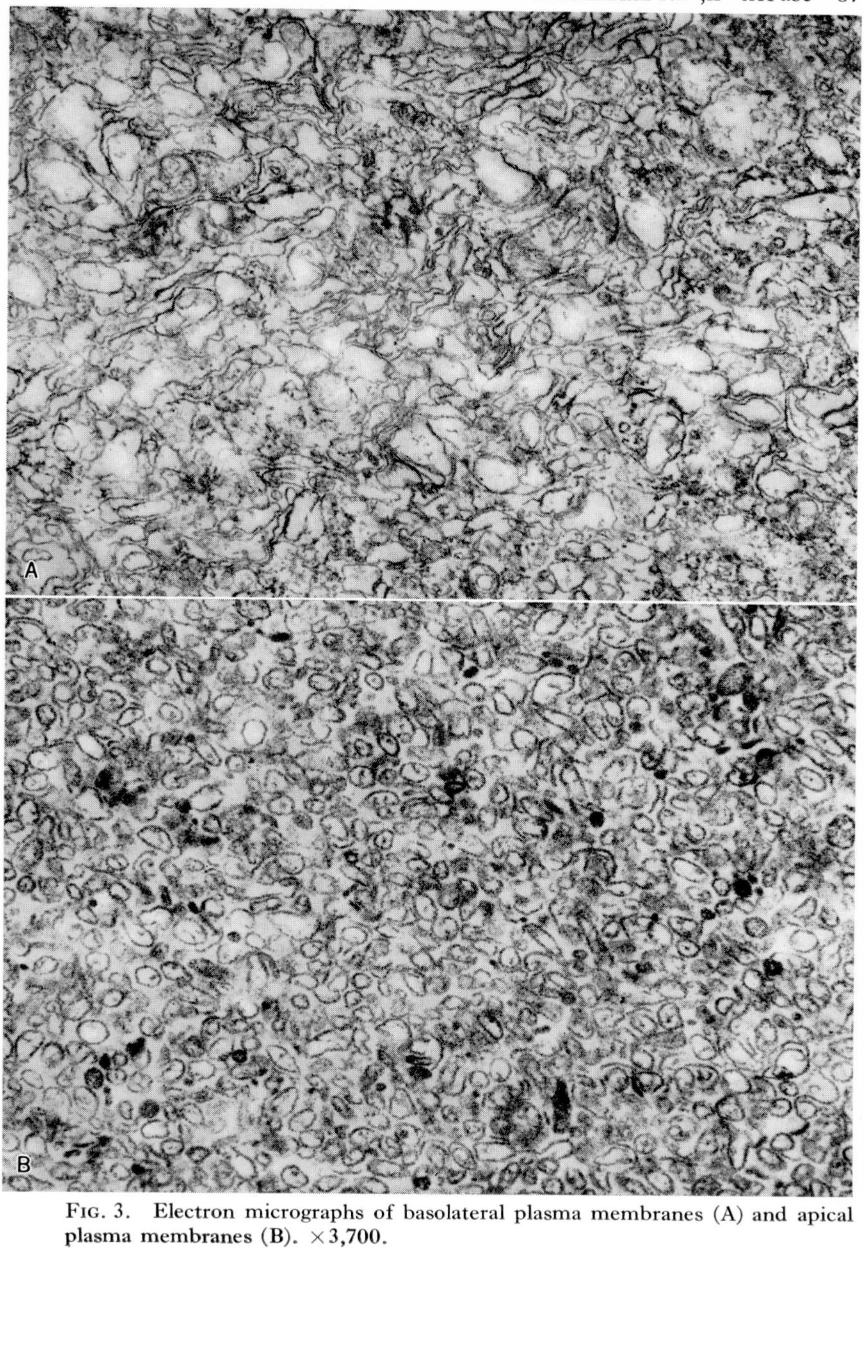

FIG. 3. Electron micrographs of basolateral plasma membranes (A) and apical plasma membranes (B). ×3,700.

TABLE I

Distribution of Ouabain-sensitive ATPase Activity and Alkaline Phosphatase Activity in Plasma Membrane Fractions of Mouse Intestinal Mucosa

Fraction	Ouabain-sensitive ATPase	Alkaline phosphatase
Homogenate	160[a] (5.1)[b]	116[a] (3.7)[b]
Basolateral membrane	27(68)	0.5(1)
Apical membrane	1(3)	19(57)
Relative specific activities:		
Homogenate	1	1
Basolateral membrane	11 ±0.6(5)[c]	0.5±0.1(7)[c]
Apical membrane	0.1±0.1(5)	18 ±1.4(7)

[a] Total activity of the fraction in μmoles of P_i (ATPase) and *p*-nitrophenol (alkaline phosphatase) liberated per hour. [b] Specific activity in μmoles per hour per mg protein. [c] Number of separate preparations.

basolateral membrane (unpublished data), whereas that for microvillous membrane showed a complexity similar to that of liver cell membrane or erythrocyte membrane.

The present findings could be summarized as follows. The digestive enzymes such as oligosaccharidase and oligopeptidase (*12*) (alkaline phosphatase may be included) are localized in apical or microvillous membrane and Na^+, K^+-ATPase activity in basolateral plasma membrane (Fig. 4).

From a physiological point of view, the differential distribution of these enzymes is very much according to the cellular function. Intermediary pro-

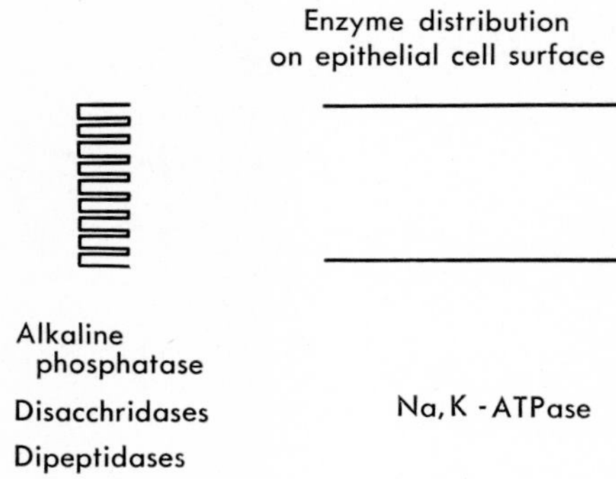

FIG. 4. Mutually exclusive distribution of digestive enzymes and ouabain-sensitive ATPase in intestinal mucosal surface membrane.

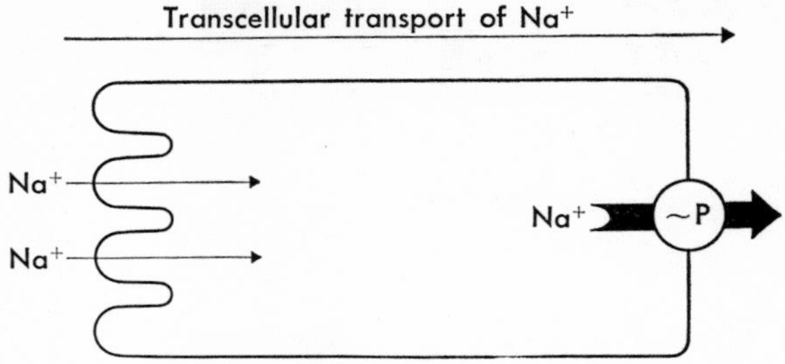

FIG. 5. Transcellular transport of Na^+ effected by asymmetric distribution of Na^+ pump activity.

ducts of intestinal digestion are further hydrolyzed to terminal products at the absorptive surface, whereas intracellular sodium ions are expelled at the opposite border of the cell (Fig. 5). The latter device indirectly drives the active absorption of nutrient end products by the mechanism of coupled transport presumably functioning at the luminal border (*3*).

REFERENCES

1. D. S. Parsons, *Brit. Med. Bull.*, **23**, 252 (1967).
2. R. T. C. Barry, *Brit. Med. Bull.*, **23**, 266 (1967).
3. S. G. Schultz and R. Zalusky, *J. Gen. Physiol.*, **47**, 1043 (1964).
4. V. Koefoed-Johnsen and H. H. Ussing, *Acta Physiol. Scand.*, **42**, 298 (1958).
5. M. Fujita, H. Matsui, K. Nagano, and M. Nakao, *Biochim. Biophys. Acta*, **233**, 404 (1971).
6. C. A. R. Boyd, D. S. Parsons, and A. V. Thomas, *Biochim. Biophys. Acta*, **150**, 723 (1968).
7. C. B. Taylor, *Biochim. Biophys. Acta*, **52**, 293 (1961).
8. G. G. Berg and B. Chapman, *J. Cellular Comp. Physiol.*, **65**, 361 (1965).
9. I. H. Rosenberg and L. E. Rosenberg, *Comp. Biochem. Physiol.*, **24**, 975 (1968).
10. D. Miller and R. K. Cane, *Biochim. Biophys. Acta*, **52**, 293 (1961).
11. M. Fujita, H. Ohta, K. Kawai, and M. Nakao, *Biochim. Biophys. Acta*, **274**, 336 (1972).
12. M. Fujita, D. S. Parsons, and F. Wojnarowska, *J. Physiol.*, **227**, 377 (1972).

Ca^{2+} TRANSPORT AND Ca^{2+} ATPase

ATPase and ATP Binding Sites in the Sarcoplasmic Reticulum Membrane*

Giuseppe Inesi, Suzanne Blanchet, and David Williams
Mellon Institute, Carnegie Mellon University, Pittsburgh, Pennsylvania, U.S.A.

Solubilization of sarcoplasmic reticulum (SR) membranes by sodium dodecylsulfate (SDS) and gel electrophoresis results in the separation of three main protein components. The major protein component displays a mobility corresponding to a molecular weight of approximately 100,000 and is identified with the membrane ATPase. Electrophoresis of membranes solubilized after aging for a few days in the absence of bacteriostatic agents reveals the presence of small protein fragments resulting from proteolytic degradation of the native proteins. Selective solubilization of ATPase from the membrane vesicles is obtained by the use of Triton X-100, a nonionic detergent. Alternatively, ATPase purification can be accomplished by solubilizing the other protein components with a prolonged incubation in 5 mM ethylenediaminetetraacetic acid (EDTA) (pH 8.6) and 0.5 mM sodium azide. This extraction leaves only the purified ATPase in the membrane fragments, which can be subsequently solubilized with Triton X-100. A 1 : 1 stoichiometric ratio is obtained between ATPase and specific ATP binding sites, which are found to be 7 moles/10^6 g membrane protein. It can be calculated that the active sites should be distributed through the plane of the membrane, with a 100–110 Å center-to-center average distance.

Vesicular fragments of sarcoplasmic reticulum (SR) display a very active and specific Ca^{2+} pump (*1, 2*). Since a tight coupling between Ca^{2+} accumulation

* This work was supported by the American Heart Association, the U.S.P.H.S., and the Muscular Distrophy Association of America, Inc.

and ATPase activity has been demonstrated, a definition of structure and function of this system requires that size and number of ATPase molecules per unit weight of membrane be determined.

With regard to the number of ATPase sites, estimates have been obtained by determining steady-state levels of enzyme phosphorylation (transfer of ATP terminal phosphate). Values of 0.1 to 6.0 moles/10^6 g of protein have thus been obtained in various laboratories, including our own (*3–6*). This variability indicates that steady-state levels depend on variants of the experimental conditions, and do not necessarily define the maximal number of ATPase sites.

On the other hand, detergent solubilization of the SR protein components has yielded ATPase preparations of high specific activity (*7–10*). However, estimates of the percentage of SR protein accounted for by ATPase vary between 18 and 90% (*7–13*).

As for the molecular weight of the enzyme, while most of the investigations have obtained figures of approximately 100,000 (*6–13*), others found that the basic protein subunits of SR have molecular weights lower than 10,000 (*14*).

In an attempt to clarify these uncertainties, and to establish frequency and size of the ATPase enzyme in the SR membrane, some of our studies on protein solubilization, ATPase purification, and ATP binding to SR are reported and discussed here.

Protein Components of the SR Membrane

Incubation of SR*[1] (0.2–1.0 mg/ml) in 2% sodium dodecyl sulfate (SDS) for 10 min at 37°C results in complete solubilization of the membrane. All of the solubilized protein enters acrylamide gels, and electrophoresis*[2] at

*[1] SR was prepared from rabbit skeletal muscle as previously described (*15*). The method includes extraction of the SR vesicles with 0.6 M KCl. *[2] Gel electrophoresis was performed in two pH systems. In system N.1 (pH 7.4) the separating gels were prepared with 7% acrylamide, 1.85% BIS (N-N'-methylene-bis acrylamide), 4.5% Tris(2-amino-2-hydroxymethyl-1,3-propanediol), 0.03% TEMED (N-N-N'-tetramethylene diamine), 0.1% SDS(sodium dodecyl sulfate), and 0.07% ammonium persulfate. The running buffer contained 4.8% Tris, 1.6% sodium acetate, and 0.1% SDS (pH 7.4). The tracking dye was Pyronine Y, and the current 7–8 mA/gel. In system N.2 (pH 8.3) the separating gels were prepared with 7% acrylamide, 1.85% BIS, 6% 1N HCl, 4.5% Tris, 0.03% (cont'd)

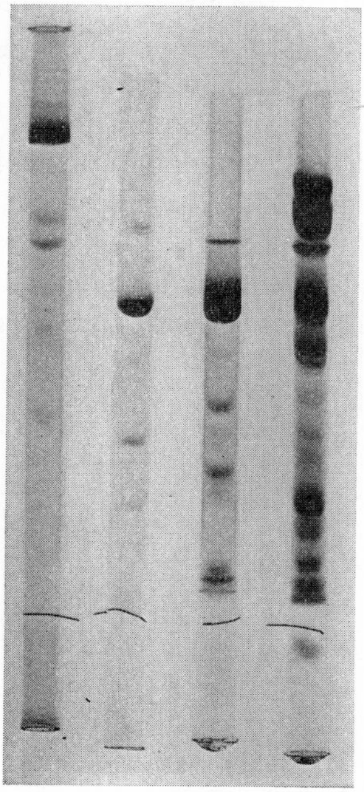

FIG. 1. Gel electrophoresis of membrane proteins solubilized in SDS. From left to right, 1st gel, 20 μg SR protein, electrophoresis in pH 7.4 system; 2nd gel, 12 μg SR protein, pH 8.3 system; 3rd gel, 48 μg SR protein, pH 8.3 system; 4th gel, 54 μg erythrocyte ghost protein, pH 8.3 system.

pH 7.4 separates two minor protein components from a major fraction which migrates with lower speed (first gel in Fig. 1).

Greater mobility and better resolution of all fractions are obtained by electrophoresis at pH 8.3. With this system, the solubilized SR protein is also resolved in one slow principal component and two faster minor bands. In addition, a variable amount of fast migrating material may be found near the buffer front (second and third gels in Fig. 1).

Identical patterns are obtained when different amounts of SR protein are

TEMED, 0.1% SDS, and 0.07% ammonium persulfate (pH 8.9). The stacking gels were prepared with 5.75% 1 N HCl, 0.75% Tris, 0.06% TEMED, 4% acrylamide, 1% BIS, 5×10^{-4}% riboflavin, 20% sucrose, 0.1% SDS (pH 6.6). The running buffer contained 0.06% Tris, 0.29% glycine, and 0.1% SDS (pH 8.3). The tracking dye was bromphenol blue, and the current 2 mA/gel.

placed on electrophoretic gels independent of the presence of reducing agents such as dithioerythritol (0.1–20 mM).

These experiments indicate that the major protein component of SR is homogeneous. On the other hand, the excellent resolving power of the pH 8.3 system is demonstrated in gel 4 of Fig. 1, where an electrophoretic pattern of solubilized erythrocyte ghosts displays great heterogeneity of protein fractions.

In both the pH 7.0 and 8.3 electrophoretic systems, a linear relationship is obtained between the electrophoretic mobilities of protein markers and the

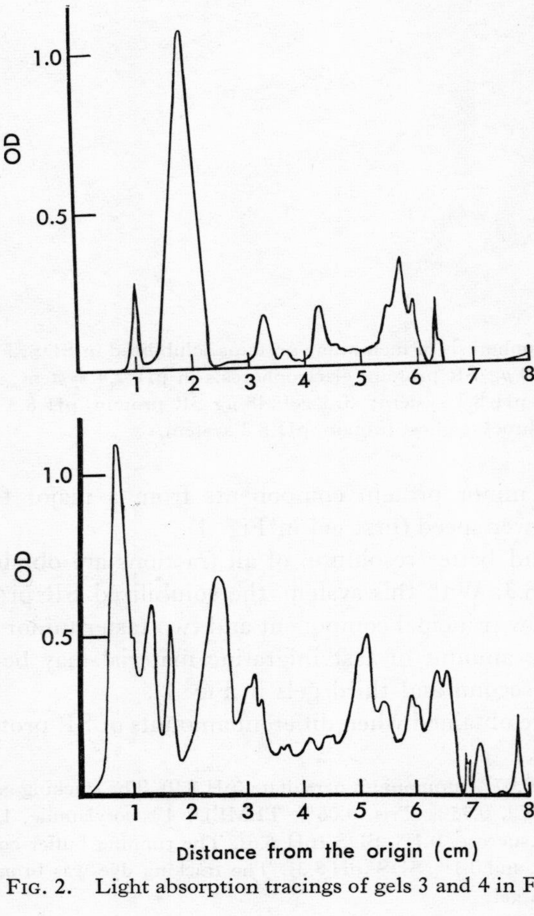

FIG. 2. Light absorption tracings of gels 3 and 4 in Fig. 1.

logarithms of their molecular weights. Based on this standardization, the apparent molecular weights derived for the protein components of SR are: 106,000 for the main component and 63,000 for the slower of the minor bands. Some discrepancy is obtained with the fast minor band in the two pH systems. In fact, this band exhibits a mobility corresponding to a molecular weight of 58,000 at 7.4, and of 48,000 at pH 8.3.

Contrary to observations made with other proteins (*16*), the presence of lipids has very little influence on the electrophoretic mobility of SR proteins. When SR lipids and protein are separated by centrifugation on density gradients in the presence of Triton X-100 (*10*), no significant changes in the electrophoretic mobility of the protein, subsequently solubilized in SDS, are observed.

Light absorption tracings of stained gels (Fig. 2) show that the main band

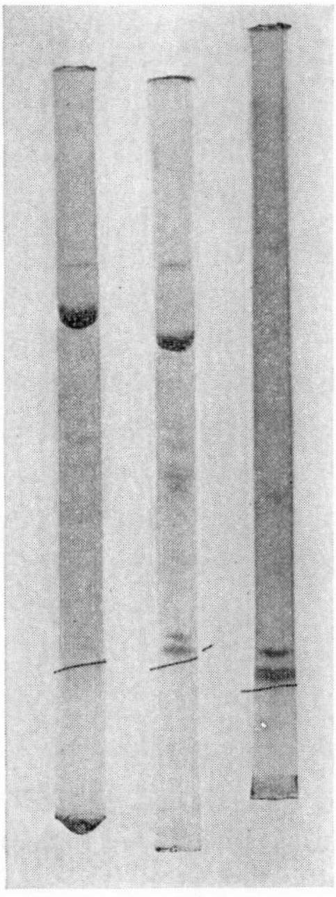

FIG. 3. Electrophoresis (pH 8.3) of SR proteins, solubilized with SDS before (gel 1) and after 4 days (gel 2) and 7 days (gel 3) incubation in 10 mM histidine (pH 7) at 4°C. If the incubation was carried out in the presence of 30% sucrose or 0.5 mM sodium azide, no change in the electrophoretic pattern was observed.

accounts for 70% of the total SR protein and exhibits a smooth profile with symmetrical slopes on both sides. For comparison, the multiple fractions obtained with erythrocyte ghost proteins are also shown in Fig. 2.

If SR vesicles are aged in a neutral buffer, extensive proteolytic degradation occurs. The gels reproduced in Fig. 3 show how the main protein fraction is progressively reduced, while the amount of rapidly migrating peptides increases. Nearly complete degradation is obtained after a week incubation at 4°C, or 4 days at 25°C. The proteolytic enzymes are of bacterial origin, since bacteriostatic agents such as sodium azide or sucrose (30%) prevent the degradation. It should be pointed out that if these proteolytic enzymes are present, proteolysis continues even after SDS solubilization.

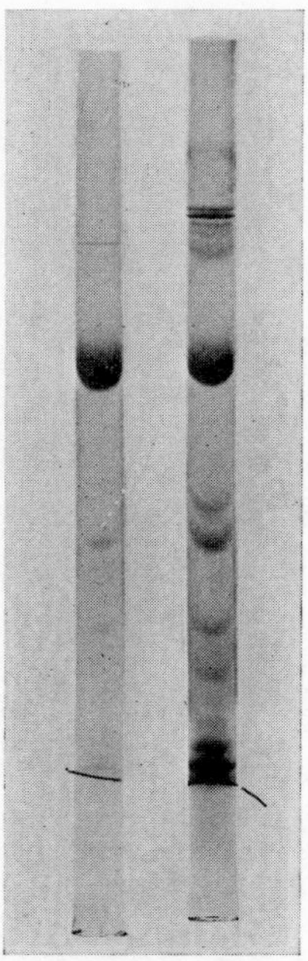

FIG. 4. SDS electrophoresis of Triton-soluble (gel 1) and Triton-insoluble (gel 2) SR proteins. Procedure: SR (8 mg protein) was incubated in 2.0 ml 0.5% (v/v) Triton for 10 min on ice. After 1-hr centrifugation at 40,000×g, 0.1 ml of the supernatant (Triton-soluble protein) was further solubilized in 1.0 ml 1% SDS (pH 8.3). The entire sediment (Triton-insoluble residue) was dissolved in 1.0 ml 1% SDS (pH 8.3); 50 μl samples were subsequently placed in acrylamide gels for electrophoresis in a pH 8.3 system.

Purification and Solubilization of SR ATPase

We (*10*) have previously reported that Triton X-100 solubilizes 80% of the SR protein, selectively removing the main protein component and the ATPase activity from the membrane. Figure 4 shows an experiment where the Triton-soluble fraction and the Triton-insoluble residue were further solubilized with SDS and permitted to migrate in electrophoretic gels. It is readily apparent that the Triton-soluble fraction contains the main protein component of SR, contaminated only by traces of the minor components. In

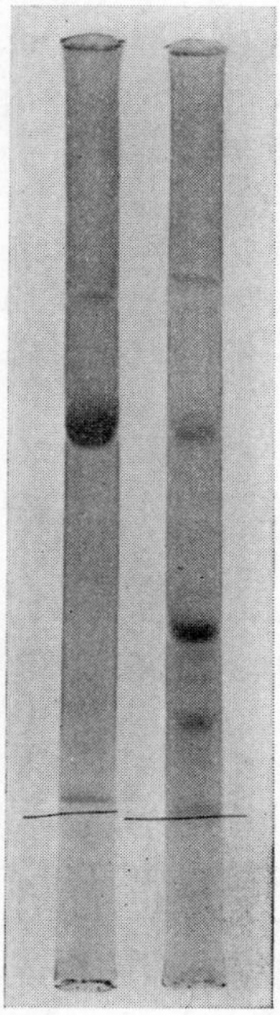

FIG. 5. SDS gel electrophoresis (pH 8.3) of purified SR ATPase (gel 1) and other SR protein components (gel 2) separated from ATPase. Procedure: SR (7.4 mg protein) was suspended in 5.0 ml 5 mM EDTA (pH 8.6) and 0.5 mM sodium azide. After 48-hr incubation at room temperature, the suspension was centrifuged at 40,000×*g* for 1 hr. Part of the supernatant (1.0 ml) was incubated for 10 min in 1.5 ml 1% SDS (pH 8.3). The sediment was resuspended in 1.0 ml 10 mM histidine (pH 7), and 25 μl of this suspension were solubilized in 1.0 ml 1% SDS (pH 8.3). 100 μl samples of either the solubilized sediment or supernatant were placed on acrylamide gels for electrophoresis in a pH 8.3 system. All of the ATPase activity was found in the sediment previous to SDS solubilization.

contrast, the Triton-insoluble residue contains a much larger proportion of the minor protein components of SR.

Another approach to ATPase purification is based on extraction of the minor protein components with low ionic strength media. In fact, Duggan and Martonosi (*17*) reported that it is possible to solubilize SR minor protein components by extraction with 1 mM ethylenediaminetetraacetic acid (EDTA). We find that extraction with 5 mM EDTA (pH 8.6) at 4°C for 48 hr in the presence of sodium azide, as described by Reynolds and Trayer (*18*) for erythrocyte ghosts, completely removes the minor protein components from the SR membrane. The main component is thus purified (Fig. 5) and remains with the membrane particles, which also retain all of the ATPase activity. These residual particles can be subsequently solubilized with Triton X-100, obtaining a solution of pure and active ATPase by a simple method and with nearly 100% yield.

The experiments reported above establish that the main protein component of SR is the ATPase enzyme, which accounts for 70% of the total protein and has a molecular weight of approximately 100,000.

ATP Binding to SR

Adenosine triphosphate binding to SR was first measured by Ebashi and Lipmann (*1*). Subsequently we performed studies with nucleotides labeled in the adenosine or phosphate moieties, to ascertain the occurrence of excess phosphate incorporation into SR, in the presence of Ca^{2+} (*20*). In addition, we have now performed experiments to determine the maximum number of specific ATP binding sites per unit weight of SR. In fact, a match of this number with that of ATPase sites, derived from protein solubilization studies, would provide an important stoichiometric ratio.

Studies of ATP binding to SR present some difficulty, due to sizable hydrolysis of ATP which is catalyzed by the ATPase, even in the absence of added divalent cations. This difficulty may be overcome by the use of a previously described method (*20*) based on rapid filtration of a small volume of an SR-ATP reaction mixture after 30-sec incubation on ice.

Our ATP binding data, displayed in a Scatchard plot, are shown in Fig. 6 and indicate the presence of 7 moles of ATP binding sites/10^6 g SR protein, with a 3×10^{-5} M dissociation constant which is very similar to the K_m determined by us for the Ca^{2+}-dependent ATPase of SR (*21*).

The plot in Fig. 6 also shows the presence of lower affinity ATP binding

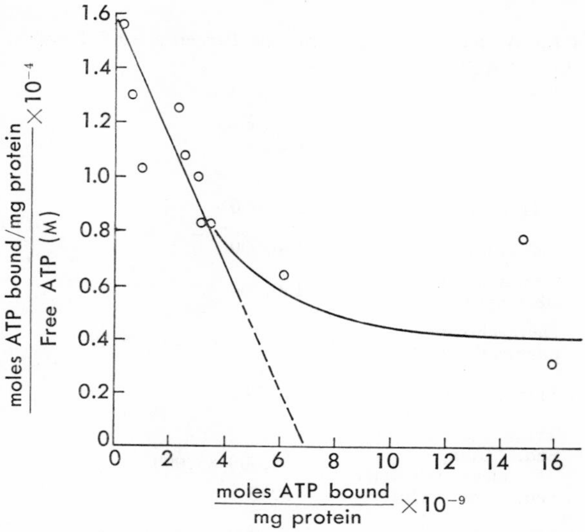

FIG. 6. Scatchard's plot of ATP binding to SR. Each point is the average of 18 determinations in 9 experiments.

in SR. However, this part of the curve cannot be evaluated due to excessive scattering of the data.

Discussion

In previous studies (*11–13*) identification of the ATPase enzyme with protein components of SR separated by electrophoresis was based on the association of a specific electrophoretic band with radioactive phosphate transfered from ATP to SR previous to membrane solubilization. The phosphorylated protein was assumed to be an intermediate form of the ATPase enzyme.

In our experiments, ATPase identification was based on the enzymatic activity displayed by a purified membrane protein, which migrates as a single band in SDS gel electrophoresis. This protein has identical electrophoretic behavior to the previously described phosphorylated protein (*11–13*).

The size of the SDS-structured ATPase enzyme is 106,000. This is in good agreement with most of the reported figures (Table I). The minor disagreement with the 85,000 M.W. reported (*10*) for the Triton-solubilized

TABLE I
Reported Estimates of the Molecular Weight and the Percentage of Total Membrane Protein Accounted for by SR ATPase

Authors	Method	Molecular weight	Total protein (%)	Ref. No.
W. Hasselbach and K. Seraydarian	-SH titration	100,000	—	(*22*)
K. Vegh *et al.*	Radiation inactivation	190,000	—	(*23*)
G. Inesi *et al.*	Triton solubilization electrophoresis	100,000	—	(*6*)
D.H. MacLennan	Deoxycholate purification	—	16	(*9*)
B.P. Yu and E.J. Masoro	SDS solubilization	10,000	90	(*14*)
B.H. McFarland and G. Inesi	Triton solubilization		80	(*10*)
	Sed. equib., electrophoresis	80,000–85,000		
	Sed. veloc. (S20).	5.3S		
D.H. MacLennan *el al.*	Electrophoresis	102,000	—	(*12*)
A. Martonosi and R.A. Halpin	SDS solubilization, electrophoresis	100,000–106,000	40–60	(*24*)
D.H. MacLennan and P.T.S. Wong	Electrophoresis	102,000	30	(*25*)
G. Meissner and S. Fleischer	Electrophoresis	115,000	50	(*13*)
G. Inesi *et al.*	SDS solubilization, electrophoresis	106,000	70	—

SR ATPase may be due to a slightly different behavior of the protein still in the active conformation (nonsolubilized in SDS) or to the lesser accuracy of methods not involving SDS solubilization, such as sedimentation equilibrium (*10*).

Much more puzzling is our failure to obtain dissociation of the SR ATPase into small subunits (10,000 M.W.) as reported by Yu and Masoro for SR (*14*), and by Laico *et al.* (*19*) for other membranes. It should be pointed out, however, that we did obtain peptide fragments when SR preparations were stored in the absence of bacteriostatic agents (Fig. 3). This effect was evidently due to proteolytic enzymes derived from contaminating bacterial growth.

In our experiments, the major protein component of SR, which contains all of the membrane ATPase activity and migrates as a single band in SDS gels, accounts for 70% of the membrane protein. This figure, although similar to our own previous estimate (*10*), is in disagreement with some other reports (Table I).

We suggest that some of the reasons for underestimating the percentage of SR protein accounted for by ATPase may be as follows:

1. Use of nonpurified SR vesicles as starting material. Our method of preparation of the vesicles begins with homogenization of muscle in low ionic strength media to avoid extraction of proteins from myofibrils and other subcellular fragments to be discarded. In addition, extraction of proteins from the SR vesicles with high ionic strength media is routinely included in our preparation. Homogenization and centrifugation are carried out in the presence of sucrose.
2. Proteolytic digestion of SR vesicles by enzymes of bacterial origin and partial degradation of ATPase to inactive protein fragments.
3. Use of electrophoretic systems that allow aggregation of native SR protein and formation of size isomers. It should be pointed out that in our SDS electrophoresis systems no protein is excluded from the gels and no significant amount of protein displays a mobility corresponding to a MW larger than 106,000.
4. Assuming that the steady-state levels of phosphorylated enzyme intermediate always correspond to the maximum number of available enzymatic sites. This assumption is kinetically unreasonable. The maximum number of specific ATP binding sites, derived from a Scatchard plot, is consistently 7 moles/10^6 g protein. This figure can only occasionally be matched by steady-state levels of phosphorylated intermediate, which vary between 4 and 6 moles/10^6 g protein.

A stoichiometric ratio between moles of ATPase and number of specific binding sites can be calculated. Since the major protein component has a molecular weight of approximately 100,000 and accounts for 70% of the total protein, 10^6 g SR protein contains 7 moles of ATPase. This figure matches the number of specific ATP binding sites (Fig. 6) and, on the assumption that each ATPase molecule contains one enzymatic site, it can be concluded that the major electrophoretic band is homogeneous.

With regard to solubilization and purification of ATPase, after early attempts, we became discouraged with the use of ionic detergents, such as deoxycholate (DOC), due to the exceedingly narrow margin between detergent concentrations causing solubilization and those producing denaturation of proteins.

In contrast, solubilization of minor protein components with EDTA is a simple and safe procedure. If desired, the residual membrane fragments containing the purified ATPase (Fig. 5), can be solubilized with Triton X-100, a nonionic detergent.

The fact that ATPase is the only SR protein requiring detergents for solubilization is consistent with hydrophobic interaction and location of this enzyme within the membrane thickness, as previously suggested (*12*, *26–28*).

Considering that 1 g of SR protein corresponds to 4.2×10^{22} Å^2 of membrane surface area (*28*), the ATPase molecules should be spaced in the membrane with an average center-to-center distance of 100–110 Å. This distance is consistent with the results obtained by Hasselbach and Elfvin (*29*) with Hg-phenyl azoferritin labels, and with previously published estimates on the structure of SR membranes (*28*).

REFERENCES

1. S. Ebashi and F. Lipmann, *J. Cell Biol.*, **14**, 389–400 (1962).
2. W. Hasselbach and M. Makinose, *Biochem. Z.*, **333**, 518–528 (1961); **339**, 94–111 (1963).
3. T. Yamamoto and Y. Tonomura, *J. Biochem.*, **62**, 558–575 (1967); *J. Biochem.*, **64**, 137–145 (1968).
4. A. Martonosi, *Biochem. Biophys. Res. Commun.*, **29**, 753–757, (1969); *J. Biol. Chem.*, **244**, 613–620.
5. M. Makinose, *Eur. J. Biochem.*, **10**, 74–82 (1969).
6. G. Inesi, E. Mering, A. J. Murphy, and B. H. McFarland, *Arch. Biochem. Biophys.*, **138**, 285–294 (1970).
7. A. Martonosi, *J. Biol. Chem.*, **243**, 71–81 (1968).
8. R. Salinger, M. Klein, and A. Amsterdam, *Biochim. Biophys. Acta*, **183**, 19–26 (1969).
9. D. H. MacLennan, *J. Biol. Chem.*, **245**, 4508–4518 (1970).
10. B. H. McFarland and G. Inesi, *Biochim. Biophys. Res. Commun.*, **41**, 239–243 (1970); *Arch. Biochem. Biophys.*, **145**, 456–464 (1971).
11. A. Martonosi and R. A. Halpin, *Arch. Biochem. Biophys.*, **144**, 66–77 (1971).
12. D. H. MacLennan, P. Seeman, G. H. Iles, and C. C. Yip, *J. Biol. Chem.*, **246**, 2702–2710 (1971).
13. G. Meissner and S. Fleischer, *Biochim. Biophys. Acta*, **241**, 356–378 (1971).
14. B. P. Yu and E. J. Masoro, *Biochemistry*, **9**, 2909–2917 (1970).
15. S. Eletr and G. Inesi, *Biochim. Biophys. Acta*, **282**, 174–178 (1972).

16. J. M. Fessenden-Roolen, *Biochem. Biophys. Res. Commun.*, **46**, 1347–1353 (1972).
17. P. F. Duggan and A. Martonosi, *J. Gen. Physiol.*, **56**, 147–167 (1970).
18. J. A. Reynolds and H. Trayer, *J. Biol. Chem.*, 7337–7342 (1971).
19. M. T. Laico, E. I. Ruoslahti, D. S. Papermaster, and W. J. Dreyer, *Proc. Natl. Acad. Sci. U.S.*, **67**, 120–128 (1970).
20. G. Inesi and J. Almendares, *Arch. Biochem. Biophys.*, **126**, 733–735 (1968).
21. G. Inesi, J. Goodman, and S. Watanabe, *J. Biol. Chem.*, **242**, 4637–4643 (1967).
22. W. Hasselbach and K. Seraydarian, *Biochem. Z.*, **345**, 159–172 (1966).
23. K. Vegh, P. Spiegler, C. Chamberlain, and W. F. H. W. Mommaerts, *Biochim. Biophys. Acta*, **163**, 266–270 (1968).
24. A. Martonosi and R. A. Halpin, *Arch. Biochem. Biophys.*, **144**, 66–77 (1971).
25. D. H. MacLennan and P. T. S. Wong, *Proc. Natl. Acad. Sci. U.S.*, **68**, 1231–1235 (1971).
26. G. Inesi and I. Asai, *Arch. Biochem. Biophys.*, **126**, 469–477 (1968).
27. D. W. Deamer and R. J. Baskin, *J. Cell Biol.*, **42**, 296–307 (1969).
28. G. Inesi, *Ann. Rev. Biophys. Bioeng.*, **1**, 191–210 (1972).
29. W. Hasselbach and L. G. Elfvin, *J. Ultrastruct.*, **17**, 598–622 (1967).

Molecular Mechanism of Ca^{2+} Transport through the Membrane of the Sarcoplasmic Reticulum*[1]

Yuji Tonomura and Shinpei Yamada*[2]

Department of Biology, Faculty of Science, Osaka University, Osaka, Japan

The partial reactions of the Ca^{2+}, Mg^{2+}-dependent ATPase of the sarcoplasmic reticulum (SR) have been studied kinetically. It has been shown that a phosphorylated intermediate (EP) is formed *via* a Michaelis complex, which is formed by binding of the enzyme with one external MgATP and 2 external Ca^{2+}. In this reaction, Sr^{2+} can replace Ca^{2+}, while Mg^{2+} inhibits the reaction by competing with Ca^{2+}. However, the ratio of binding constants of Mg^{2+} and Ca^{2+} is only $1:2\text{–}3 \times 10^4$. ATP is formed from EP by adding ADP externally, and the formation of ATP requires internal Ca^{2+}. Therefore, it is concluded that 2 Ca^{2+} in the Michaelis complex are translocated from the outside to the inside of the membrane, coupled with EP formation. The binding of 1 mole of internal Mg^{2+} to 1 mole of EP is required for EP decomposition. The EP decomposition is inhibited by internal Ca^{2+}, which competes with Mg^{2+}, and the ratio of the binding constants of Mg^{2+} and Ca^{2+} is 1: 2.5. Thus the formation of EP, *i.e.*, the translocation of the cation site, is accompanied by very large changes in the affinities for Ca^{2+} and Mg^{2+}.

Biological transport of solutes through a biological membrane usually involves 3 steps: Recognition, translocation and release. Of necessity, recognition must occur at the membrane boundary. Translocation through the

*[1] This investigation was supported by grants from the Ministry of Education of Japan and the Muscular Dystrophy Associations of America, Inc. *[2] Present address: Department of Biology, Faculty of Science, Shizuoka University, Shizuoka, Japan.

boundary, release at the other side, and, ultimately, movement across the membrane complete the transport process. The biochemical foundation of the active transport of Na^+ and K^+ was discovered by Skou (*1*), when he found an ATPase in nerve microsomes which shows high activity only when Mg^{2+}, Na^+, and K^+ are present. This ATPase has received much attention, since it probably participates in the active transport of Na^+ and K^+ (*2–8*). However, because of the complexity of the structure of the cell membrane, which contains only a small amount of the Na^+, K^+-dependent ATPase, many aspects of the molecular mechanism of active transport of Na^+ and K^+ remain to be clarified.

Fragmented sarcoplasmic reticulum (SR) isolated from skeletal muscle has several features which are very useful for investigating the molecular mechanism of cation transport. First, SR contains Ca^{2+}, Mg^{2+}-dependent ATPase (*9*) and takes up Ca^{2+} actively (*9, 10*). Its chemi-osmotic coupling is tight, and 2 moles of Ca^{2+} are actively transported coupled with hydrolysis of 1 mole of ATP by the ATPase (*11–13*). Second, the ATPase accounts for more than 50% of the total protein of SR (*14*), and ATP and P_i appear on the outside of the membrane. Third, the membrane structures can be easily destroyed by Triton-100 or deoxycholate, without impairing the ATPase activity (*15*). Fourth, the concentration of Ca^{2+} is easily controlled by glycoletherdiaminetetraacetic acid (EGTA) and that of Mg^{2+} by ethylenediaminetetraacetic acid (EDTA).

Therefore, the coupling mechanism between Ca^{2+} uptake and ATPase in SR has been studied by many workers, and the properties and physiological functions of the enzyme have already been reviewed (*8*, *16–18*). This article summarizes our kinetic studies on the Ca^{2+}, Mg^{2+}-dependent ATPase of SR, with emphasis on the coupling mechanism of the elementary steps of the ATPase with the 3 steps in cation transport mentioned above, *i.e.*, recognition, translocation and release of Ca^{2+}.

Phosphorylation of SR by ATP and Recognition of Ca^{2+}

The presence of a phosphorylated intermediate in the ATPase reaction has been suggested from the discovery of ADP-ATP exchange activity of SR (*10*, *19*, *20*), and subsequently Ca^{2+}-dependent formation of a phosphorylated protein from SR and ATP was discovered by us (*15*, *21*) and also by Makinose (*22*).

In the initial phase of the reaction of SR ATPase, the concentration of phosphorylated protein (EP) increased rapidly with time without showing any lag phase, while P_i liberation showed a definite lag phase which closely corresponded to the period when EP concentration was increasing *(23)*. Furthermore, the rates of $E^{32}P$ decomposition measured directly under various conditions on suppressing $E^{32}P$ formation by removing Ca^{2+} with EGTA or on adding a large amount of unlabeled ATP were equal to the values of $v/[EP]$, where v is the rate of P_i liberation. From these and other, more indirect results *(15, 21)* it was concluded that the phosphorylated protein is a true intermediate in the reaction of the ATPase.

Over a variety of Ca^{2+} concentrations, there was a satisfactory linear relationship between the reciprocal of the rate of EP formation, v_f^{-1}, and the reciprocal of ATP concentration. When the Ca^{2+} concentration was reduced by increasing the EGTA concentration, the maximum rate, V_f, of v_f decreased, while the Michaelis constant, K_f, was unchanged *(23)*. This indicates that the binding of ATP and Ca^{2+} to the enzyme (necessary for the formation of EP) occurs at random. There was also a satisfactory linear relationship between the reciprocal of V_f and the reciprocal of the square of the free Ca^{2+} concentration, indicating that the enzyme binds with 2 Ca^{2+}. Furthermore, the value of V_f is significantly lower when Mg^{2+} is not added to the system, probably because the substrate for EP formation is MgATP *(23)*.

It is known that EGTA cannot diffuse into the SR membrane *(12)*. When excess EGTA was added to the external medium to chelate free Ca^{2+} in the external medium, EP formation stopped promptly and EP decayed exponentially with time without any appreciable lag phase *(23)*. When unlabeled ATP was added to the medium to dilute the $AT^{32}P$ after $E^{32}P$ formation had started, subsequent $E^{32}P$ formation stopped instantaneously and $E^{32}P$ decayed exponentially without any appreciable lag phase *(23)*. These results clearly show that Ca^{2+} and substrate (MgATP) participating in EP formation bind to the enzyme on the outside of the membrane. From these results the following reaction scheme was proposed for the formation of EP:

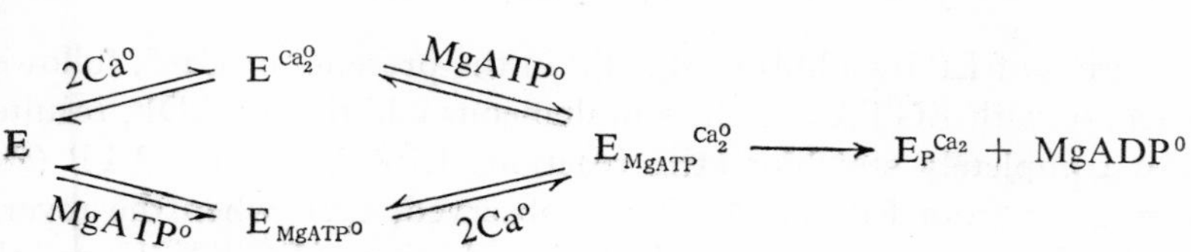

where the superscript o represents the outside of the SR membrane. The amount of active site estimated from the amount of EP formed in the presence of large amounts of Ca^{2+} and ATP was 5–7 moles per 10^6 g protein. Furthermore, the molecular weight of the ATPase protein was about 10–10.5×10^4 (*14*, *24*, *25*), and the content of the ATPase in our SR preparation was 60–70% (*14*). Thus, the results are consistent with the assumption that each molecule of the ATPase contains one phosphorylation site.

We measured the competitive inhibition of Ca^{2+}-dependent EP formation by Mg^{2+}, using SR which was treated with deoxycholate to destroy the membrane structure (*26*). The relationships between v_f and concentrations of Ca^{2+} and Mg^{2+} in the presence of a sufficiently large amount of ATP could easily be interpreted by the following reaction scheme:

$$E_{MgATP} + Mg^{2+} \underset{K_{Mg}}{\rightleftharpoons} E_{MgATP}{}^{Mg}$$

$$E_{MgATP} + 2Ca^{2+} \underset{K_{Ca}}{\rightleftharpoons} E_{MgATP}{}^{Ca_2} \longrightarrow E_P{}^{Ca_2} + MgADP\,.$$

Under the experimental conditions used, the values of the dissociation constants, K_{Ca} and K_{Mg}, were 0.35 μM and 10.6 mM, respectively, and V_f, the maximum rate of EP formation in the presence of Ca^{2+}, was 1.33 mole/10^6 g·sec (*26*). Thus, Ca^{2+} with twice the value of ε, *i.e.*, 12–14 moles per 10^6 g SR protein, should bind to the ATPase with a dissociation constant of 0.35 μM. Actually, Chevallier and Butow (*27*) have recently found by the equilibrium dialysis method that 10–20 moles of Ca^{2+} bind to 10^6 g SR with a dissociation constant of 0.4 μM. It is well known that not only Ca^{2+} but also Sr^{2+} is transported into SR. Under the same conditions, the values of V_f and K_{Sr} for Sr^{2+} were 2.73 mole/10^6 g·sec and 27.5 μM, respectively. These results are consistent with those on the transport of Sr^{2+} into SR reported by Weber *et al.* (*12*), who showed that the rate of transport is similar to that of Ca^{2+}, while the affinity for Sr^{2+} is much less than that for Ca^{2+}.

Formation of ATP from EP and ADP and Translocation of Ca^{2+}

The formation of EP by addition of ATP in the presence of Ca^{2+}, followed by quenching with EGTA and the simultaneous addition of ADP, resulted in almost completely stoichiometric reaction: EP+ADP→E+ATP (*28*). ATP formation from EP and ADP was observed, even when the reverse reaction (EP formation) was not blocked with addition of EGTA (*23*) and the amount of ADP added was of the same order of magnitude as that of ATP added initially. This clearly indicates that EP is a high-energy, phosphate-type compound. In order to study the binding of the phosphoryl group in EP, the dependence of the stability of EP denatured by trichloroacetic acid (TCA) upon pH and hydroxylamine concentration was measured (*14, 21*). The dependence suggested that P from ATP is incorporated into the protein as an acyl phosphate. We have also demonstrated the incorporation of P into the ATPase protein (*14*) and the formation of a hydroxamate by the reaction of EP with 2-hydroxyl-5-nitrobenzyl hydroxylamine (*29*).

ATP was produced when the formation of EP was stopped by applying EGTA to the outside of the membrane, and ADP was then added, as mentioned above. The formation of ATP from EP and ADP was completely independent of the concentration of external Ca^{2+} (*23*). Thus, it appears that the formation of EP+ADP from E+ATP involves external Ca^{2+}, while the reverse reaction requires the presence of internal Ca^{2+}. We confirmed this by the following 2 experiments (*23*). When SR was exposed to excess EGTA at alkaline pH after EP had been formed, ATP formation from EP and ADP was inhibited gradually, in sharp contrast with the prompt inhibition of external Ca^{2+}-dependent EP formation occurring on addition of excess EGTA. The inhibition of ATP formation was reversed by subsequent addition of $CaCl_2$ to the external medium at a high concentration. Duggan and Martonosi (*30*) showed that Ca^{2+} leaks through the SR membrane when the pH is raised and at alkaline pH almost all the Ca^{2+} in SR leaks out within 30 sec when external Ca^{2+} is chelated. The second observation confirming that ATP formation from EP and ADP requires Ca^{2+} located within the SR membrane is that ATP formation from EP and ADP was instantaneously inhibited by addition of excess EGTA, even at neutral pH, when SR had been pretreated with Triton X-100, which made the SR membrane markedly leaky. In control experiments in which unlabeled ATP was used in place of EGTA to stop $E^{32}P$ formation, $AT^{32}P$ formation from $E^{32}P$ and ADP was observed, whether SR was pretreated with Triton X-100 or not. Thus it seems certain that EP formation is activated by Ca^{2+} located outside the SR membrane but that the reverse reaction, *i.e.*, ATP formation from EP and ADP, is activated by Ca^{2+} inside the membrane:

$$E + MgATP \underset{Ca^{i}}{\overset{Ca^{o}}{\rightleftharpoons}} EP + ADP + Mg^{2+},$$

where the superscript i represents the inside of the membrane. According to the principle of microscopic reversibility, this indicates that the Ca^{2+} binding site is translocated from the outside to the inside of the SR membrane coupled with EP formation, and conversely, that it is translocated from the inside to the outside of the membrane coupled with ATP formation from EP and ADP.

With respect to the molecular mechanism of the translocation of the cation site, it should be added that inducement of conformational changes in SR by ATP has been reported by several researchers (*31–33*). In particular, Nakamura *et al.* (*33*) found that the EPR spectrum of spin-labeled SR is altered by the addition of ATP. This alteration of the EPR spectrum occurred in the presence of both Mg^{2+} and Ca^{2+}, and was reversibly abolished by the removal of external Ca^{2+}.

Release of Ca^{2+} and Decomposition of EP

The rate of EP decomposition in the steady state, measured as the ratio of the rate of P_i liberation to EP concentration (v_0/[EP]) was found to increase with increase in the concentration of $MgCl_2$. A double reciprocal plot of the value of v_0/[EP] against the Mg^{2+} concentration fitted a straight line (*23*). This suggests that Mg^{2+} stimulates EP decomposition, and activation of EP decomposition is caused by the binding of 1 mole of Mg^{2+} to 1 mole of EP. When excess EDTA was added to the reaction medium to chelate Mg^{2+} and Ca^{2+} after EP had been formed, further EP formation stopped immediately, but subsequent EP decomposition ceased almost completely after a definite lag period. The inhibited EP decomposition was reactivated by adding $MgCl_2$ at high concentrations. There was no inhibition of EP decomposition when EGTA was used to chelate Ca^{2+} instead of EDTA. These results provide strong additional evidence for the activation of EP decomposition by Mg^{2+}, and suggest that the Mg^{2+} involved in the activation of EP decomposition is located inside the SR membrane. This possibility is strengthened by the fact that the lag period for the termination of EP decomposition after addition of EDTA was greatly shortened by pretreatment of SR with Triton X-100 (*23*). Thus, it was concluded that binding of 1 mole of Mg^{2+} inside the SR membrane to the cation binding site per mole of EP is required for EP decomposition.

To clarify the changes in affinities of the cation binding site for Ca^{2+} and Mg^{2+} with the state of EP, the competitive inhibition of Mg^{2+}-dependent EP decomposition by Ca^{2+} was measured using Ca^{2+}, Mg^{2+}-dependent

ATPase from SR treated with deoxycholate (*26*). In the presence of a sufficiently large amount of Ca^{2+}, all the results obtained were easily interpreted by the following reaction scheme:

$$E_P{}^{Mg} + Ca^{2+} \underset{K_{Ca}/K_{Mg}}{\rightleftharpoons} E_P{}^{Ca} + Mg^{2+}$$

$$E_P{}^{Mg} \longrightarrow E + P_i + Mg^{2+}.$$

The value of K_{Mg}/K_{Ca} was 2.5. Thus the formation of EP, *i.e.*, the translocation of the cation site, is accompanied by very large changes in the affinities for Ca^{2+} and Mg^{2+}. After the translocation, the ratio of the concentrations of Mg^{2+} and Ca^{2+} necessary for half saturation changes from $2\text{–}3 \times 10^4 : 1$ to $2.5 : 1$.

We (*13*) have found that the acidity change in the external medium during the uptake of Ca^{2+} by SR is due to the hydrolysis of ATP itself, and that P_i released from the ATP is not transported inside the SR. Carvalho and Leo (*34*) investigated the transport of cations coupled with the uptake of Ca^{2+}, and found that the sum of the equivalents of Mg^{2+}, K^+, and Ca^{2+} within the SR vesicles is constant. This suggests that there is an efflux of Mg^{2+} and K^+ coupled with the influx of Ca^{2+}.

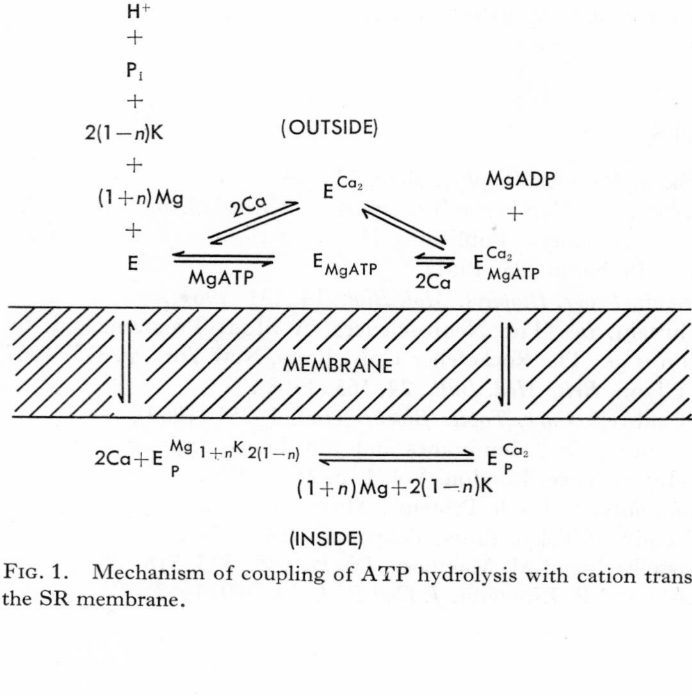

FIG. 1. Mechanism of coupling of ATP hydrolysis with cation transport across the SR membrane.

Makinose and Hasselbach (*35*) and Panet and Selinger (*36*) observed that, under conditions in which a fast ADP- and P_i-dependent release of Ca^{2+} from Ca^{2+}-loaded SR occurs, the net outward movement of 2 moles of Ca^{2+} through the membrane is stoichiometrically related to a net formation of 1 mole of ATP. More recently, Makinose (*37*) and we (*14, 38*) showed that ATPase is phosphorylated by P_i coupled with the outward translocation of Ca^{2+} through the membrane. Thus, the ATPase reaction of SR is reversible.

Figure 1 shows a simplified mechanism of coupling of hydrolysis of ATP by SR with the transport of Ca^{2+} and Mg^{2+}. Although a more elaborate mechanism is required to interpret the acceleration of EP formation by high concentrations of ATP (*23*), the phosphorylation of ATPase by P_i coupled with the outward movement of Ca^{2+} through the membrane (*14, 38*) and the exchange reaction between P_i and $H_2{}^{18}O$ catalyzed by SR (*39*), this simple mechanism is sufficient to explain all the kinetic results on the ATPase reaction mentioned in this paper. Thus the formation of 1 mole of EP causes 2 moles of Ca^{2+} to be transported from the outside to the inside of the membrane, where they replace $(1+n)$ moles of Mg^{2+} and $2(1-n)$ moles of K^+, since EP formation induces remarkably large affinity changes of the cation site for Ca^{2+} and Mg^{2+}. The binding of 1 mole of Mg^{2+} to the cation site is required for EP decomposition, which releases P_i and protons outside the membrane.

REFERENCES

1. J. C. Skou, *Biochim. Biophys. Acta*, **23**, 394 (1957).
2. J. C. Skou, *in* "Membrane Transport and Metabolism," ed. by A. Kleinzeller and A. Kotyk, Publishing House of the Czechoslovak Academy of Sciences, Praha, p. 228 (1961).
3. J. C. Skou, *Progr. Biophys. Mol. Biol.*, **14**, 131 (1964).
4. R. Whittam, *in* "The Neurosciences, A Study Program," ed. by G. C. Quarton *et al.*, The Rockefeller Univ. Press, New York, p. 313 (1967).
5. I. M. Glynn, *Brit. Med. Bull.*, **24**, 165 (1968).
6. P. C. Caldwell, *Curr. Topics Bioenergetics*, **3**, 251 (1969).
7. S. L. Bonting, *in* "Membranes and Ion Transport," ed. by E. E. Bittar, Wiley-Interscience, London, Vol. 1, p. 257 (1970).
8. Y. Tonomura, "Muscle Proteins, Muscle Contraction and Cation Transport," Univ. of Tokyo Press, Tokyo, pp. 454 (1972).
9. W. Hasselbach and M. Makinose, *Biochem. Z.*, **333**, 518 (1961).
10. S. Ebashi and F. Lipmann, *J. Cell Biol.*, **14**, 389 (1962).

11. W. Hasselbach and M. Makinose, *Biochem. Z.*, **339**, 94 (1963).
12. A. Weber, R. Herz, and I. Reiss, *Biochem. Z.*, **345**, 329 (1966).
13. S. Yamada, T. Yamamoto, and Y. Tonomura, *J. Biochem.*, **67**, 789 (1970).
14. S. Yamada, M. Sumita, and Y. Tonomura, *J. Biochem.*, **72**, 1537 (1972).
15. T. Yamamoto and Y. Tonomura, *J. Biochem.*, **62**, 558 (1967).
16. W. Hasselbach, *Progr. Biophys. Mol. Biol.*, **14**, 167 (1964).
17. A. Weber, *Curr. Topics Bioenergetics*, **1**, 203 (1966).
18. A. Martonosi, *in* "Biomembranes," ed. by L. A. Manson, Plenum Press, New York, Vol. 1, Chapter 3 (1971).
19. W. Hasselbach and M. Makinose, *Biochem. Biophys. Res. Commun.*, **7**, 132 (1962).
20. M. Makinose and W. Hasselbach, *Biochem. Z.*, **343**, 360 (1965).
21. T. Yamamoto and Y. Tonomura, *J. Biochem.*, **64**, 137 (1968).
22. M. Makinose, *Eur. J. Biochem.*, **10**, 74 (1969).
23. T. Kanazawa, S. Yamada, T. Yamamoto, and Y. Tonomura, *J. Biochem.*, **70**, 95 (1971).
24. D. H. MacLennan, P. Seeman, G. H. Iles, and C. C. Yip, *J. Biol. Chem.*, **246**, 2702 (1971).
25. A. Martonosi and R. A. Halpin, *Arch. Biochem. Biophys.*, **144**, 66 (1971).
26. S. Yamada and Y. Tonomura, *J. Biochem.*, **72**, 417 (1972).
27. J. Chevallier and R. A. Butow, *Biochemistry*, **10**, 2733 (1971).
28. T. Kanazawa, S. Yamada, and Y. Tonomura, *J. Biochem.*, **68**, 593 (1970).
29. T. Yamamoto, A. Yoda, and Y. Tonomura, *J. Biochem.*, **69**, 807 (1971).
30. P. E. Duggan and A. Martonosi, *J. Gen. Physiol.*, **56**, 147 (1970).
31. G. Inesi and W. C. Landgraf, *Bioenergetics* **1**, 355 (1970).
32. J. M. Vanderkooi and A. Martonosi, *Arch. Biochem. Biophys.*, **149**, 99 (1971).
33. H. Nakamura, H. Hori, and T. Mitsui, *J. Biochem.*, **72**, 635 (1972).
34. A. P. Carvalho and B. Leo, *J. Gen. Physiol.*, **50**, 1327 (1967).
35. M. Makinose and W. Hasselbach, *FEBS Letters*, **12**, 271 (1971).
36. R. Panet and Z. Selinger, *Biochim. Biophys. Acta*, **225**, 34 (1972).
37. M. Makinose, *FEBS Letters*, **25**, 113 (1972).
38. S. Yamada and Y. Tonomura, *J. Biochem.*, **71**, 1101 (1972).
39. T. Kanazawa and P. D. Boyer, *J. Biol. Chem.*, in press.

Comparative Studies of Fragmented Sarcoplasmic Reticulum of White Skeletal, Red Skeletal, and Cardiac Muscles

Shoichi Harigaya and Arnold Schwartz

Biological Research Laboratory, Tanabe Seiyaku Co., Toda, Saitama, Japan, and Department of Cell Biophysics, Baylor College of Medicine, Houston, Texas, U.S.A.

Comparative studies on the binding of Ca^{2+} and Sr^{2+} of isolated sarcoplasmic reticulum (SR) obtained from white skeletal, red skeletal and cardiac muscles of rabbit are described. Maximum amounts of Ca^{2+} bound by white skeletal (WSR), red skeletal (RSR), and cardiac SR (CSR) are approximately 170, 60, and 40 nmoles Ca^{2+}/mg protein, respectively, at 25°C. The rate of Ca^{2+} binding is highest in WSR and lowest in RSR. Maximum amounts of Sr^{2+} binding of RSR and CSR are 1.5 times those of Ca^{2+} binding, while WSR binds almost the same amounts of Sr^{2+} as of Ca^{2+}. The rates of Sr^{2+} binding in RSR, WSR, and CSR are 1.8 times, 1.6 times, and 1.6 times the rate of Ca^{2+} binding, respectively. Oxalate markedly stimulates Sr^{2+} uptake, but to a lesser extent than with Ca^{2+}. The amount of Sr^{2+} uptake by RSR is higher than that by WSR. Ca^{2+} uptake by these preparations in the presence of oxalate is always preceded by a lag period, but Sr^{2+} uptake takes place without any lag time. From the time course of Ca^{2+} binding and release, bound Ca^{2+} is classified into 3 categories and their physiological implications are discussed.

It is now widely accepted that the contraction-relaxation cycle in skeletal muscle is based on the movement of Ca^{2+} between 2 Ca^{2+} binding sites, namely, troponin and the sarcoplasmic reticulum (SR) (*1*, *2*).

Compared with its early establishment in skeletal muscle (*3–6*), the role of SR in cardiac muscle has long been the subject of controversial discussions. Although there is no doubt that cardiac SR (CSR) plays a substantial role in regulating the cardiac contractile processes (*7*, *8*), it is still a matter of argument whether or not the control of Ca^{2+} concentration is exclusively performed by SR, as in the case of skeletal muscle.

One of the reasons for this might be the difficulty in isolating cardiac SR preparations in the native state. The situation is somewhat similar in skeletal red muscle (*9–13*). Recently we have succeeded in preparing improved preparations of SR from these muscles (*10*, *14*). Due to this success, we are now able to make comparative studies of SR from white skeletal (WSR), red skeletal (RSR), and cardiac muscles.

Comparative Aspects of SR of White Skeletal, Red Skeletal, and Cardiac Muscles

The properties of fragmented SR preparations (FSR) of white skeletal, red skeletal and cardiac muscles of the rabbit are summarized in Table I. The rate constant of Ca^{2+} binding* is highest in WSR and lowest in RSR. The rate constant of RSR (1,300 $\text{M}^{-1}\text{sec}^{-1}$) (*10*) and that of CSR (2,375 $\text{M}^{-1}\text{sec}^{-1}$) (*14*) are almost one-third and half that of WSR (4,000 $\text{M}^{-1}\text{sec}^{-1}$) (*14*), respectively. These values may be reasonable if we consider the slower rates of the contraction-relaxation cycle in red and cardiac muscles than in white muscle. Langer (*16*) has suggested that the relaxation time for cardiac muscle is not less than 200 msec at 70 to 80 beats/min at 20°C. The amount of Ca^{2+} bound by CSR in 200 msec is estimated to be 3 to 8 nmoles Ca^{2+}/mg protein, based on the above rate constant (*14*). Since the yield of CSR from rabbit heart muscle is about 1.5 mg/*g* muscle, CSR should bind Ca^{2+} at a rate of 4.5 to 12 nmoles Ca^{2+}/g muscle per 200 msec at room temperature. According to Ebashi *et al.* (*1*), skeletal troponin binds about 4 moles of exchangeable Ca^{2+} per 10^5 g; the amount of Ca^{2+} bound by cardiac troponin is about half that of skeletal troponin, and the troponin content in skeletal muscle would be about 3 mg/ml. If the assumption is made that the content of cardiac troponin is approximately the same as in skeletal muscle, the amount of Ca^{2+} required for full activation of the cardiac muscle contractile system is estimated to be about 60 nmoles/g muscle. Since cardiac muscle attains only about one-fourth to one-third the tension of maximum contrac-

* The terms " Ca^{2+} binding," which means Ca^{2+} accumulation in the absence of oxalate, and " Ca^{2+} uptake," which means Ca^{2+} accumulation in the presence of oxalate, are used for convenience following the suggestion of Katz and Repke (*15*), although these terms might not represent the true mechanism of Ca^{2+} accumulation by FSR.

TABLE I
Comparison of Properties of FSR

Property	FSR		
	WSR	RSR	CSR
a. In the absence of oxalate:			
Rate of Ca^{2+}binding			
Rate constant ($M^{-1}sec^{-1}$)	4,000	1,300	2,400
Half-time for maximum Ca^{2+}binding (sec)	<3	20	10
Amount of maximum			
Ca^{2+}binding (nmoles/mg)			
At 25°C	170	58	40
At 37°C	160	83	42
K_m of ATP for Ca^{2+}binding	2×10^{-7}	2×10^{-6}	8×10^{-6}
Energy of activation for Ca^{2+}binding (kcal/mole)	11–16	11.2	10.5
b. In the presence of oxalate:			
Amount of Ca^{2+}uptake (nmoles/mg)	4,800	2,000–3,000	2,000–3,000

tion under physiological conditions, the actual amount of Ca^{2+} required is about 20 nmoles. If we consider the low yield and deterioration during preparation of such a labile material as FSR, the rate of Ca^{2+} binding derived from experiments with isolated preparations may not be far below that required physiologically.

As to the capacity for Ca^{2+} binding, RSR is superior to CSR. The Ca^{2+} binding capacity of CSR varies from species to species. CSR of dog (*14*), hamster (*17*) or chick (*8*) has high Ca^{2+} binding capacities such as 50 to 70 nmoles/mg. The apparent Ca^{2+} binding constant of rabbit CSR (2×10^6 M^{-1}) is almost one-third that of rabbit WSR (7×10^6 M^{-1}) (*14*). At present it is not clear whether this low binding constant represents the real state of CSR or reflects its denaturation during the preparation process.

Ca^{2+} binding by FSR is temperature-dependent. The apparent activation energies for Ca^{2+} accumulation by the 3 types of FSR are almost equal in the range between O° and 25°C (*12, 14, 18*). The maximum amount of Ca^{2+} bound by RSR is greater at 37°C than at 25°C, while those by WSR and CSR change only slightly with increase in temperature from 25° to 37°C. Since the Ca^{2+} bound to FSR is released very rapidly in WSR and CSR at a high temperature such as 37°C, the apparent capacities of WSR or CSR may not increase with increase in temperature from 25° to 37°C. On the other hand, RSR is different from WSR and CSR in that the Ca^{2+} bound to RSR is hardly released, as will be described later.

The K_m values of ATP for Ca^{2+} binding by FSR varies with muscle types, being 2×10^{-7}, 2×10^{-6}, and 8×10^{-6} in WSR, RSR, and CSR, respectively.

The effects of ITP and GTP on Ca^{2+} binding by RSR are almost the same as that of ATP in supporting the amount and rate of Ca^{2+} binding, but these nucleotides are less effective on WSR and CSR.

In RSR the amount of Ca^{2+} binding with ADP is the same as with ATP, but the rate is much slower. It is not yet clear whether this effect is due to ADP itself or due to the presence of adenylkinase in RSR preparations.

The amount of Ca^{2+} uptake in the presence of oxalate is also highest in WSR; the capacities of RSR and CSR are almost the same, about half that of WSR.

Sr^{2+} Binding by SR of White Skeletal, Red Skeletal, and Cardiac Muscles

It is well known that Sr^{2+} can replace Ca^{2+} in various biological functions. Weber *et al.* (*19*), and Yamada and Tonomura (*20*) have shown that Sr^{2+} is also bound by WSR with less affinity but with a faster rate than Ca^{2+}.

Ebashi *et al.* (*1*) have shown that the different sensitivities of myosin B to Sr^{2+} and Ca^{2+} between the cardiac and skeletal muscles can be explained by different properties of troponin in these muscle types. Therefore it is of interest to compare Sr^{2+} binding by FSR from various muscle types with Ca^{2+} binding (Sr^{2+} binding by FSR could be measured by double-beam spectrophotometry using the murexide method of Ohnishi and Ebashi (*21*) as in the case of Ca^{2+}).

Ca^{2+} and Sr^{2+} binding by FSR obtained from white skeletal, red skeletal and cardiac muscles of rabbit is shown in Fig. 1 and Table II. It should be pointed out that the binding of Ca^{2+} or Sr^{2+} is immediately followed by a fairly rapid releasing phase, though this is not so fast as the binding phase. All 3 FSR preparations bind and release Sr^{2+} faster than Ca^{2+}. The rate of Sr^{2+} binding by FSR is 1.6 times that of Ca^{2+} binding in white and cardiac muscle, and 1.8 times that in red muscle. The maximum amount of Sr^{2+} bound by FSR is 1.5 times that of Ca^{2+} in red and cardiac muscle, and about the same as that of Ca^{2+} in white muscle. In white and cardiac muscle,

Sr^{2+} bound to FSR is released faster than Ca^{2+}, whereas no significant release of Ca^{2+} or Sr^{2+} takes place in red muscle in an hour.

Oxalate, which is known to increase Ca^{2+} uptake by FSR, also enhanced

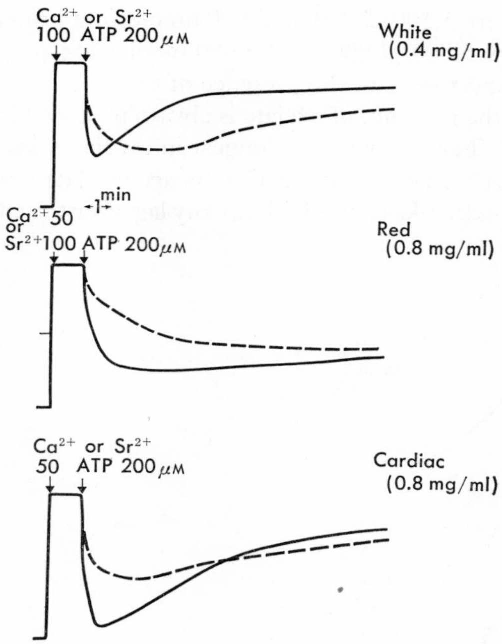

FIG. 1. Ca^{2+} or Sr^{2+} binding by FSR from various muscles of rabbit, measured by double-beam spectrophotometry. The reaction mixture consisted of 100 mM KCl, 10 mM $MgCl_2$, 20 mM Tris-maleate buffer (pH 6.8), 0.2 mM murexide, and 0.4 mg/ml (WSR) or 0.8 mg/ml (RSR, CSR) protein. Ca^{2+} or Sr^{2+}, and ATP were added at the point indicated by the arrow in the figure. The change in transmission difference between 507 and 542 nm was recorded with respect to time. Reaction was carried out at 25°C. —— Sr^{2+}; ---- Ca^{2+}.

TABLE II
Comparison of Ca^{2+} and Sr^{2+} Binding of FSR

FSR	Rate (Sr^{2+}/Ca^{2+})	Max. binding (Sr^{2+}/Ca^{2+})
WSR	1.6 ± 0.12	1.03 ± 0.06
RSR	1.8 ± 0.17	1.49 ± 0.10
CSR	1.6 ± 0.13	1.51 ± 0.11

the uptake of Sr^{2+} but to a lesser extent. Nagai *et al.* (*22*) have reported that the amount of Sr^{2+} uptake in the presence of oxalate by WSR is almost the same as that of Ca^{2+} uptake. This discrepancy may have been caused by the limited amount of cations used in their assay. In our experiments, the amounts of Sr^{2+} uptake were 1,500, 2,000, and 600 nmoles/mg protein and those of Ca^{2+} uptake were 5,000, 3,000, and 3,000 nmoles/mg protein in WSR, RSR, and CSR, respectively, in the presence of oxalate.

Ca^{2+} uptake by FSR in the presence of oxalate is always preceded by a lag period, as shown in Fig. 2. The lag period is longest in cardiac muscle and shortest in white muscle. After the lag phase, Ca^{2+} is accumulated rapidly. On the other hand, Sr^{2+} uptake takes place without any lag after the addition

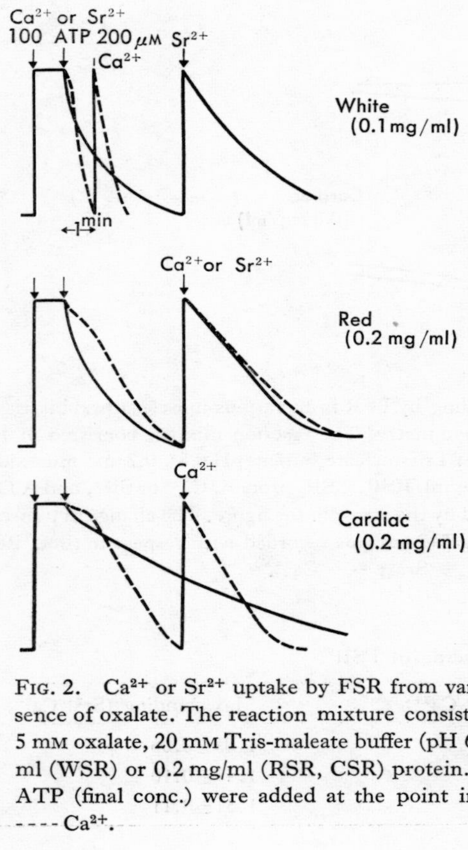

FIG. 2. Ca^{2+} or Sr^{2+} uptake by FSR from various muscles of rabbit in the presence of oxalate. The reaction mixture consisted of 100 mM KCl, 10 mM $MgCl_2$, 5 mM oxalate, 20 mM Tris-maleate buffer (pH 6.8), 0.2 mM murexide, and 0.1 mg/ml (WSR) or 0.2 mg/ml (RSR, CSR) protein. 100 mM Ca^{2+} or Sr^{2+} and 0.2 mM ATP (final conc.) were added at the point indicated by the arrow. —— Sr^{2+}; ---- Ca^{2+}.

TABLE III
Effect of Ca^{2+} and Sr^{2+} on the ATPase Activity of FSR

FSR	ATPase activity (P_i μmoles/min/mg)		
	Basic ATPase	Ca^{2+}-ATPase	Sr^{2+}-ATPase
WSR	0.049	0.050	0.070
RSR	0.160	0.010	0.045
CSR	0.300	0.050	0.075

The reaction mixture consisted of 100 mM KCl, 4 mM $MgCl_2$, 20 mM Tris-maleate buffer (pH 6.8), 0.1 mM GEDTA, 0.1 mM Ca^{2+} (or Sr^{2+}), 0.05 mg/ml protein, and 1 mM ATP. The activity of basic ATPase was measured in the absence of Ca^{2+} (or Sr^{2+}). The activity of Ca^{2+} (or Sr^{2+})-activated ATPase was estimated as the difference between the activity of total ATPase and that of basic ATPase. Reaction was carried out at 25°C for 15 min.

of ATP in any type of FSR. Ca^{2+} uptake, once started, proceeds faster than Sr^{2+} uptake and the amount accumulated eventually surpasses that of Sr^{2+} in WSR and CSR, whereas in RSR, the amount of Ca^{2+} taken up reaches the same level as that of Sr^{2+} in due course and both proceed in the same manner thereafter, at least up to 1,000 nmoles/mg protein. The presence of a lag period in Ca^{2+} uptake may indicate an inhibitory effect of Ca^{2+} itself on the Ca^{2+} uptake mechanism by FSR.

The effect of Ca^{2+} and Sr^{2+} on ATPase activity of FSR is shown in Table III. Sr^{2+} activates ATPase of FSR to a greater extent than Ca^{2+}, and this effect is most remarkable in red muscle. The high Sr^{2+} uptake by RSR may be related to the high activity of Sr^{2+}-activated ATPase of RSR.

Discussion of the Properties of FSR

The results presented above show that differences in the properties of FSR are greatest between red and white muscles and that the properties of CSR are situated between those of the other 2 FSR preparations. The absolute values of Ca^{2+} and Sr^{2+} binding by CSR are low compared with those of WSR, but are similar to those of RSR. With respect to the ratio of Sr^{2+} binding to Ca^{2+} binding, the ratio for CSR is similar to that for WSR, while the ratio of binding rate in CSR is similar to that in RSR. Ca^{2+} and Sr^{2+} bound to RSR are hardly released, while those bound to WSR or CSR are released rapidly.

In the presence of an ATP-regenerating system (phosphoenol pyruvate and pyruvate kinase), Ca^{2+} or Sr^{2+} bound to FSR is released following a

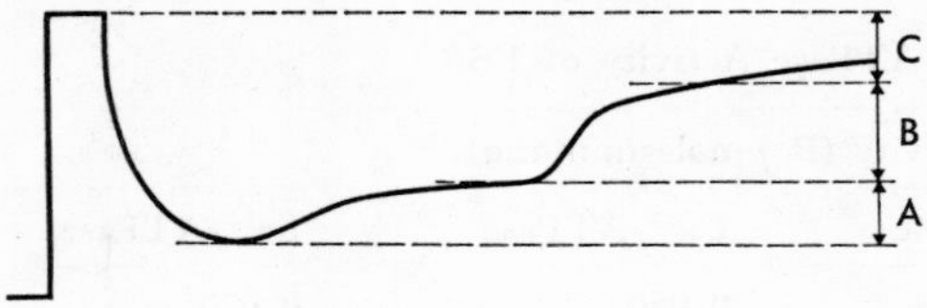

% of fraction

Muscle	Ca			Sr		
	A	B	C	A	B	C
White	33	45	22	33	45	22
Cardiac	26	30	44	45	37	18
Red	22		78	30		70

FIG. 3. Ca^{2+} or Sr^{2+} binding and release by FSR in the presence of ATP regenerating system. The upper figure shows a schematic pattern of Ca^{2+} or Sr^{2+} binding and release by FSR in the presence of an ATP-regenerating system (1 mM phosphoenol pyruvate and 0.05 mg/ml pyruvate kinase). Other reaction mixtures were the same as in Fig. 1. The lower table shows the estimated ratio of Ca^{2+} or Sr^{2+} binding capacity of each proposed fraction of FSR calculated from 10 to 20 experimental results.

particular pattern, as shown schematically in Fig. 3 (unpublished data). From this pattern we may tentatively divide the bound Ca^{2+} into 3 fractions. Fraction A becomes apparent only in the presence of an ATP-regenerating system and occurs in a similar ratio in each muscle. It is attractive to assume that the Ca^{2+} releasing tendency of SR, which is the basis of the contraction-triggering action (*23*) of SR, is reflected in this phenomenon.

Fraction B is dominant in WSR and lacking in RSR. The replenishment of ATP or phosphoenol pyruvate prevents this rapid release of Ca^{2+} from FSR. It is possible that this release may be related to the result by Weber *et al.* (*24*) that the efflux of Ca^{2+} from FSR is influenced by the ratio of ATP to ADP.

Fraction C is dominant in RSR. The nature of this fraction is not yet clear.

It is interesting that the Ca^{2+} binding process of CSR prepared from failing human heart (*14*) is rather similar to that of RSR, exhibiting a slow Ca^{2+} binding rate and slow or no Ca^{2+} release. The mean value of Ca^{2+} binding of these preparation is 33 nmoles/mg protein (*14*, *25*) which is about 70% of the normal amount. Similar results are also obtained from CSR of genetic cardiac myopatic hamsters (*17*). In this connection, it is interesting to investigate whether or not SR of white skeletal muscle which has been converted to red-type muscle by cross-innervation might possess fraction B.

REFERENCES

1. S. Ebashi, A. Kodama, and F. Ebashi, *J. Biochem.*, **64**, 465–477 (1968).
2. S. Ebashi and M. Endo, *in* "Progr. Biophys. Mol. Biol.," ed. by J. A. V. Butler and D. Noble, Pergamon Press, Oxford and New York, pp. 123–183 (1968).
3. S. Ebashi, *J. Biochem.*, **50**, 236–244 (1961).
4. W. Hasselbach and M. Makinose, *Biochem. Z.*, **333**, 518–528 (1961).
5. S. Ebashi and F. Lipmann, *J. Cell Biol.*, **14**, 389–400 (1962).
6. A. Weber, R. Herz, and I. Reiss, *J. Gen. Physiol.*, **46**, 679–702 (1963).
7. G. Inesi, S. Ebashi, and S. Watanabe, *Am. J. Physiol.*, **207**, 1339–1344 (1964).
8. A. Weber, R. Herz, and I. Reiss, *Biochim. Biophys. Acta*, **131**, 188–194 (1967).
9. J. Gergely, D. Pragay, A. F. Scholz, J. C. Seidel, F. A. Sreter, and M. M. Thompson, *in* "Molecular Biology of Muscular Contraction," ed. by S. Ebashi *et al.*, Igaku Shoin, Tokyo and Elsevier Publishing Co., Amsterdam, pp. 145–159 (1965).
10. S. Harigaya, Y. Ogawa, and H. Sugita, *J. Biochem.*, **63**, 324–331 (1968).
11. M. Takauji, T. Yamamoto, and T. Nagai, *Jap. J. Physiol.*, **17**, 111–121 (1967).
12. F. A. Sreter, *Arch. Biochem. Biophys.*, **134**, 25–33 (1969).
13. W. Fiehn and J. B. Peter, *J. Clin. Invest.*, **50**, 570–573 (1971).
14. S. Harigaya and A. Schwartz, *Circ. Res.*, **25**, 781–794 (1969).
15. A. M. Katz and D. I. Repke, *Circ. Res.*, **21**, 153–162 (1967).
16. G. A. Langer, *Physiol. Rev.*, **48**, 708–757 (1968).
17. W. B. McCollum, C. Crow, S. Harigaya, E. Bajusz, and A. Schwartz, *J. Mol. Cell. Cardiology*, **1**, 445–457 (1970).
18. G. Inesi and S. Watanabe, *Arch. Biochem. Biophys.*, **121**, 665–671 (1967).
19. C. Edwards, H. Lorkovic, and A. Weber, *J. Physiol.*, **186**, 295–306 (1966).
20. S. Yamada and Y. Tonomura, *J. Biochem.*, **72**, 417–425 (1972).
21. T. Ohnishi and S. Ebashi, *J. Biochem.*, **54**, 506–511 (1963).
22. T. Nagai, H. Takahashi, and M. Takauji, *in* "Molecular Biology of Muscular Contraction," ed. by S. Ebashi *et al.*, Igaku Shoin, Tokyo and Elsevier Publishing Co., Amsterdam, pp. 169–176 (1965).
23. M. Endo, M. Tanaka, and Y. Ogawa, *Nature*, **228**, 34–36 (1970).

24. A. Weber, R. Herz, and I. Reiss, *Biochem. Z.*, **345**, 329–369 (1966).
25. G. E. Lindenmayer, L. A. Sordahl, S. Harigaya, J. C. Allen, H. R. Besh, Jr., and A. Schwartz, *Am. J. Cardiology*, **27**, 277–283 (1971).

Ca^{2+} Uptake and Release by Fragmented Sarcoplasmic Reticulum with Special Reference to the Effect of β,γ-Methylene Adenosine Triphosphate

Yasuo Ogawa and Setsuro Ebashi

Department of Pharmacology, Faculty of Medicine, University of Tokyo, Tokyo, Japan

β,γ-Methylene adenosine triphosphate depresses E-P formation and Ca^{2+} uptake by the SR with various kinds of substrates other than ATP and at low concentrations of ATP. AMPOPCP also shows a Ca^{2+}-releasing effect on Ca^{2+}-filled SR with a concomitant decrease in the amount of E-P. The releasing effect is markedly accelerated by the presence of caffeine; the Ca^{2+}-releasing effect of the latter is also augmented by AMPOPCP. Caffeine has no influence on the amount of E-P under any circumstances. The mechanism of the Ca^{2+}-releasing action of AMPOPCP is dicussed in relation to the connection between the mechanisms of Ca^{2+} uptake and release.

Since the discovery of the ATP-dependent Ca^{2+} uptake of SR by Ebashi and Lipmann (*1*) and Hasselbach and Makinose (*2*), the mechanism of Ca^{2+} uptake has been worked out by many investigators, particularly Weber (*3*), Tonomura (*4*), Inesi (*5*) and Makinose and Hasselbach (*6*, *7*). Compared with this, biochemical studies on the mechanism of Ca^{2+} release from SR, which is of primary importance from a physiological standpoint, are still in a preliminary stage. One of the reasons for this might be

Abbreviations used: SR, sarcoplasmic reticulum; E-P, phosphorylated intermediate; AMPOPCP, β,γ-methylene adenosine triphosphate; AMPCPOP, α,β-methylene adenosine triphosphate; AcP, acetyl phosphate; CarbP, carbamyl phosphate; pNPP, *p*-nitrophenyl phosphate; GEDTA, glycoletherdiaminetetraacetic acid (or EGTA); EDTA, ethylenediaminetetraacetic acid; PEP, phosphoenol-pyruvate; PK, pyruvate kinase EC 2.7.1.40.

that the mechanism of Ca^{2+} release may be similar to the excitation mechanism of the surface membrane and, therefore, it would be very difficult to find a direct biochemical approach, as has been the case with the excitation mechanism.

Following suggestions by Dr. Y. Kaziro, we started studies of the effect of AMPOPCP on the Ca^{2+} uptake of SR in view of the interesting action of β,γ-methylene guanosine triphosphate on the ribosome dependent guanosine triphosphatase (*8*). Unexpectedly, it was found that the reagent possessed a remarkable effect on the process of Ca^{2+} release from SR.

Ca^{2+} Uptake Mechanism and AMPOPCP

It has been shown that not only nucleoside triphosphates other than ATP (*9*), but also agents entirely different from nucleotides in their chemical structure, such as acetyl phosphate (*10*; Fig. 1) or CarbP (*11*, *12*; Fig. 2), can serve as the energy source of the Ca^{2+} uptake mechanism. One might argue that the site of action of these non-nucleotides might not be identical with that of ATP. AMPOPCP provides reasonable counterevidence against this argument. This nucleotide, which itself cannot be hydrolyzed by SR and cannot induce Ca^{2+} uptake, inhibits the initial rates of Ca^{2+} uptake induced by non-nucleotides* (Fig. 3) as well as by low concentrations of ATP (the competitive nature of this inhibition has been proved with CarbP). In correspondence to this inhibition, Ca^{2+}-dependent E-P formation due to CarbP (Fig. 4) is also depressed (Table I).

Thus we can deal with the results with non-nucleotides on the same level as those with ATP and other nucleotides. If the affinity of SR for Ca^{2+} is represented by the apparent binding constant, *i.e.*, the reciprocal of the free Ca^{2+} concentration which gives half-maximum Ca^{2+} uptake, it is less than $5 \times 10^5\ \text{M}^{-1}$ in the case of AcP as the substrate, in contrast to a high value, $3 \times 10^6\ \text{M}^{-1}$, in the case of ATP (Fig. 1,A). In parallel with this low affinity, the maximum rate of Ca^{2+} uptake with AcP is very low, nearly one-tenth that of ATP. This is also the case with CarbP or AMPCPOP (Figs. 2 and 4).

* Although to a much lesser extent, the total amount of Ca^{2+} uptake is also depressed by AMPOPCP at higher concentrations of Ca^{2+}. However, this inhibition is not shown at lower concentrations of Ca^{2+}; as a result, the apparent binding constant appears to be somewhat increased in the presence of AMPOPCP (Fig. 3,B). This fact may indicate the presence of a heterogeneity in the SR preparation as regards its affinity for the substrate.

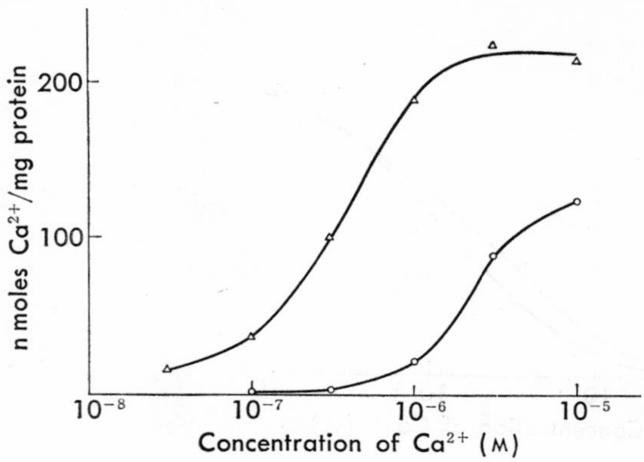

A

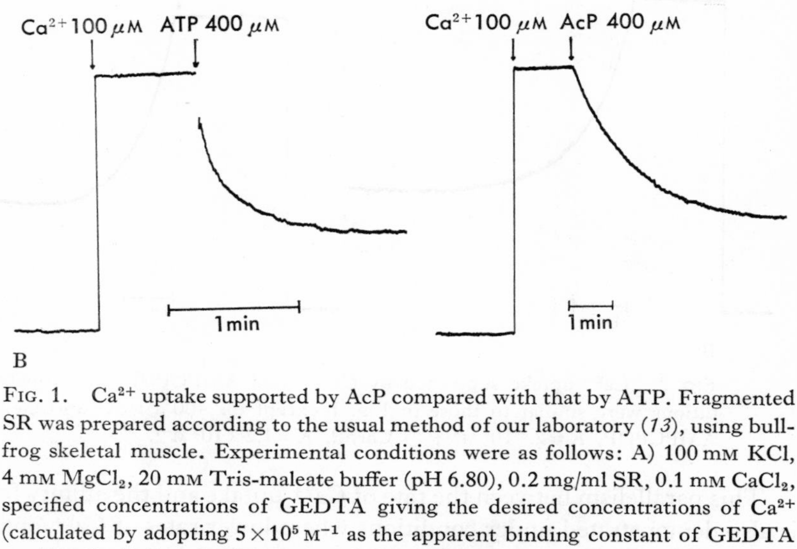

B

FIG. 1. Ca^{2+} uptake supported by AcP compared with that by ATP. Fragmented SR was prepared according to the usual method of our laboratory (*13*), using bull-frog skeletal muscle. Experimental conditions were as follows: A) 100 mM KCl, 4 mM $MgCl_2$, 20 mM Tris-maleate buffer (pH 6.80), 0.2 mg/ml SR, 0.1 mM $CaCl_2$, specified concentrations of GEDTA giving the desired concentrations of Ca^{2+} (calculated by adopting 5×10^5 M^{-1} as the apparent binding constant of GEDTA at pH 6.80 (*14*)) and substrates. △ 2.7 mM ATP, $K=3\times10^6 M^{-1}$; ○ 2.7 mM AcP, $K=5\times10^5 M^{-1}$. B) 100 mM KCl, 4 mM $MgCl_2$, 20 mM Tris-maleate buffer (pH 6.80), 0.4 mg/ml SR, 0.25 mM murexide. Total volume was 3 ml. To this medium, 10 μl of 30 mM $CaCl_2$ and 30 μl of 40 mM substrates were added in turn. The change of free Ca^{2+} in the medium was determined by recording the change of absorbance difference between 507 and 542 nm using a dual-wavelength spectrophotometer (*13*).

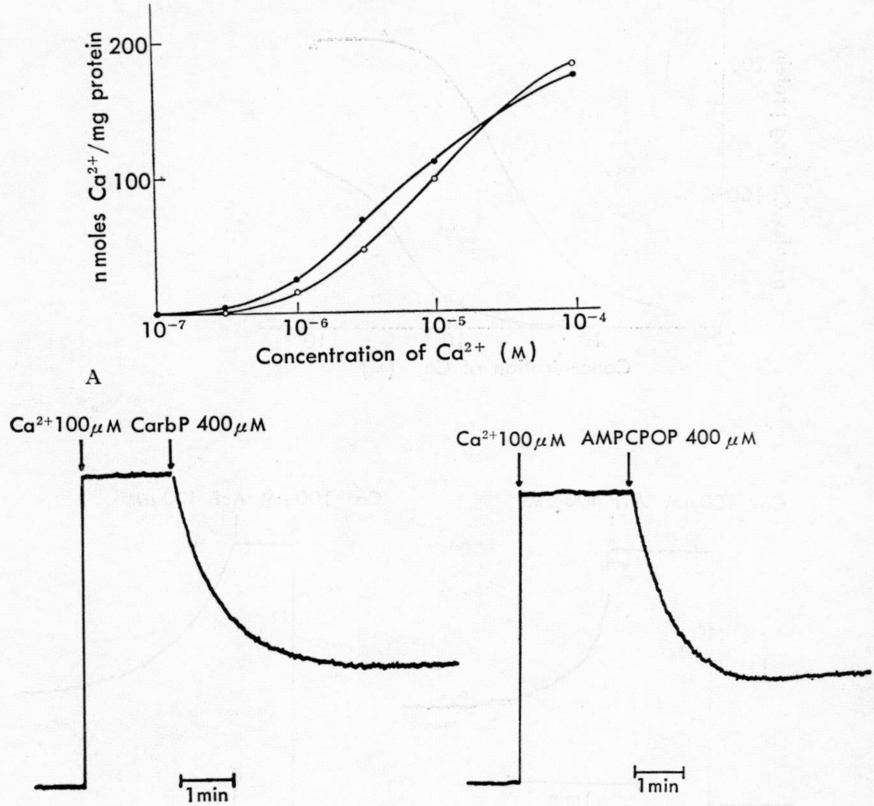

FIG. 2. Ca^{2+} uptake supported by CarbP and AMPCPOP. Experimental conditions were similar to those in Fig. 1 except for 400 μM of substrate. A: ● AMPCPOP, $K=2\times10^5$ M^{-1}; ○ CarbP, $K=1.2\times10^5$ M^{-1}.

This parallelism between the rate of Ca^{2+} uptake and the affinity for Ca^{2+} is also demonstrated under conditions where lower rates of Ca^{2+} uptake are produced by lowering the concentration of ATP, the capacity being kept constant (Fig. 5). If we plot the affinities for Ca^{2+} against the rates of Ca^{2+} uptake irrespective of the type and concentrations of substrate, they fall on the same line (Fig. 6).

This linear relationship between the affinities for Ca^{2+} and the rates of Ca^{2+} uptake does not mean, however, that the absolute value for the rate of

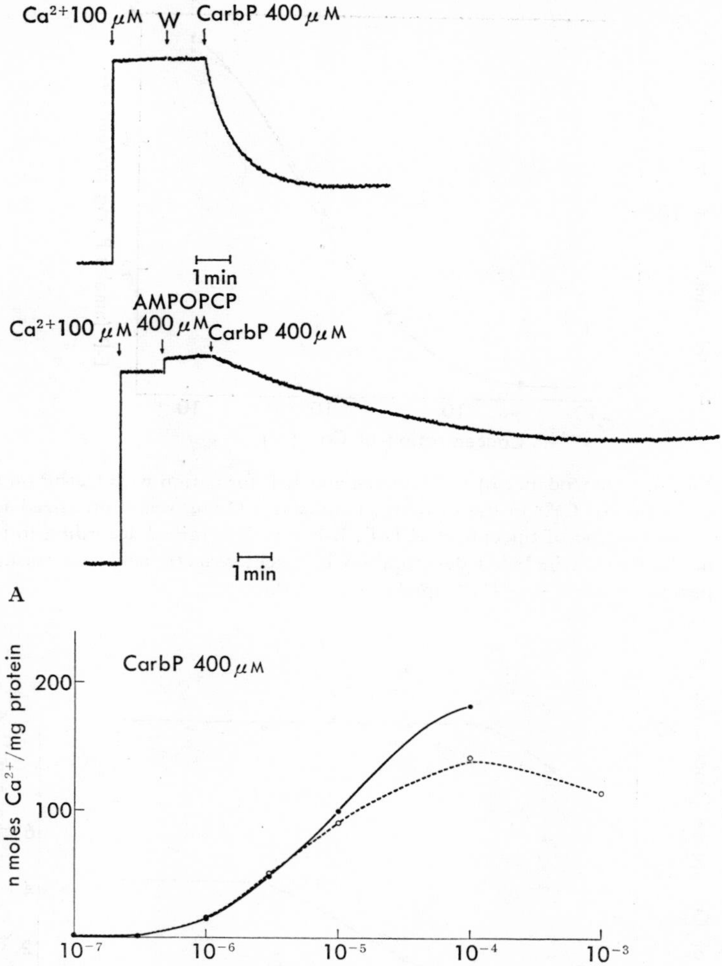

FIG. 3. Effect of AMPOPCP on Ca^{2+} uptake supported by CarbP. Experimental conditions were similar to those in Fig. 1. B: ● without AMPOPCP, $K=1.2 \times 10^5$ M^{-1}; ○ with 400 μM AMPOPCP, $K=2.0\times10^5$ M^{-1}.

Ca^{2+} uptake determines the affinity for Ca^{2+}. If the same substrate or the same concentration of ATP is used, the affinity is independent of temperature and, therefore, of the absolute rate of reaction of each step. One plausi-

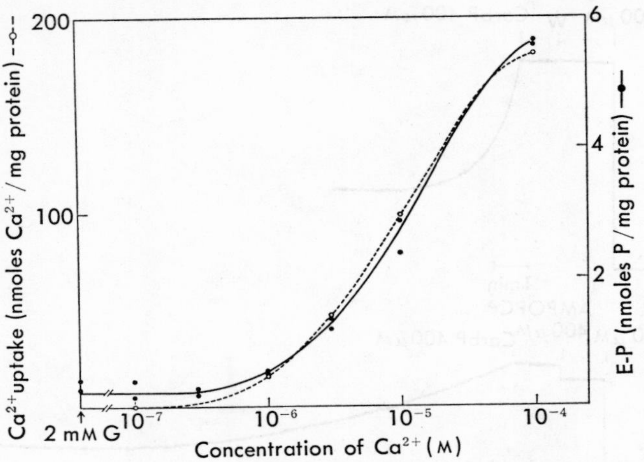

FIG. 4. Dependence of Ca^{2+} uptake and E-P formation with CarbP on the concentration of Ca^{2+} in the medium. Radioactive CarbP was synthesized according to the method of Spector *et al.* (*15*). E-P was determined according to the usual method which included denaturation by trichloroacetic acid and washing with perchloric acid. —— Ca^{2+} uptake; ······ E-P.

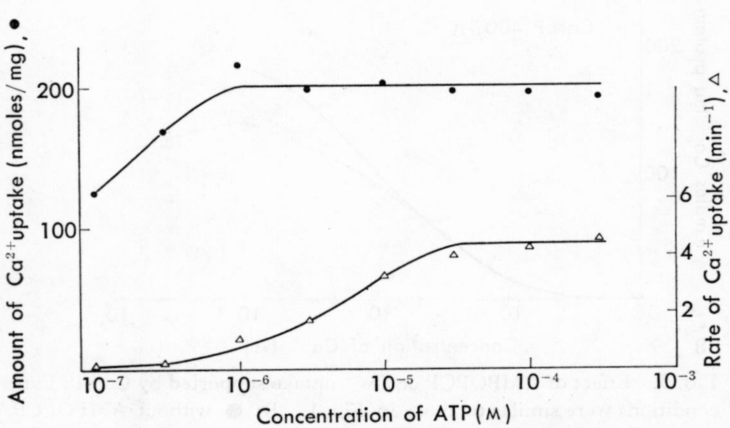

FIG. 5. Dependence of the amount and rate of Ca^{2+} uptake on the concentration of ATP. Experiments were done under similar conditions to those in Fig. 1 with an ATP-regenerating system of PEP+PK, using a dual-wavelength spectrophotometer. The rate was expressed as the reciprocal of the time required for half-maximum uptake. For details refer to (*17*).

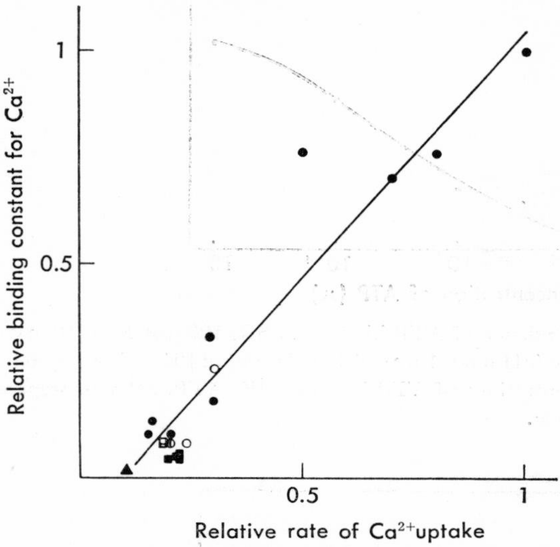

FIG. 6. Relation between rate of Ca^{2+} uptake and affinity of SR for Ca^{2+}. Experimental conditions were similar to those in Fig. 1. For details refer to (*13*). ● ATP (0.3–30 μM; 2.7 mM); ◐ AMPCPOP (0.4 mM); ○ ITP (0.4, 2.7 mM); □ AcP (0.4, 2.7, 4 mM); ■ CarbP (0.4, 2.7, 4 mM); ▲ pNPP (4 mM).

ble explanation compatible with these observations is that the affinity of the uptake mechanism of SR for Ca^{2+} is largely determined by the affinity of the enzyme-substrate complex for Ca^{2+}, and that the rate of uptake is mainly dependent on the rate of Ca^{2+} binding to the enzyme-substrate complex.

There is a problem which deserves further attention, *i.e.*, whether or not the binding of ATP to SR shown by Ebashi and Lipmann (*1*) represents the enzyme-substrate complex. To discuss this, it is necessary to enumerate the properties of this binding: 1) Binding is observed only in the presence of sufficient ATP-regenerating system, being quite different from the stable nature of E-P. 2) The amount of bound ATP (Fig. 7) is comparable with that of E-P so far reported (*4–6*, *16*) (according to Ebashi and Lipmann, the amount of bound ATP reaches a plateau at about 10^{-5} M ATP and then it increases again with increase in ATP concentration, but in our studies, this plateau is hardly seen (Fig. 7)). 3) Binding is not influenced by EDTA, indicating no requirement for Mg^{2+}. 4) Depression of ATP binding by various agents is accompanied by a parallel decrease in Ca^{2+} uptake

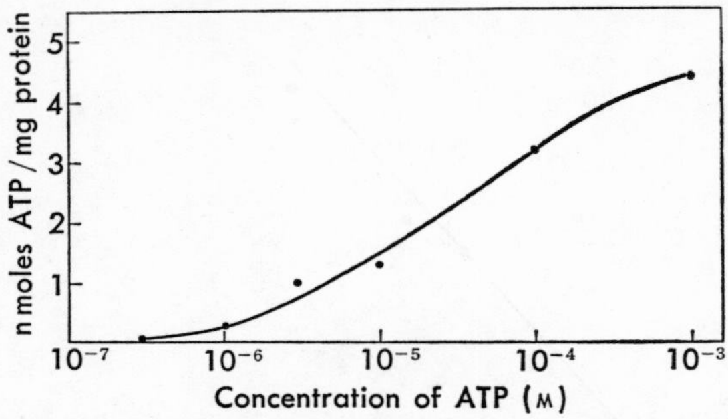

FIG. 7. ATP dependence of ATP binding to SR. 100 mM KCl, 10 mM $MgCl_2$, 20 mM Tris-maleate (pH 6.80) 1 mM PEP, 0.35 mg/ml PK, 0.2 mg protein SR/ml and specified concentrations of ATP including ^{14}C-ATP. SR was prepared from rabbit skeletal muscle.

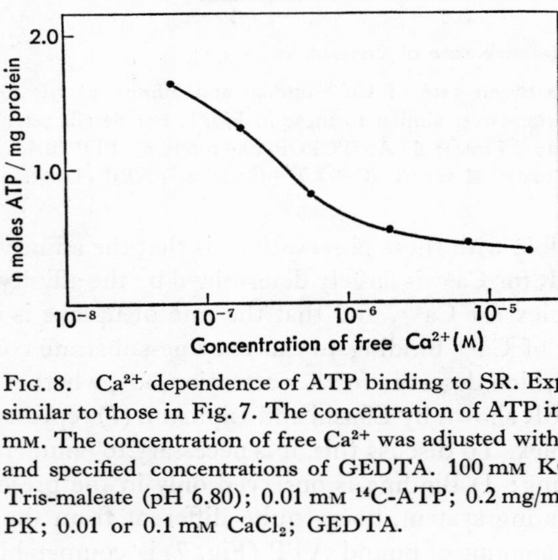

FIG. 8. Ca^{2+} dependence of ATP binding to SR. Experimental conditions were similar to those in Fig. 7. The concentration of ATP including ^{14}C-ATP was 0.01 mM. The concentration of free Ca^{2+} was adjusted with 0.01 mM or 0.1 mM $CaCl_2$ and specified concentrations of GEDTA. 100 mM KCl; 10 mM $MgCl_2$; 20 mM Tris-maleate (pH 6.80); 0.01 mM ^{14}C-ATP; 0.2 mg/ml SR (rabbit); 1 mM PEP; PK; 0.01 or 0.1 mM $CaCl_2$; GEDTA.

(*17*). 5) In agreement with the results of Weber *et al.* (*3*), the amount of bound ATP is reduced in the presence of Ca^{2+}. It follows a mirror image of the amount of E-P and Ca^{2+} uptake at varied concentrations of Ca^{2+} (Fig. 8).

These findings may favor an affirmative answer to the above question. This is further substantiated by the fact that the amount of bound ATP is lowered by AMPOPCP to the extent expected from its inhibitory effect on E-P formation or Ca^{2+} uptake.

Although not directly related to the present subject, it should be noted that the stoichiometric decrease in the Mg^{2+} content of SR concomitant with the increase in Ca^{2+} uptake (*18*) is not found in frog SR. While Ca^{2+} content markedly increases on addition of ATP, *e.g.*, from 58 to 235 nmoles/mg protein, Mg^{2+} content decrease only slightly, *e.g.*, from 93 to 55 nmoles/mg protein. This raises the problem of which ion may exchange or accompany Ca^{2+} movement. Since SO_4^{2-}, which is considered to be practically impermeable, can induce a higher rate of Ca^{2+} uptake than Cl^-, the usual anions may not accompany Ca^{2+}. At present we cannot answer this problem.

AMPOPCP and the Ca^{2+} Release Mechanism

With the background of the above findings, we would like to refer to the effect of AMPOPCP on the Ca^{2+} release mechanism.

When AMPOPCP is added immediately after saturation of SR with Ca^{2+} in the presence of low concentrations of ATP, reversible Ca^{2+} release, *i.e.*, immediate release of Ca^{2+} and subsequent reuptake of Ca^{2+}, even in the presence of the agent, was observed (Fig. 9; this release cannot be observed at high concentrations of ATP). This Ca^{2+}-releasing action is also observed

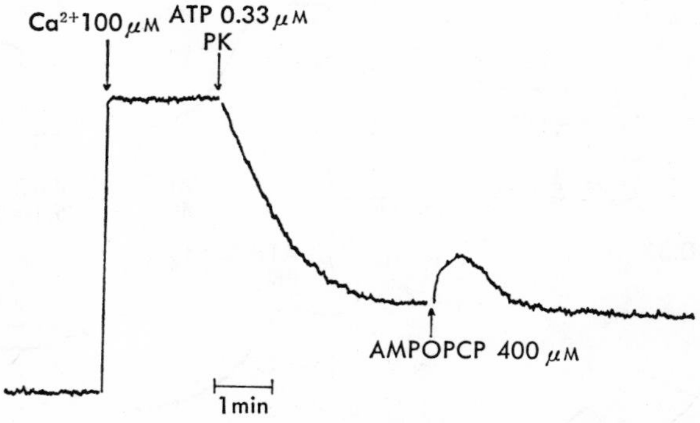

FIG. 9. Ca^{2+}-releasing effect of AMPOPCP. Experiments were made using a dual-wavelength spectrophotometer. 100 mM KCl, 4 mM $MgCl_2$, 20 mM Tris-maleate (pH 6.80), 2 mM PEP, 0.25 mM murexide, 0.4 mg SR/ml. SR was 2,000–8,000 × *g* fraction from bullfrog skeletal muscle. Total volume was 3 ml. Temperature was 15°C. To this medium, 10 μl of 30 mM $CaCl_2$, 10 μl of 0.1 mM ATP and 30 μl of PK, and 30 μl of 40 mM AMPOPCP were added in this order.

when CarbP is used as substrate (Fig. 11, upper figure; although the rate of release is somewhat lower, the amount reaches to the same level as with ATP). The releasing action of AMPOPCP is apparently similar to that of caffeine (*13, 19–21*) in some respects: 1) H-fraction, the fraction obtained between 2,000× *g* and 8,000× *g*, is more sensitive to AMPOPCP than the usual microsome fraction. 2) The releasing action is inhibited by procaine. 3) The amount of released Ca^{2+} increases as the temperature falls. 4) The amount of released Ca^{2+} decreases with decrease in the concentrations of free Ca^{2+}.

However, these similarities do not indicate that the sites of action of AMPOPCP and caffeine are the same. If SR is once subjected to caffeine, no Ca^{2+} release or only a slight and sluggish release is observed on further addition of caffeine. AMPOPCP behaves in a similar way. However, when AMPOPCP is added to SR filled with Ca^{2+} in the presence of caffeine, instantaneous release of an enormous amount of Ca^{2+} can be observed (Fig. 10). In turn, when caffeine is added in the presence of AMPOPCP, the releasing action of caffeine is also augmented markedly (Fig. 10; this Ca^{2+} release is so rapid compared with the ^{40}Ca-^{45}Ca exchange rate (*3*), that it

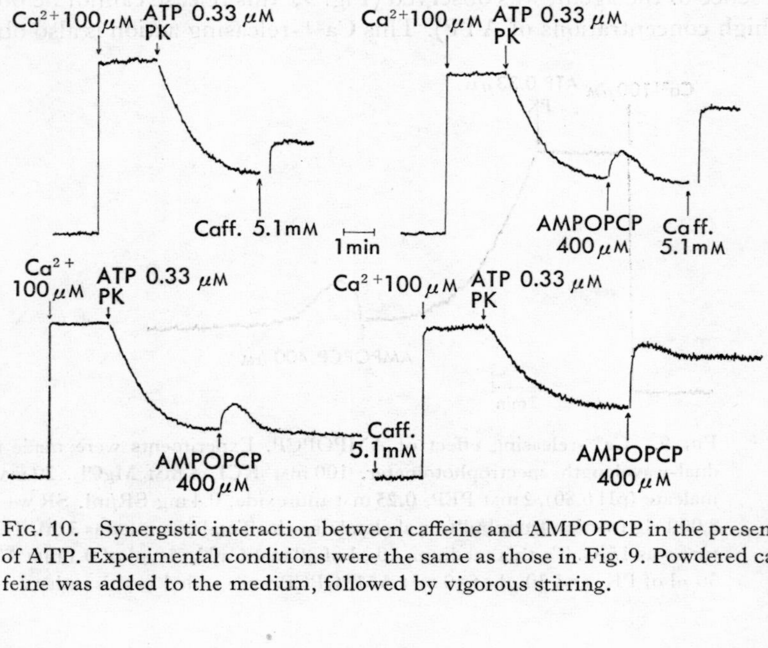

FIG. 10. Synergistic interaction between caffeine and AMPOPCP in the presence of ATP. Experimental conditions were the same as those in Fig. 9. Powdered caffeine was added to the medium, followed by vigorous stirring.

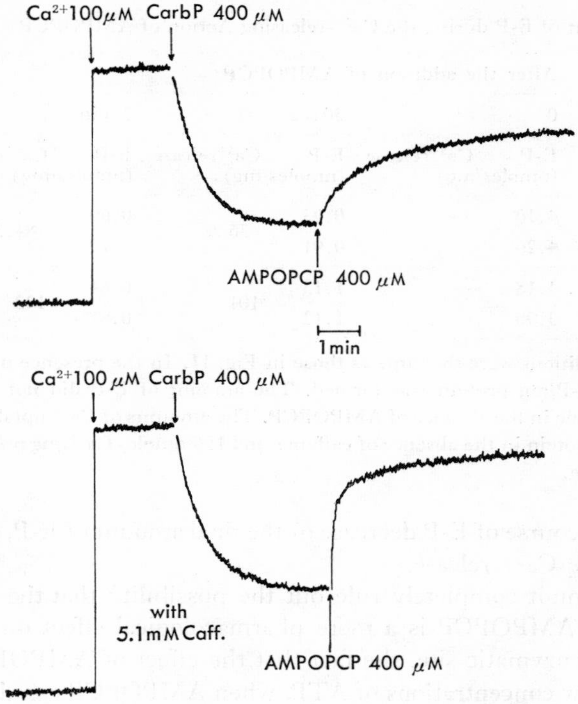

FIG. 11. Synergistic interaction between caffeine and AMPOPCP in the presence of CarbP. Experimental conditions were the same as those in Fig. 10 except for CarbP instead of ATP and ATP-regenerating system. If higher concentrations of CarbP, *i.e.*, 2 mM, were used, reversible Ca^{2+} release could be observed as was the case with ATP.

cannot be a simple reverse reaction of Ca^{2+} uptake). Essentially the same phenomenon is observed when ATP is replaced by CarbP (Fig. 11 and Table I). Thus, caffeine and AMPOPCP have synergistic action, suggesting that the site for AMPOPCP should be different from that for caffeine.

The difference in the mode of action of caffeine and AMPOPCP is also noted in their actions on E-P. While caffeine does not show any effect on the amount of E-P in spite of its remarkable Ca^{2+}-releasing action, the amount of E-P decreases very rapidly and nearly reaches a minimum 30 sec after addition of AMPOPCP, when the release of Ca^{2+} is still at an intermediate stage (Table I). Addition of caffeine at this stage does not exert any in-

TABLE I
Changes in the Amount of E-P during the Ca^{2+}-releasing Action of AMPOPCP

	After the addition of AMPOPCP					
	0		30 sec		3 min	
	E-P (nmoles/mg)	Ca^{2+}release	E-P (nmoles/mg)	Ca^{2+}release	E-P (nmoles/mg)	Ca^{2+}release
Without caffeine	4.10	—	0.93	36.9	0.60	84.3
	4.26	—	0.91		0.79	
With 5 mM caffeine	4.18	—	1.13	101	0.84	123
	3.90	—	1.12		0.81	

The experimental conditions were the same as those in Fig. 11. In the presence of 2 mM GEDTA, 0.4 nmole E-P/mg protein was formed. The amount of E-P did not change through the reaction time in the absence of AMPOPCP. The amounts of Ca^{2+} uptake were 168 nmoles Ca^{2+}/mg protein in the absence of caffeine, and 156 nmoles Ca^{2+}/mg protein in the presence of caffeine.

fluence on the time course of E-P decrease or the final amount of E-P, giving rise to a remarkable Ca^{2+} release.

Although we cannot completely rule out the possibility that the Ca^{2+}-releasing action of AMPOPCP is a mere pharmacological effect on a site different from the enzymatic site, the fact that the effect of AMPOPCP is exhibited only at low concentrations of ATP, when AMPOPCP can depress E-P formation and Ca^{2+} uptake, strongly indicates that the agent exerts its effect through the enzymatic site.*

It has been tacitly admitted that the mechanism of Sr^{2+} uptake should be the same as that of Ca^{2+} uptake. Indeed, AMPOPCP inhibits the rate of Sr^{2+} uptake as in the case of Ca^{2+} uptake. However, AMPOPCP alone cannot release Sr^{2+} from SR filled with Sr^{2+} by CarbP; it can only release a minute amount of Sr^{2+} in the presence of caffeine. We do not have a plausible explanation for this at present, but there is an observation which might reflect different behavior of accumulated Ca^{2+} and Sr^{2+}.

Sr^{2+} uptake is stimulated by the presence of oxalate, as is Ca^{2+} uptake (Table II). The apparent solubility product of oxalate with Sr^{2+} is about 30 times as high as that with Ca^{2+}. Unexpectedly, however, the minimum con-

* After the conference, we found that AMPOPNP, kindly supplied by Dr. R.G. Yount to whom our cordial thanks are due, had essentially the same effect as AMPOPCP; this is additional evidence for the above assumption.

TABLE II
Sr^{2+} Uptake and Its Stimulation by Oxalate

	Ca^{2+}	Sr^{2+}
Apparent solubility products with oxalate	2×10^{-7} M^2 (4×10^{-9} M^2)[a]	6.4×10^{-6} M^2 (1.6×10^{-7} M^2)[a]
Minimum concentrations of oxalate to stimulate uptake	0.3 mM	0.8 mM
Estimated concentrations of free cations inside the vesicles	0.7 mM	8 mM

[a] Solubility products when the activities of ions are taken as unity.

centration of oxalate which was required for stimulating Sr^{2+} uptake was not much different from that for Ca^{2+} uptake. A simple calculation using apparent solubility products shows that the concentrations of free cations inside the vesicles should be 0.7 mM in the case of Ca^{2+} and 8 mM in the case of Sr^{2+}. Since either the maximum amount of Ca^{2+} uptake or Sr^{2+} uptake is about 200 nmoles/mg protein in the absence of oxalate, the concentration of divalent cations inside the vesicles is estimated to be around 20 mM, provided that the accumulated divalent cations exist as free ions (under the assumption that 1 mg protein of SR occupies a volume of 10 μl). These considerations indicate that a considerable part of the accumulated Sr^{2+} may exist as free ions in the vesicles, in contrast to the case of Ca^{2+} which may exist mostly in a bound form. It is interesting to suppose that this difference between Ca^{2+} and Sr^{2+} may be the cause of the different responses to AMPOPCP of Ca^{2+} and Sr^{2+}, *i.e.*, the released cation would be derived mainly from the bound form so that Sr^{2+} is more difficult to release than Ca^{2+}.

Conclusions

The uptake and release mechanisms of Ca^{2+} from SR are often compared with the active and passive transport of Na^+ and K^+ at the surface membrane. This implies that the Ca^{2+} release mechansim under physiological conditions may be independent of the Ca^{2+} uptake mechanism. It is certain that Ca^{2+} release is not the reverse reaction of the uptake process. However, the effect of AMPOPCP on the Ca^{2+} uptake and release mechanisms indicates the presence of a connection between these two mechanisms. This agent may be considered as a tool for a biochemical approach to the release mechanism.

REFERENCES

1. S. Ebashi and F. Lipmann, *J. Cell Biol.*, **14**, 389–400 (1962).
2. W. Hasselbach and M. Makinose, *Biochem. Z.*, **333**, 518–528 (1961).
3. A. Weber, R. Herz, and I. Reiss, *Biochem. Z.*, **345**, 329–369 (1966).
4. T. Kanazawa, S. Yamada, T. Yamamoto, and Y. Tonomura, *J. Biochem.*, **70**, 95–123 (1971).
5. G. Inesi, E. Maring, A. J. Murphy, and B. H. McFarland, *Arch. Biochem. Biophys.*, **138**, 285–294 (1970).
6. M. Makinose, *Eur. J. Biochem.*, **10**, 74–82 (1969).
7. M. Makinose and W. Hasselbach, *FEBS Letters*, **12**, 271–272 (1971).
8. F. Lipmann, *Science*, **164**, 1024–1031 (1969).
9. A. Martonosi and R. Feretos, *J. Biol. Chem.*, **239**, 648–658 (1964).
10. L. DeMeis, *J. Biol. Chem.*, **244**, 3733–3739 (1969).
11. A. Pucell and A. Martonosi, *J. Biol. Chem.*, **246**, 3389–3397 (1971).
12. Y. Ogawa, *J. Biochem.*, **71**, 571–573 (1972).
13. Y. Ogawa, *J. Biochem.*, **67**, 667–683 (1970).
14. Y. Ogawa, *J. Biochem.*, **64**, 255–257 (1968).
15. L. Spector, M. E. Jones, and F. Lipmann, *in* "Methods in Enzymology," ed. by S. P. Colowick and N. O. Kaplan, Academic Press, New York, Vol. III, pp. 653–655 (1957).
16. T. Yamamoto and Y. Tonomura, *J. Biochem.*, **64**, 137 (1968).
17. S. Ebashi and M. Endo, *in* "Biochemistry of Muscle Contraction," ed. by J. Gergely, Little Brown & Co., Boston, pp. 199–206 (1964).
18. A. P. Carvalho and B. Leo, *J. Gen. Physiol.*, **50**, 1327–1352 (1967).
19. A. Weber and R. Herz, *J. Gen. Physiol.*, **52**, 750–759 (1968).
20. A. Weber, *J. Gen. Physiol.*, **52**, 760–772 (1968).
21. M. Endo, M. Tanaka, and Y. Ogawa, *Nature*, **228**, 34–36 (1970).

CONFORMATIONAL CHANGE IN ATPase AND SIMILAR PHENOMENA

Use of ATP Analogs to Analyze ATPase Catalysis*[1]

Joseph A. Duke and Manuel F. Morales*[2]

Cardiovascular Research Institute, University of California, San Francisco, California, U.S.A.

This article treats several chemical compounds which are structurally and functionally related to ATP. Two classes of such compounds are considered, those which can serve as enzymatic substrates and those which do not exhibit this function. The first class comprises those compounds in which the amino group at position 6 of the adenine portion of the ATP is replaced by a sulfhydryl group. The resulting compound, under appropriate conditions, can react in either of two modes, as a substrate or as an affinity label for myosin. In the latter mode the symmetrical disulfide appears to be the operative species. Further structural modifications of this analog are considered which involve the attaching of spin labels or fluorescent labels to the sulfhydryl group. Other related thiol compounds are discussed, one of which (an analog of adenosine monophosphate) forms a thioether linkage with phosphorylase and a second of which has a sulfhydryl group on the terminal phosphorus atom of ATP. The second class of compounds, those which are not able to be hydrolyzed, are those which replace the oxygen bridge between the beta and gamma phosphorus atoms of ATP with a methylene or an imido group. These analogs have been used to study the mechanism of the binding of nucleotides to enzymes and the incorporation of nucleotides as prosthetic groups. The use of the analogs as affinity labels is discussed and the results thus far obtained are evaluated.

*[1] This research was supported by grants from the N.S.F., the P.H.S., and the A.H.A., of the United States of America. *[2] Career Investigator of the American Heart Association.

There is an unspecificity in the title of this paper which it is necessary to appreciate. Analogs of ATP of a certain kind are not at all new. Tonomura and his former colleague Ikehara in a series of papers (*1*) reported a dazzling array of analogs from which they were able to go some distance in mapping the active sites in various contractile processes. Their analogs may be called "intellectual" analogs. In contrast, what we shall consider here are "utilitarian" analogs designed to accomplish a particular job in a particular experiment. The literature on utilitarian analogs of ATP is very sparse and most of it is very new.

In the last two years our laboratory and Yount's have developed analogs of ATP for the express purpose of probing the ATPases of the contractile system. It seems likely, however, that these substances will find wider application and on this account they may be worth discussing. Still more recently other laboratories have produced interesting compounds with similar potential. Even though we have as yet no experience with these newer compounds, we shall indicate here the manner in which we suppose they can be used. Our plan for this brief paper is to say a little about the analogs themselves and then to present a catalog of uses.

Thio Analogs of ATP

Following a prior suggestion of H. Asai, Murphy *et al.* in our laboratory first synthesized the 6-sulfhydryl analog of ATP (*2*).

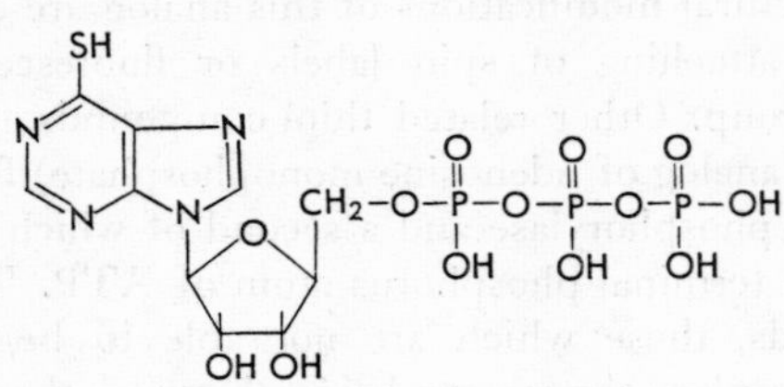

6-Mercapto-purine-9-β-D-ribofuranosyl-5′-triphosphate (SH-TP)

Abbreviations used: SH-TP, 6-mercapto-purine-9-β-D-ribofuranosyl-5′-triphosphate; $(S\text{-}TP)_2$, 6-6′-dithio-bis-purine-riboside triphosphate; NBA-S-MP, 6-(4-nitro-3-carboxyphenyl)thiopurine-9-β-D-ribofuranosyl-5′-monophosphate; ATP-SH, adenosine-5′-O-(3-thiotriphosphate); ε-ATP, 1-N^6-ethenoadenosine triphosphate; ADP-CH_2P, adenyloyl methylenediphosphonate; ADP-NP, adenyloyl imidodiphosphate; I-AEDANS, N-iodoacetyl-N′-1-(5-sulfonaphthyl)ethylene diamine; SH-DP-NP, 6-thio-inosinoyl imidodiphosphate.

Murphy started with the monophosphate, successfully coupled orthophosphate with dicyclohexylcarbodiimide and isolated the desired product on a Dowex-1 column. Later, one of us (J.A.D.) developed a superior method, now used routinely, *viz.*, pyrophosphate is directly coupled to the organic monophosphate with carbonyldiimidazole and the desired product is isolated on DEAE-cellulose. The starting compound may be refluxed with elemental ^{35}S to incorporate a radioactive atom when desired (*3*). We originally intended to produce a chromophoric analog (absorption peak at 322 nm), and in fact did so, but Murphy found that under slightly alkaline conditions (pH 8.2) the compound was also an affinity label of ATPase sites, although the labeling required some 60 hr at 0°C. While using this compound as an affinity label, Tokiwa and Morales (*4*) noted that the incubation of the analog at pH 8.2 prior to labeling accelerated the subsequent reaction with protein and suggested that labeling proceeded through the intermediate formation of the disulfide. Yount and his colleagues (*5*) synthesized the disulfide of a closely similar analog and showed it to be a much faster labeling agent. Yount not only confirmed Tokiwa's suggestion but produced the best labeling scheme (labeling time about 15 min).

Letting SH-TP stand for the 6-SH analog of ATP and $(S\text{-}TP)_2$ for the corresponding disulfide [$(S\text{-}DP)_2$ and $(S\text{-}MP)_2$ have also been synthesized and used] we may sum up: SH-TP is capable of substituting for ATP as a substrate; its ring ionization pK is 8.00 at 25°C; its acid form has a strong absorbance ($\varepsilon = 2.2 \times 10^4$) which is decreased on passing from an aqueous to a hydrophobic environment. Under conditions in which thiols are partly ionized, (pH 8.2), $(S\text{-}TP)_2$ rapidly and specifically labels such enzymes as myosin ATPase, creatine kinase and the F-1 particle of mitochondria according to the reaction,

$$\text{Prot-XH} + (\text{S-TP})_2 \rightarrow \text{Prot-XS-TP} + \text{SHTP}$$
(in myosin X=S; in creatine kinase it may be N).

Prot-XS-TP is sufficiently stable so that the ATPase sites are stoichiometrically inhibited but, in our experience, it is insufficiently stable to withstand the maneuvers incident to peptide mapping.

Since our work, others have prepared related analogs. Hulla and Fasold (*6*) reported the synthesis of the thioether of the monophosphate and 2-nitro-4-mercapto-benzoic acid,

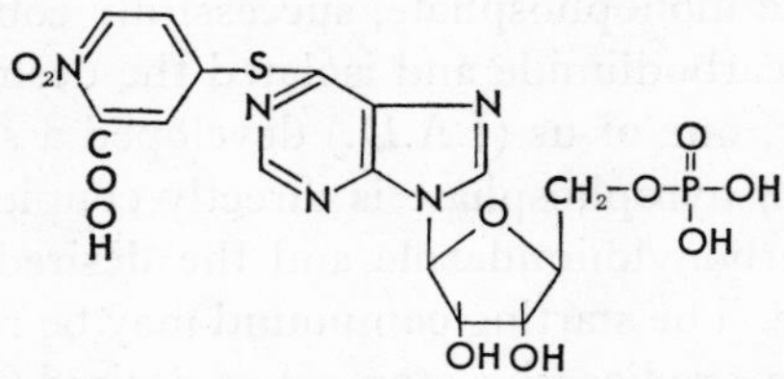

6-(4-Nitro-3-carboxyphenyl) thiopurine - 9-β-D - ribofuranosyl-5′-monophosphate (NBA - S - MP)

This compound specifically labels phosphorylase:

$$\text{Prot-SH}+\text{NBA-S-MP}\rightarrow\text{Prot-S-MP}+\text{NBA-SH}\ .$$

The authors report that Prot-S-MP is very stable. The triphosphate analog has not been reported but presumably can be synthesized similarly. Possibly the thioether linkage would be sufficiently stable for peptide mapping.

The labeling reaction we employed, and presumably that of Hulla and Fasold, depends for its primary specificity on the fact that its structure is complementary to that of adenine. Goody and Eckstein (7), however, have produced a thiophosphate analog of ATP.

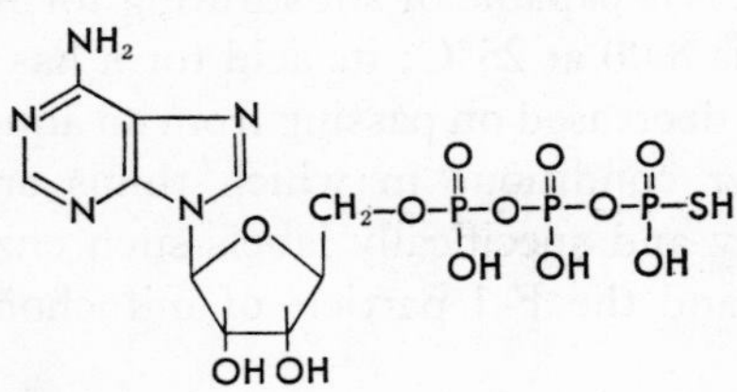

Adenine - 5′ - O - (3 - thiotriphosphate) (ATP - SH)

The schematic labeling reaction for this analog is

$$\text{Prot-SH}+\text{ATP-SH}\rightarrow\text{Prot-S-S-ATP}+2\text{H}^{+}+2e^{-}\ .$$

In this case the primary specificity must arise from the polyphosphate chain. For any particular ATPase the specificity of this analog must differ from that of the SH-TP, so comparative studies on the same enzyme with two analogs should be very rewarding. These are not yet available although several laboratories are working with the ATP-SH analog.

Chromophoric and affinity labels are familiar from work with other enzymes but it is certain that analysis of ATPase will also profit from fluorescent as well as spin labels. The compound SH-TP is a convenient starting material for some of these newer labels. Commercially available nitroxyl spin labels with iodoacetamide or N-ethyl-maleimide reactivities, commercially available dansyl cystine, or Hudson's (*8*) I-AEDANS

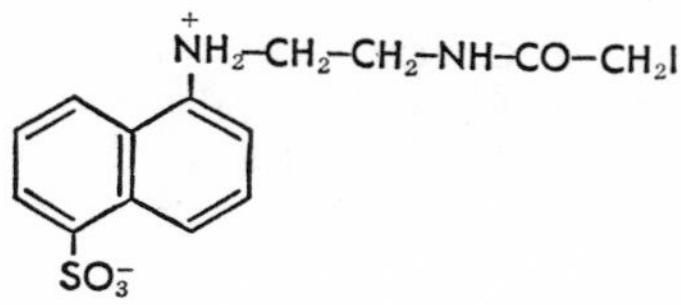

N - Iodoacetyl - N′ - 1 - (5-sulfonaphthyl)ethylene diamine (I - AEDANS)

readily react with SH-TP to form either spin labeled or fluorescent labeled analogs of ATP. Our laboratory has already reported on the use of the spin label (*9*). In solution, derivatives of SH-TP probably exhibit more or less free rotation between the label and the ATP moiety. After attachment this feature is likely to be unimportant since ring structures tend to be immobilized on hydrophobic regions of proteins. It is noteworthy, nevertheless, that an analog of ATP in which the nitroxyl radical is rigidly attached to the nucleotide ring has been synthesized by McConnell and Hamilton (*10*) and also Secrist *et al.* (*11*) have synthesized the fluorescent compound ε-ATP:

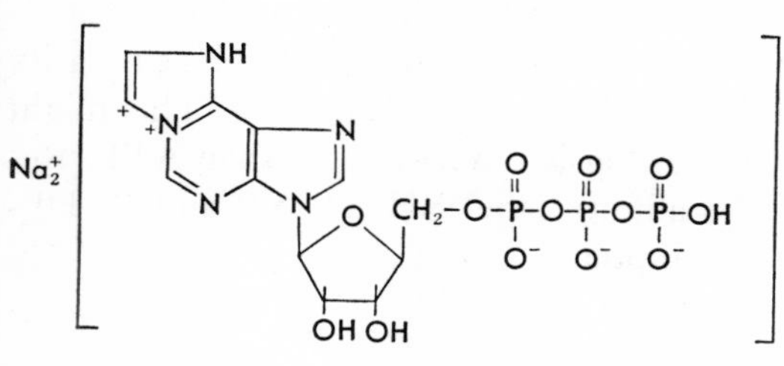

1 - N^6- Ethenoadenosine triphosphate (ε - ATP)

Unsplittable Analogs of ATP

The energetic consequences of ATP hydrolysis are widely appreciated but, under physiological conditions, ATP is also interesting because it is a quadrivalent anion with chelating capability. Unsplittable analogs suppress the first property but sustain the second nearly as well as ATP. On this account they can be used to test whether, in a given process, ATP functions either as an energy donor or by virtue of its binding or chelating properties. Two such analogs are available:

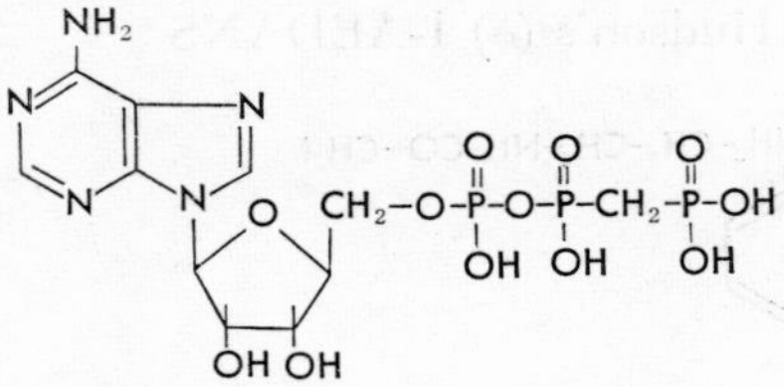

Adenyloyl methylenediphosphonate (ADP - CH_2P) Myers *et al.* (*12*)

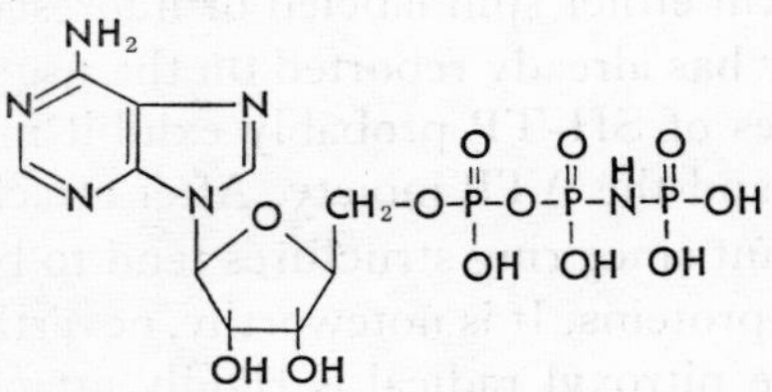

Adenyloyl imidodiphosphate (ADP - NP) Yount *et al.* (*13*)

According to the measurements reported by Yount, the affinity constants of these compounds for H^+, Ca^{2+}, and Mg^{2+} are all of the same order of magnitude as those of ATP. It is our experience with ADP-NP that its similarity in effect to ATP is greater at pH 7.5–8.0 than it is at pH 6.5–7.0. This might be expected from the fact that the analog is a weaker acid than ATP. Recently, Yount has synthesized the affinity label SH-DP-NP and (S-DP-NP)$_2$, the disulfide of the former compound.

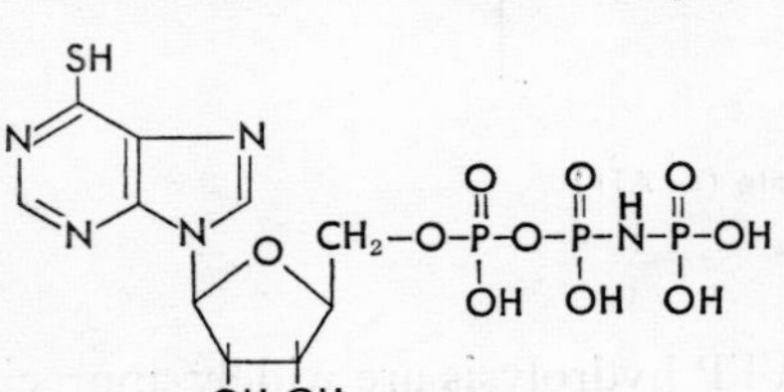

6 - Thio - inosinoyl imidodiphosphate (SH - DP - NP)

The use of these compounds guarantees that the bound moiety will be the triphosphate.

We now turn to cataloging some of the uses to which the foregoing analogs have been put. While most of our examples are drawn from applications to the contractile system, we have sought to abstract what we think is the essence of the application rather than detail the application itself.

1) If an instantaneous indicator of binding of the nucleotide to an enzyme is desired: The 322 nm absorbance of SH-TP decreases when the purine ring transfers from an aqueous to a hydrophobic environment (*14*). With sufficiently sensitive apparatus this effect can be used to characterize equilibrium (*14*) or transient (*15*) distributions of unbound and enzyme-bound nucleotides. If only the un-ionized analog is bound, and the ionized species is not, then a much more sensitive indication can probably be obtained by working at a pH near the p*K* of the ring ionization (8.0) and taking advantage of the very large difference in absorbance between the two species of the analog. Leonard's analog (*11*) has not been tested in this way. It is very possible that it will turn out to be an excellent probe to use in this mode.

2) If one desires to ascertain whether a process depends only on ATP binding (as distinct from hydrolysis) or on the formation of a "Michaelis complex": The behavior of ADP-NP may be tested provided that the concentrations of hydrogen ions and metal ions make the simulation of ATP exact. Such a use of ADP-NP was made by Yount *et al.* (*16*) to show that the binding of a nucleotide triphosphate is insufficient to cause contraction of glycerol-extracted psoas fibers or superprecipitation of actomyosin. It has also been used by us (*17*) to show that it is sufficient to produce relaxation of fibers.

3) If one desires to simulate the incorporation of ATP or ADP as a prosthetic group: It was shown early that SH-TP will exchange readily with the "prosthetic" ATP of monomeric (G-) actin (*14*). Once incorporated, the ring moiety of the analog is a chromophoric probe since the p*K* of the prosthetic molecule can be compared with that of the free analog. In G-actin this difference is very small, which suggests that at least a part of the prosthetic purine ring is in full contact with solvent.

More recently, Murphy (*18*) has studied SH-TP and SH-DP incorporation into G- and F-actin respectively for the purpose of creating Cotton effects in the visible spectrum thereby enabling studies of circular dichroism uncomplicated by contributions from the protein. By this approach Murphy was able to show that the protein interacts much more strongly with the nucleotide when it is in the F- form than when it is in the G- form.

A very different application of a prosthetic analog in actin was made by Cooke and Duke (*9*). The EPR spectrum of the "spin labeled ATP" made by these authors shows the three sharp bands characteristic of the nitroxyl radical tumbling in solution with a correlation time of 0.01 nsec or less. When the analog is incorporated into either G-actin of F-actin, however, those peaks become very broadened and distorted due to the asymmetric environment which the protein provides for the nitroxyl structure. Cooke and Duke therefore used the progressive appearance of sharp spectral bands to detect release of prosthetic nucleotide. By this means they revealed the considerable mechanical fragility of F-actin.

4) If one desires to affinity-label ATPase sites of enzymes: Until now only studies employing SH-TP (acting through a disulfide intermediate) or $(S\text{-}TP)_2$, $(S\text{-}DP)_2$, $(S\text{-}MP)_2$ (acting by disulfide exchange) have been reported. In the first of these studies (*14*) it was shown that two moles of reagent stoichiometrically inhibited one mole of myosin. Tokiwa later (unpublished) showed that this reaction was completely reversed and that activity was completely restored by treatment with β-mercaptoethanol. Myosin happens to be a duplex molecule and has two ATPase sites. Suitable plots of SH-TP binding indicate that these two sites have the same affinity constant for SH-TP. It is therefore simplest to assume that the two sites bind the nucleotides independently. On this basis, Tokiwa and Morales (*4*) partially reacted populations of myosin molecules in which the label was Bernoulli-distributed among halves of the myosin molecules. He was then able to show that certain functions could only be performed by molecules in which both halves were unlabeled (suggesting that in such processes both halves in some way cooperated), whereas in other processes any unlabeled half appeared to function independently of its mate.

In one way SH-TP and $(S\text{-}TP)_2$ have been disappointing as the label is instabilized by the maneuvers normally used to dissociate myosin into subunits or to prepare peptide maps, *e.g.*, by high concentrations of acid or base or guanidine hydrochloride. This instability led Murphy and Morales (and us) (*14*) to think mistakenly that it was the so-called "light" chains of myosin which were labeled. Our recent experiments show that in all probability it is on the "heavy" chains. This instability also has prevented thus far the determination of the amino acid sequence around the ATPase site. We would expect that a similar instability would plague enzyme complexes with the Goody-Eckstein analog. On the other hand the superior stability of the thioether linkage should make the Hulla-Fasold analog able to stand subunit dissociation or hydrolytic enzyme cleavage.

An ambitious use of the disulfide analogs of the adenosine mono-, di-, and triphosphates has been to synthesize nucleotide-enzyme complexes resembling those which are intermediate in ATPase activity. In the case of myosin ATPase we (*19*) and other investigators have thought that an essential intermediate (in contrast to the case of membrane ATPase) is a special complex of myosin and ADP which is different from that obtainable simply by equilibrating myosin and ADP. We have called this complex myosin *ADP (as distinct from myosin•ADP). When a myosin system is labeled with $(S\text{-}TP)_2$ the nucleotide later detached by β-mercaptoethanol is the diphosphate. Because the enzyme is not permanently damaged by labeling it is uncertain whether hydrolysis of the covalently bound label occurs upon binding or whether hydrolysis occurs at the instant of β-mercaptoethanol treatment. The same state of the myosin system is achieved, however, by labeling with $(S\text{-}TP)_2$. In our view (*17*) the disulfide-bound complexes may be equivalent to myosin *ADP. Of course these inferences may be of little relevance to students of membrane ATPase wherein the enzyme appears to be phosphorylated rather than "ADP-ized." Nonetheless, the parallel experiments with membranes may be great interest, *e.g.*, to affinity-label with the Goody-Eckstein analog and then to study the ADP-ATP exchange to see whether one of the enzyme intermediates has been synthesized.

In closing we express our intuitive feeling that experiments on membranes using analogs in some of the ways we have described will prove very rewarding.

REFERENCES

1. Y. Tonomura, K. Imamura, M. Ikehara, H. Uno, and F. Harada, *J. Biochem.*, **61**, 460 (1967).
2. A. J. Murphy, J. A. Duke, and L. Stowring, *Arch. Biochem. Biophys.*, **137**, 297 (1970).
3. J. A. Duke, L. Stowring, and A. J. Murphy, *Abstr. 14th Meet. Biophys. Soc.*, Baltimore, 82a (1970).
4. T. Tokiwa and M. F. Morales, *Biochemistry*, **10**, 1722 (1971).
5. R. G. Yount, J. S. Frye, and K. R. O'Keefe, unpublished.
6. F. W. Hulla and H. Fasold, *Biochemistry*, **11**, 1056 (1972).

7. R. S. Goody and F. Eckstein, *J. Am. Chem. Soc.*, **93**, 6252 (1971).
8. B. Hudson, Dissertation, University of Illinois, 39 (1970).
9. R. Cooke and J. A. Duke, *J. Biol. Chem.*, **246**, 6360 (1971).
10. H. M. McConnell and C. L. Hamilton, *Proc. Natl. Acad. Sci. U.S.*, **60**, 776 (1968).
11. J. A. Secrist, J. R. Barrio, and N. J. Leonard, *Science*, **175**, 646 (1972).
12. T. C. Myers, K. Nakamura, and J. W. Flesher, *J. Am. Chem. Soc.*, **85**, 3294 (1963).
13. R. G. Yount, D. Babcock, W. Ballantyne, and D. Ojala, *Biochemistry*, **10**, 2482 (1971).
14. A. J. Murphy and M. F. Morales, *Biochemistry*, **9**, 1528 (1970).
15. D. R. Trentham, R. G. Bardsley, J. F. Eccleston, and A. G. Weeds, *Biochem. J.*, **126**, 635 (1972).
16. R. G. Yount, D. Ojala, and D. Babcock, *Biochemistry*, **10**, 2490 (1971).
17. C. G. dos Remedios, R. G. Yount, and M. F. Morales, *Abstr. 16th Meet. Biophys. Soc.*, Toronto, 281a (1972).
18. A. J. Murphy, *Biochemistry*, **10**, 3723 (1971).
19. G. Viniegra and M. F. Morales, *J. Bioenergetics*, **3**, 101 (1972).

Subfragment-1: The Enzymatically Active Portion of Myosin

Koichi Yagi, Yoichi Yazawa, Fumio Otani, and Yoh Okamoto

Department of Chemistry, Faculty of Science, Hokkaido University, Sapporo, Japan

Physical and enzymatic properties of subfragment-1 prepared using different proteolytic enzymes such as trypsin, chymotrypsin, and Nagarse were compared. Size, shape, and ATPase activities of these subfragments were all similar. However, the cleavage of intramolecular peptide bond, which was analyzed by dinitrophenylation, was limited in subfragment-1 prepared by chymotryptic digestion, while it was clearly seen in the others. This result was accordant with the difference in viscosities of the subfragments in 5 M guanidine-HCl. Ca^{2+}-activated ATPase activity of myosin solution did not change but EDTA, K^+-activated ATPase activity decreased on incubation with Nagarse for 10 min. The enzymatically active components found in the incubation mixture after 10 min appeared to be subfragment-1 alone. It was assumed from this result that the subfragment-1 produced from myosin was continuously decomposed but the decrease in amount was counterbalanced by the activation of Ca^{2+}-activated ATPase accompanied by the change from myosin to 2 monomers of subfragment-1. Subfragment-1 prepared by chymotryptic digestion of myosin was fractionated into 2 components. One contained a small component (probably g_1 chain) and showed lower ATPase activity, but the other lacked this component and showed higher ATPase activity.

It has long been known that the ATPase activity of myosin is suddenly and completely lost on increasing the pH to 10.5–11.0. Trotta *et al.* (*1*) suggested the involvement of low-molecular weight components (g-chains) in this pH

TABLE I
Reconstitution of Myosin from Separated g-Chains and f-Chains

Expt. No.	ATPase activity of original myosin prep. (U)	ATPase activity in % of original activity			
		Separated g-chains	Separated f-chains	After reconstitution	
				Calc.	Obs.
1	0.46	47.8	1.9	6.5	72.5
2	0.37	25.8	2.8	5.1	77.9
3	0.39	76.8	47.4	50.1	84.6
4	0.27	52.0	19.0	22.3	110.5
5	0.29	0	14.4	12.9	11.0
6	0.44	58.5	64.3	63.7	54.4
7	0.39	9.9	24.8	23.4	29.8
8	0.45	38.6	41.4	41.1	52.8

effect. Gershman and Dreizen (*2*) succeeded in the reconstitution of myosin from separated g-chains and main chains (f-chains). Reconstitution of subfragment (S-1(T)) was then reported by Stracher (*3*). They emphasized from these results that g-chains are indispensable factors of myosin not only for ATP hydrolysis but also for actin binding.

We have repeated the renaturation experiment with myosin carried out by Frederiksen and Holtzer (*4*). A complete recovery of myosin ATPase which was denatured in a solution of pH 11.45 for 5 min at 1±0.5°C, was also achieved after reversing the pH to neutral. However, reconstitution experiments with myosin from separated f-chains and g-chains by LiCl treatment (*1*, *2*) were rather difficult and results were variable, as shown in Table I. As has generally been obtained, 3 components were found in the g-chain preparations by sodium dodecyl sulfate (SDS) disc electrophoresis (*5*), and they are called g_1-, g_2-, and g_3-chains in decreasing order of molecular weight. The function of these 3 g-chains still remains ambiguous and it is necessary to investigate whether g-chains are really essential for myosin and then what their functions are.

Abbreviations used: EDTA, ethylenediaminetetraacetic acid; S-1(T), subfragment-1 prepared by tryptic digestion of HMM(T); S-1(N), S-1(CT), and S-1(P), subfragment-1 prepared from myosin by using Nagarse, chymotrypsin, and papain, respectively; S-n, subfragment-1 prepared from S-1(T) by Nagarse digestion; HMM(T), heavy meromyosin preparedf rom myosin by tryptic digestion.

We have been working on subfragments (S-1) prepared by using different proteolytic enzymes such as trypsin, chymotrypsin and Nagarse. Similarities and also some differences were found in properties among these subfragments. Therefore, information on the function of g-chains may be obtained from investigations of each S-1.

Comparison of Physical and Enzymatic Properties of S-1 Prepared Using Different Proteolytic Enzymes (*6*, *7*)

Table II shows the physical parameters of S-n, S-1(T), S-1(N), S-1(CT) and also S-1 (P) (*8*), measured at neutral pH and in 0.15 M KCl unless otherwise stated. The sedimentation coefficients and intrinsic viscosities were in the ranges from 5.0S to 6.0S and from 0.036 to 0.064 dl/g, respectively. An extremely similar molecular weight was obtained (0.84–1.25×10^5) for these subfragments using different methods.

It is suggested from these results that the size and shape of all these sub-

TABLE II
Physical and Enzymatic Properties of Subfragments(S-1) Prepared by Proteolysis of Myosin by Different Proteinases

Subfragments	Sedimentation coefficient (s)	Intrinsic viscosity (dl/g)	b_0	m_{238} [f]	Molecular weight ($\times 10^{-5}$)
S-n (*10*)	5.60	0.036	−130[g]	−37,000[g]	0.84,[a] 1.02[b] (*15*)
S-1(T) (*7*)	5.74	0.052	−158[g]	−44,000[g]	1.05,[a] 1.2[b] (*15*), 1.2[c]
S-1(T) (*11*)	5.95				
S-1(T) (*12*)	5.7–6.0	0.08–0.11[h]			
S-1(T) (*13*)	5.7	0.08[h]			
S-1(T) (*14*)					0.95,[d] 1.10,[d] 1.19[d]
S-1(N) (*7*)	5.63	0.044			0.95,[a] 1.2[c]
S-1(CT) (*9*)	5.0	0.06			0.93,[a] 1.25,[c] 1.1,[e] 1.04[b] (*15*)
S-1(P) (*8*)	5.8	0.064			1.15[d]

[a] Scheraga-Mandelkern equation using s^{0}_{20w} and $[\eta]$. [b] Archibald approach to sedimentation equilibrium. [c] Gel filtration using Sephadex G-200. [d] High-speed sedimentation equilibrium of Yphantis. [e] Light scattering. [f] Solvent: 20 mM Tris-HCl, pH 7.6. [g] Unpublished data. [h] These values might be higher than the real one due to contamination by heavy meromyosin or other larger molecules, because we obtained similar values when purification was incomplete.

fragments are very similar. This may result from the proteolytic cleavage of a definite region of myosin by different proteolytic enzymes.

The steady-state ATPase activity in the presence of Ca^{2+} (1.0–1.25 U) was similar among the subfragments except for S-n. The magnitude of the initial rapid liberation of P_i was measured with S-1(T) (*16*), S-1(N) and S-1(CT) and was in the range between 0.3 and 0.5 moles P_i per 1×10^5 g. No initial burst was observed, however, with S-n, whose steady-state ATPase activity in the presence of Ca^{2+} (1.4 U) was a little higher than that of others (*16*).

In contrast to the similar physical properties and ATPase activities shown above, a remarkable difference was detected in the reactivity with actin between S-1(CT) (and S-1(P) (*8*)) and other subfragments. This was examined in 2 ways—firstly by actin activation of Mg^{2+}-activated ATPase and secondly by an acceleration effect of subfragments on actin polymerization. As shown in Figs. 1 and 2, higher activities were obtained with S-1(CT) than with S-1(T) and S-1(N). From the 2 different experimental results, it appeared that the reactivity of S-1(CT) towards actin in the presence of ATP was much higher than those of S-1(T) and S-1(N). It now remained to determine what structural differences might be responsible for the considerable difference in the reaction with actin.

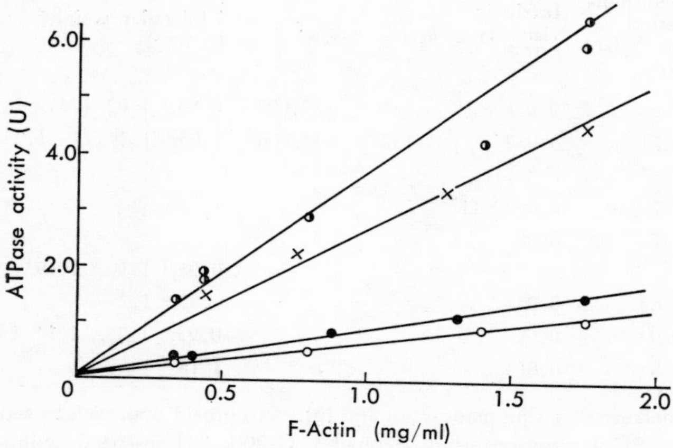

FIG. 1. Activation effect of F-actin on Mg^{2+}-activated ATPase activity of subfragments. The reaction mixture contained 1 mM $MgCl_2$, 20 mM Tris-KOH buffer (pH 7.0), 2 mM ATP, 0.1 mg of S-1 or HMM(T), and variable amount of F-actin. × HMM(T); ◑ S-1(CT); ● S-1(T); ○ S-1(N).

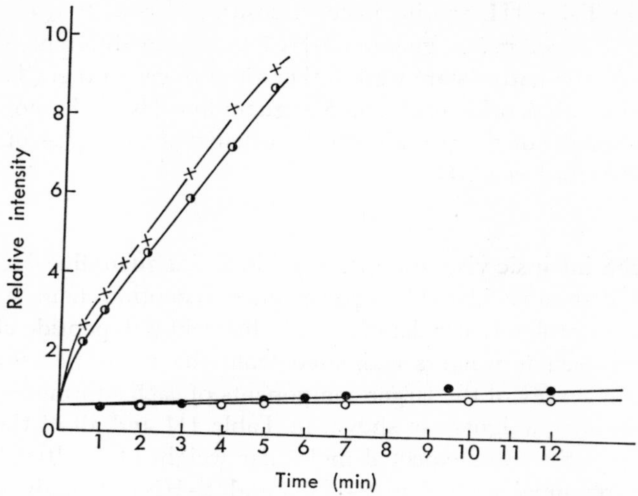

FIG. 2. Acceleration effect of subfragments on actin polymerization. Actin polymerization was followed by measuring the light intensity observed at right angles to the incident light. Protein concentration: G-actin, 0.5 mg/ml: S-1 and HMM(T), 0.4 mg/ml. Reaction was started by adding 0.2 ml of S-1 or HMM(T) to 2.8 ml of G-actin solution in 5 mM Tris-maleate buffer (pH 7.0). Same symbols as in Fig. 1 are used.

Subfragments in the Denatured State (7)

Since the subfragments were prepared using different proteolytic enzymes, differences can be expected in the peptide bonds cleaved. In 5 M guanidine-HCl, the conformation of proteins is expected to be random coil and all of the peptide chains will become free in this solution. Therefore, the differences in peptide bond cleavage can be detected by measuring the viscosity in concentrated guanidine-HCl.

TABLE III
Viscosities and Molecular Weights of Subfragments in 5 M Guanidine-HCl

	S-1(N)	S-1(T)	S-n	S-1(CT)
Intrinsic viscosity (dl/g)	0.25	0.25	0.20	0.48
Molecular weight estimated from viscosity ($\times 10^{-4}$)	2.4	2.4	1.6	6.5

As shown in Table III, the intrinsic viscosities of S-n, S-1(T), S-1(N) and S-1(CT) measured in 5 M guanidine-HCl were quite different, although the viscosities in the native state were rather close to each other (Table II). If the size of the polypeptide chains in 5 M guanidine-HCl is homogeneous, the molecular weight of the protein can be calculated by means of Eq. (1) proposed by Tanford *et al.* (*17*).

$$[\eta] = 0.716 n^{0.66}, \qquad (1)$$

where $[\eta]$ is the intrinsic viscosity measured in 5–7 M guanidine-HCl and n is the number of amino acid residues per single polypeptide chain. Although the subfragments probably consist of several different polypeptide chains as regards size, molecular weights estimated from the viscosities should be useful to make a survey of the larger components of each subfragment. The calculated molecular weights are shown in Table III and all of the values were much lower than the expected molecular weight of 1×10^5. The low value (2×10^4) obtained with S-n, S-1(T), and S-1(N) strongly suggests that several peptide bonds per mole were cleaved in these subfragments. The small polypeptides produced may be bonded together in the particle in the native state. N-Terminal amino acid analysis of each subfragment was performed by dinitrophenylation in order to estimate the number of intramolecular peptide bonds cleft. Only 0.47 mole of 2, 4-dinitrophenol (DNP)-amino acid was found per 1×10^5 g of S-1(CT), but 2–3 moles were found with S-1(T) and S-1(N) and about 5 moles with S-n.

Components of these subfragments were further investigated by gel filtration using a Sepharose 6B column with 5 M guanidine-HCl. A large component, which occupied 88.7±1.5% by weight, was obtained with S-1(CT) and the molecular weight estimated from the elution volume was 6.9×10^4. This value was very close to 6.5×10^4, which was estimated from the viscosity, indicating that intramolecular peptide bond cleavage is limited in S-1(CT) and that it contains a g-chain as another component. Molecular weight of the larger component was calculated as 4.3×10^4 for S-1(T), S-1 (N), and S-n, but it occupied about 35% in S-1(T) and S-1(N) and less than 20% in S-n. Other components were found at an elution position corresponding to a molecular weight of 1.8×10^4.

As described above, S-1(CT) was distinguished from other subfragments in reactivity to actin and a large difference was found in structure between the 2 groups.

Possibility of Myosin-ATPase Being in the Inhibited State

For the preparation of S-1(N), the weight ratio of Nagarse to myosin largely affected the yield of S-1(N) from myosin. At a fixed ratio of 1/200 at 25°C, products appeared to be mostly S-1(N) after 10-min incubation. Myosin was decomposed nearly completely and HMM(N) found in the reaction mixture was less than 5% of added myosin. The ATPase activity of the original reaction mixture was that of myosin itself, but after 10-min incubation it can be regarded as that of S-1(N). The Ca^{2+}-activated ATPase activity was followed as a function of digestion time and, as shown in Fig. 3,

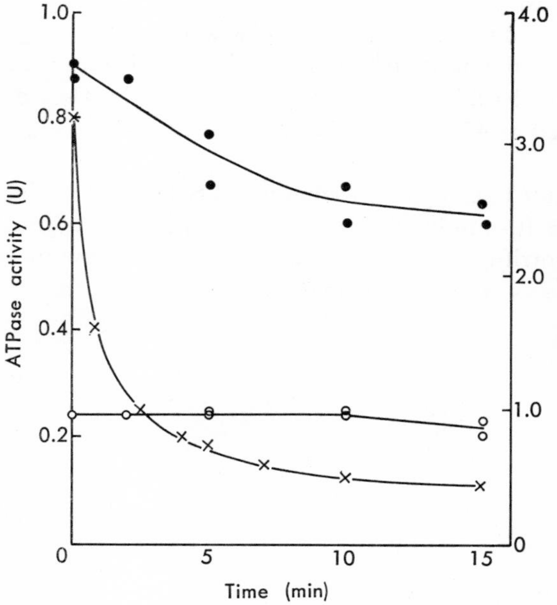

FIG. 3. Changes in ATPase activity and viscosity of myosin solution during Nagarse digestion. Myosin (10 mg/ml) was incubated with Nagarse 1/200 by weight of total myosin. After an appropriate period of incubation, the reaction was stopped by adding the incubation mixture to 0.1 M DFP. Final concentration of DFP was adjusted to 0.1 mM. Ca^{2+}-activated ATPase activity (○) was measured in 5 mM $CaCl_2$, 0.5 M KCl, 2 mM ATP at pH 7.0, and at 25°C. EDTA, K^+-activated ATPase (●) activity was measured in 5 mM EDTA, 0.5 M KCl, 2 mM ATP at pH 7.6, and at 25°C. For the viscosity measurements (×), incubation was performed in a viscometer at 25°C.

it was found that the activity of reaction mixture did not change for about 10 min.

The result could be explained in 2 ways.

1) If S-1(N) produced was not hydrolyzed further, the sum of Ca^{2+}-activated ATPase activities of 2 moles of S-1(N) is the same as the activity of 1 mole of myosin and no substantial change in the active site appears in the change from myosin to S-1(N).

2) If S-1(N) produced was continuously decomposed to inactive components, the amount of S-1(N) at 10-min incubation should be lower than the amount expected from the myosin consumed. It has to be assumed in this case that the sum of ATPase activities of 2 moles of S-1(N) was higher than that of 1 mole of myosin.

It is generally accepted that myosin (molecular weight 4.65×10^5) is composed of 2 subunits and the head portion of each subunit is separated as S-1 (molecular weight 1.0×10^5). The weight fraction of S-1(2 moles) per mole of myosin is calculated as 43%, which is the yield of S-1(N) expected in case (1).

An attempt was made to compare the amount of S-1(N) remaining in the reaction mixture after 10-min incubation with the calculated value of 43%. After dialysis and centrifugation of the Nagarse digest of myosin, the resulting supernatant was found to contain about 60–80% of the initial

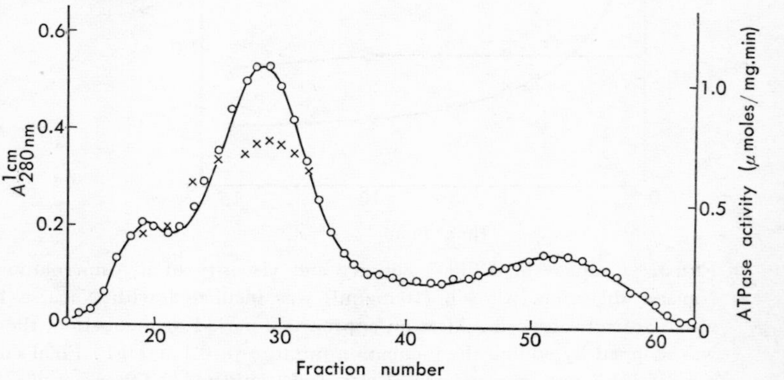

FIG. 4. Gel filtration pattern of Nagarse digest of myosin. The digest was dialyzed and the precipitate formed was discarded by centrifugation. The supernatant was applied to a Sephadex G-200 column. Three-ml fractions were collected. ×, Ca^{2+}-activated ATPase activity.

amount of protein. The supernatant was applied to a Sephadex G-200 column and 3 peaks were obtained, as shown in Fig. 4. The second peak contained S-1(N) and the amount of protein estimated from the area under the peak corresponded to 46% of added myosin. The Ca^{2+}-activated ATPase activity of this preparation was 0.75–0.77 U. Fractions in the second peak were collected and further purified by salting-out precipitation and DEAE-cellulose chromatography. Final yield of S-1(N) became 30% of added myosin and the Ca^{2+}-activated ATPase activity increased to 1.25 U. Rechromatography of this preparation did not increase the ATPase activity.

$A_{280\,nm}{}^{1\,cm}$ of 1% protein solution was 0.5 for myosin, but was 0.8 for purified S-1(N). Therefore, the amount of S-1(N) was actually 19% of added myosin in weight ratio.

We may thus suppose that nearly one-half of the S-1(N) produced was decomposed after 10 min of Nagarse digestion. In order to keep the ATPase activity unchanged during Nagarse digestion, the decrease in the amount of S-1 (N) should be counterbalanced by the activation, as assumed in case (2). This was clearly shown in the difference between Ca^{2+}-activated ATPase activities of myosin (0.24 U) and of purified S-1(N)(1.25 U). In other words, myosin ATPase may be in the inhibited state.

In contrast to the results of Ca^{2+}-activated ATPase activity, EDTA, K^+-activated ATPase activity decreased with digestion time (Fig. 3). (In order to obtain the decrease in the latter ATPase activity, fresh myosin preparations should be used for the digestion experiments.) Assuming that EDTA released myosin from the postulated inhibitory state and the EDTA, K^+-activated ATPase activity of S-1(N) in myosin is the same as that in the free state, the results obtained in EDTA might indicate the gradual decomposition (or inactivation) of S-1(N) as it was released from myosin.

If the 2 f-chains of myosin are not identical and if only 1 of the 2 constitutionally has the catalytic activity of myosin ATPase, it is not necessary to assume that myosin is in the inhibited state. The inactive f-chain might be sensitive to Nagarse-digestion and the active one alone would be obtained as pure S-1(N) after the purification. This view is, however, opposed to the economical viewpoint that protein molecules are synthesized from a small amount of genetic information and the complex formed by association is biologically active. The latter viewpoint may be particularly applicable to a large molecule such as myosin.

If the 2 f-chains of myosin are identical, they may still have different properties, as suggested by Morita (*18*) and Tonomura's group (*19–21*). It has been shown that the content of each g-chain was 1 mole per mole of myosin (*22*), and this might bring about a discrimination between the functions of the 2 f-chains. When the first substrate (or modifier) binds to 1 active site, the g-chain might modify either of the f-chains, which discriminates the second substrate (or modifier) binding from the first.

Two Different States of S-1 (CT)

S-1(CT) was prepared by chymotryptic digestion of myosin and purified carefully by ion exchange chromatography, salting-out precipitation and gel filtration. The acrylamide gel electrophoresis pattern of purified S-1 (CT) showed 2 main bands and a slightly colored band between them. This

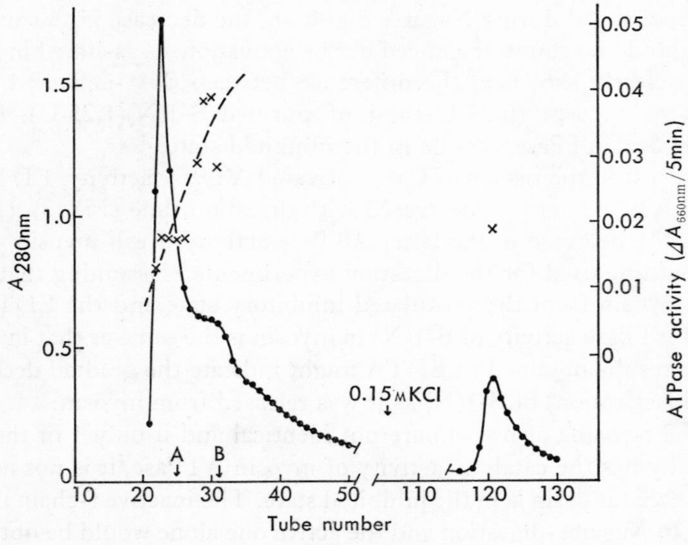

FIG. 5. DE-32 chromatography of S-1(CT). S-1(CT) was dialyzed against 0.08 M KCl-20 mM TES buffer (pH 7.5) and 1 ml of the solution (26.7 mg/ml) was applied to a column (0.9×60 cm). One-step elution was performed with 0.08 M KCl-20 mM TES buffer (pH 7.5) and the remaining protein was eluted out with 0.15 M KCl-20 mM TES buffer (pH 7.5) as indicated by the arrow. One-ml fractions were collected. ATPase activity was measured in a reaction mixture containing 5 mM $CaCl_2$, 0.064 M KCl, 16 mM TES buffer (pH 7.5), 2 mM ATP, and 0.038 mg/ml of enzyme. ×, Ca^{2+}-activated ATPase activity.

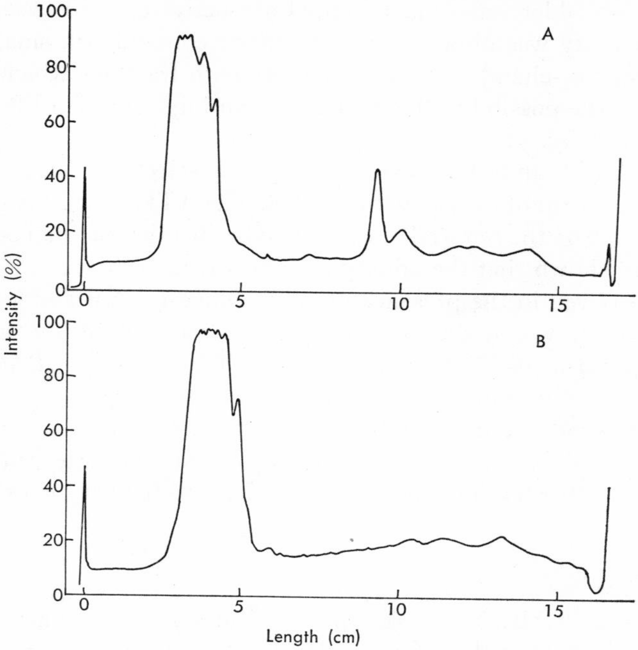

FIG. 6. Difference in components of 2 S-1(CT) preparations fractionated by DE-32 chromatography. Densitometer traces are shown from SDS gels run of samples given as A and B in Fig. 5.

sharp separation of the main band could be observed neither with S-1(T) nor with S-1(N). Figure 5 shows a preliminary fractionation experiment of purified S-1(CT) by DE-32 chromatography. The 2 main components were roughly separated. The polypeptide composition of each fraction was analyzed by SDS disc electrophoresis (*23*). In addition to large components, a small component (probably g_1-chain) was detected in the early fractions (Fig. 6,A). It gradually faded in later fractions and was hardly detected in fractions at the edge of the shoulder of the fractionation pattern (Fig. 6,B).

The Ca^{2+}-activated ATPase activity of each fraction was also measured and the results are shown in Fig. 5. The activity of early fractions was very low and that at the shoulder was very high. The ATPase activity of original S-1(CT) added to this column was the same as the average value of the activities of total fractions. EDTA, K^+-activated ATPase activity of the fraction at the shoulder was similar to the Ca^{2+}-activated ATPase activity. Since lower activity was obtained with fractions containing a small component (probably g_1-chain) and higher activity with fractions which lacked the component, the possibility that g_1-chain is an inhibitor of ATPase may be worth considering.

Gazith *et al.* (*22*) and Weeds and Lowey (*24*) selectively extracted g_2-chain by the treatment of myosin with DTNB. The ATPase activity of myosin did not change on the removal of g_2-chain. On the other hand, Tonomura *et al.* (*25*) have shown that the affinity of actin for myosin pretreated with CMB increased even in the presence of ATP. Since DTNB and CMB are both SH-reagents, we might assume that the removal of the g_2-chain stabilizes the actin-myosin interaction in ATP. The role of g_2-chain might be an interaction inhibitor between actin and myosin.

Although it is rather speculative at present, the function of g_1-chain and g_2-chain may be inhibitory based upon findings so far available, but lack of correlation among the findings should be rectified and further investigations are needed.

Acknowledgments
We are indebted to Mr. N. Sasaki for technical assistance. This study is supported by grants from the Muscular Dystrophy Associations of America, Inc. and from the Ministry of Education of Japan.

REFERENCES

1. P. P. Trotta, P. Dreizen, and A. Stracher, *Proc. Natl. Acad. Sci. U.S.*, **61**, 659 (1968).
2. L. C. Gershman and P. Dreizen, *Biophys. J.*, **9**, A235 (1969).
3. A. Stracher, *Biochem. Biophys. Res. Commun.*, **35**, 519 (1969).
4. D. W. Frederiksen and A. Holtzer, *Biochemistry*, **7**, 3935 (1968).
5. E. Gaetjens, K. Barany, G. Bailin, H. Oppenheimer, and M. Barany, *Arch. Biochem. Biophys.*, **123**, 82 (1968).
6. K. Yagi, *Seikagaku* (*J. Jap. Biochem. Soc.*), **43**, 66 (1971) (in Japanese).
7. Y. Yazawa and K. Yagi, *J. Biochem.*, **73**, 567 (1973).
8. S. Lowey, H. S. Slayter, A. G. Weeds, and H. Baker, *J. Molec. Biol.*, **42**, 1 (1969).
9. M. Onodera and K. Yagi, *J. Biochem.*, **69**, 145 (1971).
10. K. Yagi, Y. Yazawa, and T. Yasui, *Biochem. Biophys. Res. Commun.*, **29**, 331 (1967).

11. H. Mueller and S. V. Perry, *Biochem. J.*, **85**, 431 (1962).
12. H. Mueller, *J. Biol. Chem.*, **240**, 3816 (1965).
13. D. M. Young, S. Himmerfarb, and W. F. Harrington, *J. Biol. Chem.*, **240**, 2428 (1965).
14. K. M. Nauss, S. Kitagawa, and J. Gergely, *J. Biol. Chem.*, **244**, 755 (1969).
15. H. Kawakami, J. Morita, K. Takahashi, and T. Yasui, *J. Biochem.*, **70**, 635 (1971).
16. Y. Yazawa and K. Yagi, *Biochim. Biophys. Acta*, **180**, 190 (1969).
17. C. Tanford, K. Kawahara, and L. Lapanje, *J. Am. Chem. Soc.*, **89**, 729 (1967).
18. F. Morita, *J. Biochem.*, **69**, 517 (1971).
19. T. Shimada, *J. Biochem.*, **67**, 185 (1970).
20. M. Ohe, B. K. Seon, K. Titani, and Y. Tonomura, *J. Biochem.*, **67**, 513 (1970).
21. Y. Hayashi, *J. Biochem.*, **72**, 83 (1972).
22. J. Gazith, S. Himmerfarb, and W. F. Harrington, *J. Biol. Chem.*, **245**, 15 (1970).
23. K. Weber and M. J. Osborn, *J. Biol. Chem.*, **244**, 4406 (1967).
24. A. G. Weeds, *Nature*, **233**, 1362 (1969); A. G. Weeds and S. Lowey, *J. Molec. Biol.*, **61**, 701 (1971).
25. Y. Tonomura, J. Yoshimura, and S. Kitagawa, *J. Biol. Chem.*, **236**, 1968 (1961).

Change of Tryptophanyl Residue in Myosin Induced by ATP

Fumi Morita, Hidenori Yoshino, and Michio Yazawa

Department of Chemistry, Faculty of Science, Hokkaido University, Sapporo, Japan

Adenosine triphosphate-induced movement of a tryptophanyl residue in myosin was investigated. From measurement of the fluorescence emission spectrum of heavy meromyosin excited at 293 nm, our previous observations made by UV absorption difference spectrum measurements were confirmed. These indicated that a tryptophanyl residue was buried in the protein interior on adding ATP and recovered after hydrolysis to ADP. A charge-transfer interaction between the adenine moiety of ATP and the tryptophanyl indole group may enlarge the side-chain movement induced by ATP. A difference between 2 moles of substrate binding to heavy meromyosin was demonstrated distinctly by the binding of ADP in the presence of $MnCl_2$. Only 1 mole of the stronger binding ADP was responsible for inducing the difference spectrum. Heavy meromyosin was modified with 2-hydroxy-5-nitrobenzyl bromide. About 2 moles of this reagent were specifically bound to heavy meromyosin. The tryptophanyl movement induced by ATP was suppressed by only 1 mole of the weaker binding 2-hydroxy-5-nitrobenzyl group. For substrate binding, either 2 nonidentical sites or 2 sites with an interaction between them may be considered. Divalent cations are required for the ATP-induced movement of the tryptophanyl residue. The possible molecular mechanism of the movement was discussed based on the difference in the coordination states of ATP and ADP to the divalent cation. The tryptophanyl movement was not affected by actin binding.

We have reported that some tyrosyl and tryptophanyl residues in heavy meromyosin are buried in the protein interior on addition of ATP as determined by means of UV absorption (*1*) and fluorescence (*2, 3*) spectroscopy. Detailed investigation showed that the tyrosyl residues are also buried by ADP, pyrophosphate (PP_i) or other ATP analogs. Changes in the tryptophanyl residue are induced by ATP or other nucleoside triphosphates specifically. With preparations of myosin (*4*) or of subfragment-1 (*5*), the same movement of the tryptophanyl residue is also observed. The movement of the tryptophanyl residue, therefore, may be considered to play some role in the function of myosin. In this paper, our studies will be described, focusing on the elucidation of the mechanism and function of the ATP-induced movement of the tryptophanyl residue in myosin. Almost all experiments presented in this paper, however, were carried out with heavy meromyosin since it is easily purified and separated from myokinase and deaminase activities to a degree suitable for our experiments (*6, 7*). Heavy meromyosin is also convenient for the measurement of small changes in the electronic spectra since turbidity changes, as observed in myosin (*4*), do not accompany the addition of ATP.

Interaction between Adenine Moiety of ATP and Tryptophanyl Indole Group

From a comparison of the UV absorption difference spectra of heavy meromyosin induced by ATP and by ADP, it was shown that a tryptophanyl residue buried in the protein interior by ATP recovered after hydrolysis to ADP (*1*). This was also observed in the fluorescence emission spectrum of heavy meromyosin. The fluorescence emission spectrum of a heavy meromyosin solution excited at 293 nm showed a maximum at 338 nm (*3*). When ATP was added, the emission spectrum increased in intensity and shifted toward blue slightly (*3*). After hydrolysis of ATP it returned to a state very similar to that obtained before the addition of ATP. Figure 1 shows the pH dependence of the quantum yield of heavy meromyosin in the 3 states. Since only the tryptophanyl chromophore is excited at 293 nm, the results in Fig. 1 indicate that the tryptophanyl movement is induced by ATP, and very slightly by ADP.

Similar movement of the tryptophanyl residue was indicated from the UV absorption difference spectrum of heavy meromyosin induced by other nucleoside triphosphates (*8*). It was not caused, however, by nucleoside

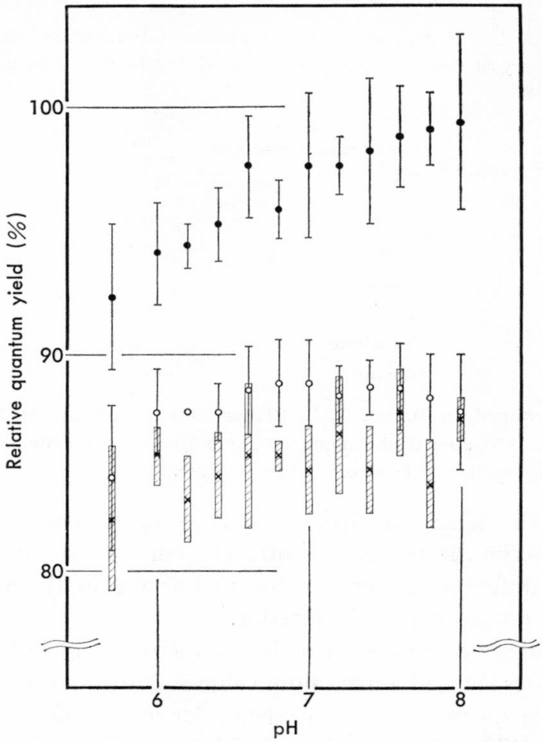

FIG. 1. Relative quantum yield of heavy meromyosin solution. Excitation was at 293 nm. Conditions were 0.6 M KCl, 1 mM $MgCl_2$, and 20 mM phosphate buffer at 20°C. Concentration of heavy meromyosin was 0.1 mg/ml. (●) Immediately after the addition of 2.22 μM ATP, (○) after the hydrolysis of ATP to ADP, and (×) in the absence of ATP or ADP.

diphosphates, ribose triphosphate or inorganic tripolyphosphate (*8*). We inferred that there is a nonspecific interaction between the base moiety of nucleoside triphosphate molecule and the tryptophanyl residue in heavy meromyosin (*3, 8*). The interaction may be a charge-transfer type. Tryptophan may be considered to be an electron donor. Using N-acetyltryptophan amide as a model compound, a new absorption band suggesting charge transfer was observed around 296 nm on addition of ATP (*3*). The molar absorption coefficient around 296 nm was only 150 M^{-1} cm^{-1}, which is less than one-tenth of $\Delta\varepsilon$ of the ATP-induced difference spectrum of heavy

TABLE I
Comparison between the Value of $\Delta\varepsilon_{289}$ of Heavy Meromyosin Induced by Nucleoside Triphosphate and the Energy of the Lowest Empty Molecular Orbital of the Constituent Base or Base Analog

Nucleoside triphosphate	$\Delta\varepsilon_{289}$ [a] ($M^{-1}cm^{-1}$)$\times 10^{-3}$	Constituent base or analog	Energy of lowest empty molecular orbital reported by Pullman (in β units)[b]
CTP[c]	6.9	Cytosine	−0.80
TTP[d]	6.0	Thymine	−0.96
ATP	5.8	Adenine	−0.87
UTP[e]	4.5	Uracil	−0.96
ITP[f]	3.1	Xanthine	−1.01
GTP[g]	1.7	Guanine	−1.05

[a] The values were determined in 0.25 M KCl, 10 mM $MgCl_2$, and 20 mM Tris-HCl (pH 8). [b] From Ref. (*9*). [c] Cytidine triphosphate. [d] Thymidine triphosphate. [e] Uridine triphosphate. [f] Inosine triphosphate. [g] Guanosine triphosphate.

meromyosin around the same wavelength. Therefore, even if there is a charge-transfer interaction at the active site, the band may not be distinguished from the difference spectrum due to burying of the aromatic chromophores in the nonpolar protein interior.

The absorption difference spectrum of heavy meromyosin induced by nucleoside triphosphates showed a maximum value at 289 nm, although the shape of the difference spectrum varied in detail depending on the type of constituent base (*8*). When the $\Delta\varepsilon$ values at 289 nm induced by various nucleoside triphosphates are compared with each other, their order of magnitude agrees well with that of the energy of the lowest empty molecular orbital of each constituent base as reported by Pullman (*9*) (Table I). The only exception is that the order of thymine and adenine is reversed. Bases having higher energy have higher electron affinity (*9*) and may interact with indole more easily. Therefore, the charge-transfer interaction between the base moiety of nucleoside triphosphate and the tryptophayl residue appears to contribute to the change in the side-chain groups in heavy meromyosin reflected in the difference spectrum.

Apparent Heterogeneity between Two Moles of Substrate Binding

As shown previously, in the presence of 10 mM $MgCl_2$ and 0.08 M KCl, a Scatchard plot of the binding of ADP to heavy meromyosin, which was measured directly by the gel filtration method, showed a concave downward curvature (*7*). Comparison between the value of ΔA at 288 nm, the maximum of the ADP-induced difference spectrum, and the number of bound ADP molecules obtained from gel filtration experiments showed that the maximum value of ΔA was attained at 1 mole of ADP binding out of 2 moles maximum binding (*7*). A more distinct difference between the 2 moles of ADP binding was observed in the presence of $MnCl_2$ (*10*), as shown in the Scatchard plot in Fig. 2. The upward curvature of the plot is different from that obtained in the presence of $MgCl_2$. The plot obviously indicates 1 mole of stronger binding of ADP and another weaker one. The dissociation constants evaluated from the Scatchard plot assuming 2 independent first-order bindings were 1.1×10^{-5} M and less than 3×10^{-8} M. Figure 3 represents a comparison between the value of ΔA at 288 nm of heavy meromyosin

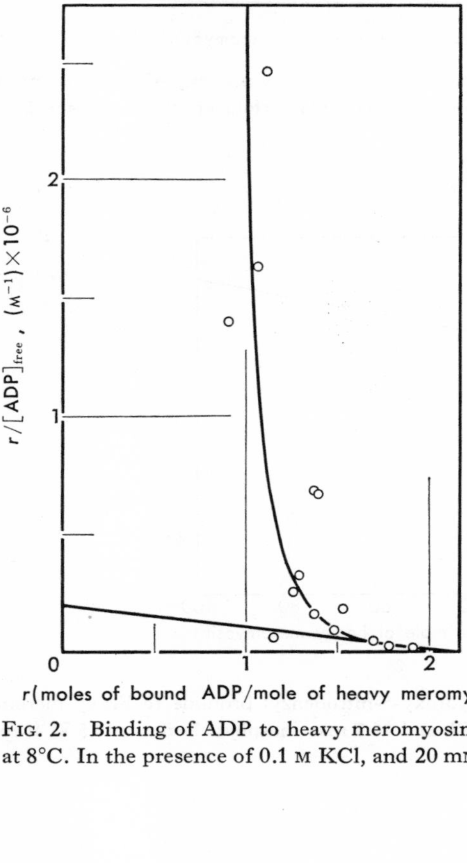

FIG. 2. Binding of ADP to heavy meromyosin in the presence of 1 mM $MnCl_2$ at 8°C. In the presence of 0.1 M KCl, and 20 mM Tris-HCl (pH 8).

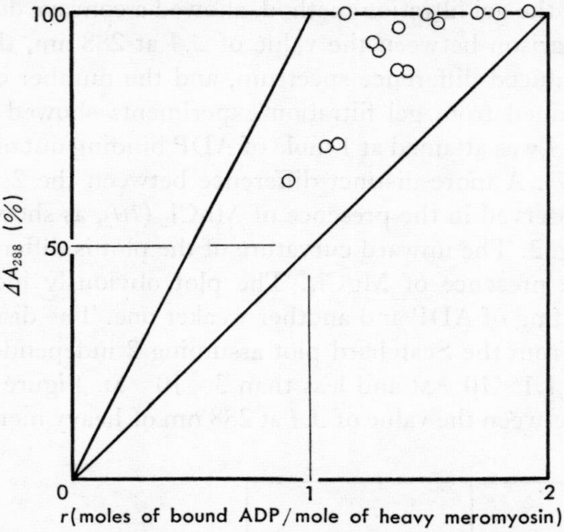

FIG. 3. Comparison between ΔA_{288} of heavy meromyosin induced by ADP and the number of bound ADP molecules. Conditions are the same as in Fig. 2.

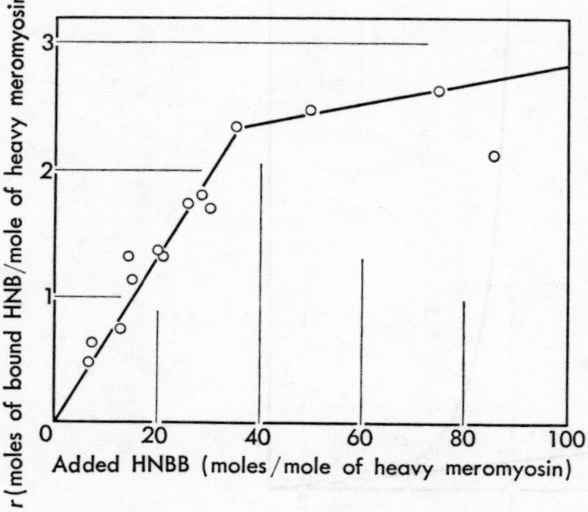

FIG. 4. Binding of 2-hydroxy-5-nitrobenzyl bromide to heavy meromyosin in the presence of 0.6 M KCl, and 16.7 mM phosphate buffer (pH 5.8) at room temperature.

induced by ADP, and bound ADP as evaluated from the data of Fig. 2. The maximum value of ΔA is attained at 1 mole of ADP binding, which is very similar to the result obtained in the presence of $MgCl_2$. These results indicate that the 2 moles of ADP binding in heavy meromyosin are apparently heterogeneous.

On the other hand, the binding of ATP can hardly be measured by direct methods. Using the stopped-flow method, we have shown that the minimum concentration of ATP needed to induce the maximum change in ΔA was twice the concentration of heavy meromyosin, indicating that 2 moles of ATP are bound per mole of heavy meromyosin (*11, 12*). However, it was not really clear whether both bound ATP molecules induce the difference spectrum or whether only 1 mole of ATP binding does. Heavy meromyosin was treated with 2-hydroxy-5-nitrobenzyl bromide (HNBB) under suitable conditions to modify the tryptophanyl residue selectively (*8*). As shown in Fig. 4, about 2 moles of HNBB reagent specifically reacted with heavy

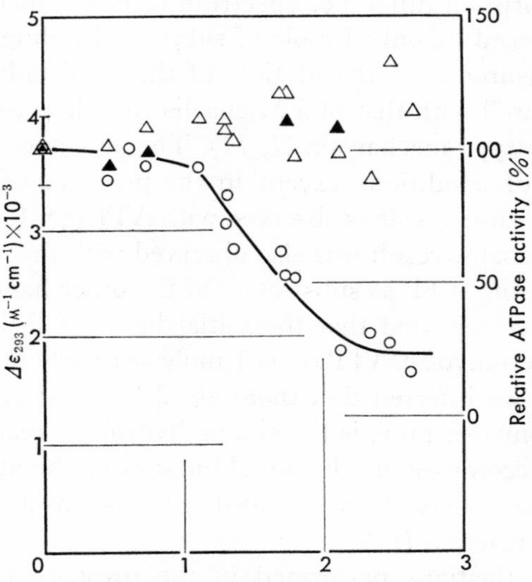

FIG. 5. Effect of 2-hydroxy-5-nitrobenzyl group binding on $\Delta\varepsilon_{293}$ of heavy meromyosin induced by ATP (○). The measurements were made in the presence of 0.25 M KCl, 10 mM $MgCl_2$, and 20 mM Tris-HCl (pH 8) at 25°C. Relative ATPase activity was determined from the time course of P_i liberation at the steady state (△) and from the lifetime of the ATP-induced difference spectrum (▲).

meromyosin. Using this modified heavy meromyosin the difference spectra induced by ATP were measured. As shown in Fig. 5, the value of $\Delta\varepsilon$ at 293 nm decreases to one-half or less with increase in HNB binding from 1 to 2 moles, although it does not change until 1 mole of HNB is bound. These results indicate that the tryptophanyl residue buried by ATP probably comprises 1 mole per mole of heavy meromyosin. As in the case of ADP, only 1 out of 2 moles of bound ATP appears to be responsible for inducing the difference spectrum of heavy meromyosin.

The apparent heterogeneity of substrate bindings may be caused either by the 2 sites for substrate binding being nonidentical, or if they are identical, there may be an interaction between them. Which is correct may be indicated after structural identity is proved between the two subfragment-1 parts constituting myosin.

Molecular Mechanism of the Movement of Tryptophanyl Residue

As stated above, the absorption difference spectrum of heavy meromyosin was considered to be induced by only 1 mole of substrate binding out of 2 moles. From kinetic measurements, the lifetime of the ATP-induced difference spectrum agreed well with that of a Michaelis complex, assuming a simple Michaelis-Menten type mechanism (*1*, *11*). The agreement is quite good over a wide range of conditions except in the presence of Ca^{2+} or Mn^{2+}. CTP also gave the same result as observed with ATP (*8*). In addition to heavy meromyosin, the same result was also observed with myosin (*4*) or with subfragment-1 (*5*) using ATP as substrate. On the other hand, Tonomura and co-workers have reported that the initial burst of P_i liberation from myosin or heavy meromyosin ATPase is 1 mole per mole of the protein (*13*, *14*). Therefore, we inferred that there are 2 kinds of ATPase in myosin or in heavy meromyosin; one is the simple hydrolysis reaction and the other is the reaction accompanying the initial burst of P_i liberation. The difference spectrum was considered to be induced in the Michaelis complex of the simple hydrolysis reaction (*11*).

However, recent investigations performed in the presence of $MnCl_2$ (*10*) showed that the difference spectrum should occur in molecular species of the reaction relating to the initial burst of P_i liberation rather than that of the simple hydrolysis reaction. The spectrum induced by ATP may be associated with the ES complex relating to the initial burst together with the succeeding intermediate containing trichloroacetic acid-labile P_i, or it

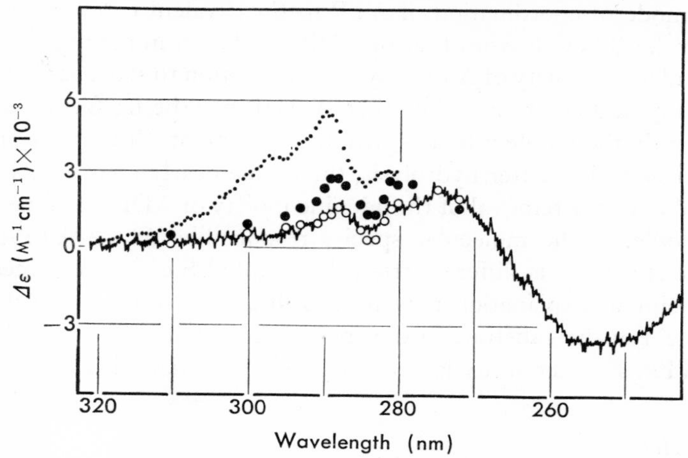

FIG. 6. Absorption difference spectra of heavy meromyosin after the addition of ATP in the absence of divalent cations. Conditions were 0.1 M KCl, and 20 mM Tris-HCl (pH 8) at 8°C. Circles were determined by the stopped-flow method before (●) and after (○) the decay of the difference spectrum induced by 5.14×10^{-5} M ATP. The recording trace is the difference spectrum induced by the addition of 9.59×10^{-5} M ADP. The dotted line represents the ATP-induced difference spectrum in the presence of 1 mM $MgCl_2$.

may be associated with the latter species alone. In any case, the substrate in the latter species may not be ATP, but may be an intermediate state from which P_i can be liberated by trichloroacetic acid, as stated by Tonomura and co-workers (*15*) and Lymn and Taylor (*16*). The shape of the difference spectrum due to this species is obviously different from that induced by the addition of ADP to heavy meromyosin (*10*). We call the difference spectrum induced by addition of ATP before decay the "ATP-induced difference spectrum" disregarding whether or not the bond between the γ and β phosphates of ATP in the species is actually cleaved.

As shown in Fig. 6, the ATP-induced difference spectrum was devoid of a shoulder around 300 nm in the absence of $MgCl_2$ (*12*). Mg^{2+} is required for ATP-induced movement of the tryptophanyl residue. Mg^{2+} was replaceable by other divalent cations with smaller ionic radius, such as Ca^{2+}, Mn^{2+} or Co^{2+} (*12*). The divalent cation may be bound to heavy meromyosin through the coordination of 2 imidazole groups (*12*). Based on the NMR results reported by Cohn and Hughes (*17*), it was considered, therefore, that the mode of coordination of ATP to the divalent cation at the active site might be different from that of ADP, as shown in the previous paper (*3*). The adenine moiety of ATP may be in a position to stack easily with the tryptophanyl indole group, while that of ADP may be far from the indole. Consequently the indole interacts with the adenine of ATP and is buried in the protein interior. After hydrolysis, the indole nearly recovers its original position due to the removal of the adenine moiety of ADP from the vicinity of the indole. If the molecular species responsible for the ATP-induced difference spectrum is different from the simple ES complex as mentioned above, a similar explanation may be possible by assuming that the steric arrangement of the substrate in the species discussed is still very similar to that of ATP. Further study is necessary to substantiate this.

Effect of Actin

In order to evaluate the role of the tryptophanyl movement, the effect of actin on the ATP-induced difference spectrum of heavy meromyosin was studied. As already reported, heavy meromyosin ATPase is activated by

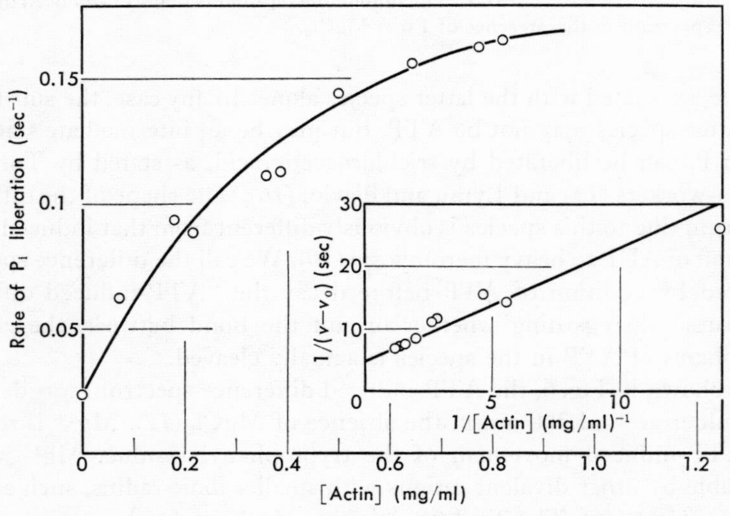

FIG. 7. Activation of heavy meromyosin ATPase by actin. Conditions were 15 mM KCl, 1 mM $MgCl_2$, and 20 mM Tris-HCl (pH 8) at 3°C. Concentration of heavy meromyosin was 1.36 mg/ml. Concentration of ATP was 2.97×10^{-4} M.

actin (*18–23*). The activation was measured in the presence of 15 mM KCl, 1 mM $MgCl_2$ and 20 mM Tris-HCl (pH 8) at 3°C (Fig. 7). The apparent dissociation constant of actin-heavy meromyosin complex in the presence of 0.3 mM ATP was estimated as 0.38 mg/ml from a similar plot made by Eisenberg and Moos (*21*) (inset of Fig. 7). Under the same conditions, the ATP-induced difference spectrum of heavy meromyosin was measured in the presence of 0 to 1 mg/ml actin. In order to avoid turbidity change due to the presence of actin, 2 cells containing heavy meromyosin, and actin plus ATP, respectively, were used for reference. The typical shape of the ATP-induced difference spectrum was not changed essentially even in the presence of 1 mg/ml actin, although a small modification of the shape was observed due to the unavoidable turbidity difference between the reference and the sample cells. The value of ΔA at 289 nm of the difference spectrum was plotted against the concentration of actin as shown in Fig. 8. In the presence of ATP the mode of interaction between actin and heavy meromyosin should be different from that in the absence of ATP, as typically shown by light scattering or viscosity changes. There is also some ambiguity about the stoichiometry of actin to myosin (*24*) or heavy meromyosin in the presence of ATP, or about the molecular species responsible for the activation of ATPase (*15*, *25*). In any case, as stated by Perry *et al.* (*19*), there should be some interaction between actin and heavy meromyosin when the

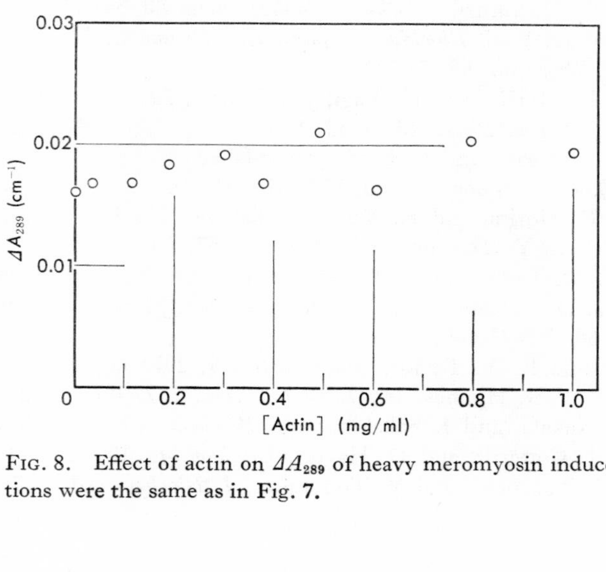

FIG. 8. Effect of actin on ΔA_{289} of heavy meromyosin induced by ATP. Conditions were the same as in Fig. 7.

ATPase is activated by actin. Therefore, the results in Fig. 8 indicating no change in ΔA seem to suggest that the tryptophanyl and tyrosyl residues in heavy meromyosin as reflected in the difference spectrum are located in an area which is not directly related to the interaction with actin in the presence of ATP. It would appear that the tryptophanyl movement may not be directly related to the interaction between actin and myosin. We cannot yet determine what the role of the ATP-induced tryptophanyl movement in myosin may be.

Addendum

After this paper was submitted a paper has appeared (M. M. Werber, A. G. Szent-Györgyi, and G. D. Fasman, *Biochemistry*, **11**, 2872 (1972)) which reports the fluorescence change in heavy meromyosin induced by substrate. Their results agree very closely with those obtained by us (*2*, *3*).

REFERENCES

1. F. Morita, *J. Biol. Chem.*, **242**, 4501 (1967).
2. F. Morita and K. Yagi, 19th Symposium on Enzyme Chemistry, Kanazawa, p. 117 (1968).
3. F. Morita, *in* " Molecular Mechanisms of Enzyme Action," ed. by Y. Ogura *et al.*, University of Tokyo Press, Tokyo, p. 281 (1972).
4. H. Yoshino, F. Morita, and K. Yagi, *J. Biochem.*, **71**, 351 (1972).
5. F. Morita and T. Shimizu, *Biochim. Biophys. Acta*, **80**, 545 (1969).
6. F. Morita and K. Yagi, *Biochim. Biophys. Res. Commun.*, **22**, 297 (1966).
7. F. Morita, *J. Biochem.*, **69**, 517 (1971).
8. H. Yoshino, F. Morita, and K. Yagi, *J. Biochem.*, **72**, 1227 (1972).
9. B. Pullman, *J. Chem. Phys.*, **43**, S233 (1965).
10. M. Yazawa, F. Morita, and K. Yagi, unpublished.
11. F. Morita, *Biochim. Biophys. Acta*, **172**, 319 (1969).
12. M. Yazawa, F. Morita, and K. Yagi, *J. Biochem.*, **71**, 301 (1972).
13. T. Kanazawa and Y. Tonomura, *J. Biochem.*, **57**, 604 (1965).
14. K. Imamura, M. Tada, and Y. Tonomura, *J. Biochem.*, **59**, 280 (1966).
15. Y. Tonomura, H. Nakamura, N. Kinoshita, H. Onishi, and M. Shigekawa, *J. Biochem.*, **66**, 599 (1969).
16. R. W. Lymn and E. W. Taylor, *Biochemistry*, **9**, 2975 (1970).
17. M. Cohn and T. R. Hughes, Jr., *J. Biol. Chem.*, **237**, 176 (1962).
18. K. Yagi, T. Nakata, and I. Sakakibara, *J. Biochem.*, **61**, 567 (1967).
19. S. V. Perry, J. Cotterill, and D. Hayter, *Biochem. J.*, **100**, 280 (1966).
20. K. Sekiya, K. Takeuchi, and Y. Tonomura, *J. Biochem.*, **61**, 567 (1967).

21. E. Eisenberg and C. Moos, *Biochemistry*, **7**, 1486 (1968).
22. E. M. Szentkiralyi and A. Oplatka, *J. Mol. Biol.*, **43**, 551 (1969).
23. E. Eisenberg and C. Moos, *J. Biol. Chem.*, **245**, 2451 (1970).
24. E. Eisenberg, L. Dobkin, and W. W. Kielley, *Proc. Natl. Acad. Sci. U.S.*, **69**, 667 (1972).
25. R. W. Lymn and E. W. Taylor, *Biochemistry*, **10**, 4617 (1971).

Velocities of ATP Hydrolysis and Superprecipitation of Actomyosin

Takamitsu Sekine and Masahiro Yamaguchi

Department of Biochemistry, School of Medicine, Juntendo University, Tokyo, Japan

A relation between the ATP splitting and the superprecipitation rate of actomyosin was investigated to find out what factor controls the rate of energy transfer from chemical (ATP) to mechanical one (contraction). The actomyosin, which was made from smooth muscle myosin (myosin S) and skeletal F-actin, superprecipitated much more slowly than skeletal actomyosin did. Myosin S incidentally lacks the so-called "S_1 group," a fast reactive thiol located in the active site of skeletal myosin A. On the other hand, the marked decrease of superprecipitation was associated with the blocking of S_1 with N-ethylmaleimide (NEM) in myosin A. In addition, a quantitative relationship was found between the rate of superprecipitation and the ratio of actomyosin ATPase to myosin ATPase in the three types of actomyosin reconstituted from myosin A, S_1-blocked myosin A, and myosin S, respectively. Therefore, the region containing this specific thiol, S_1, in the active site of myosin ATPase was suspected to be responsible for regulating the speed of muscle contraction.

Among many "functional" ATPases in energy-transducing systems in living cells, myosin ATPase, a mechanochemical transducer of energy in muscle tissues, is notable for passing actin filaments along its own filaments under the influence of ATP.

Energy-transducing processes, in general, are closely regulated by intrinsic or extrinsic control systems. In the actin-myosin-ATP system, such

TABLE I
ATPase Activities of Myosin S and Myosin A (Horse)

	Temp. (°C)	ATPase activity (ΔP_i μmoles/min per mg protein)			
		EDTA-	Ca^{2+}-	Ca^{2+}-	Mg^{2+}-
		0.5 M KCl		0.05 M KCl	
Myosin S	37	1.37	0.80	0.39	0.030
	25	0.61	0.28	0.13	0.0042
Myosin A	37	3.80	0.73	1.67	0.032
	25	2.42	0.54	1.25	0.0080

ATPase activity was measured in a system containing 1 mM ATP and histidine buffer (pH 7.6). For details see Ref. (*2*).

regulation can be analyzed in terms of the rate of muscle contraction, superprecipitation of actomyosin being the simplest measure *in vitro*.

Efforts have been made to find a correlation between ATPase activity (rate of energy production) and contraction velocity (*1*) (rate of energy consumption). At present, however, there is no clear conclusion. As can be seen in Table I, the Ca^{2+}-ATPase and Mg^{2+}-ATPase activities of smooth muscle myosin (referred to as myosin S) at 37°C are almost the same as those of skeletal myosin (referred to as myosin A), although smooth muscle contracts much more slowly than skeletal muscle does. In fact, superprecipitation of skeletal muscle actomyosin is 40 times faster than that of smooth muscle actomyosin (*2*), as shown in Fig. 1.

This remarkable difference in the contraction rates of the two types of myosin cannot be accounted for yet, but might be due to the following factors; the affinity of myosin to actin, the degree and speed of conformational changes of myosin induced by actin and/or ATP, probably based on some subtle differences in the tertial structure of the active sites of skeletal and smooth muscle myosins. These might reflect the following marked differences in enzymatic properties; strong activation of the Ca^{2+}-ATPase activity of myosin S by tryptic digestion (*3*), the reverse dependence on KCl concentration of the catalytic activity of both myosins (*2, 4*), activation of the Ca^{2+}-ATPase of myosin A, but not of myosin S, by N-ethylmaleimide (NEM) treatment (*2*) (Fig. 2). The last item means that myosin S lacks the specific sulfhydryl group referred to as S_1 in its active site, which is responsible for the Ca^{2+}-ATPase activity of myosin A (*5*).

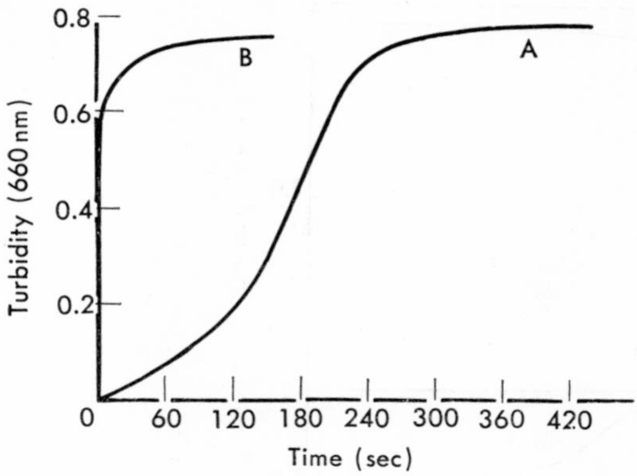

FIG. 1. Superprecipitation of reconstituted actomyosins of smooth and skeletal muscle myosins with smooth muscle F-actin at 25°C. The reaction mixture (*2*) contained 0.14 mg/ml myosin (A, horse esophagus myosin S (*2*); B, horse skeletal myosin A (*10*)) and 0.07 mg/ml esophagus smooth muscle F-actin in the presence of 0.05 M KCl, 1 mM $MgCl_2$, 20 mM Tris-maleate buffer (pH 6.5), and 0.1 mM ATP.

On the other hand, myosin S is very similar to myosin A as regards molecular weight (5.8×10^5), subunit molecular weight (2×10^5), axis ratio (*80–90*) (*6*) and aggregate-forming ability (*7*). Furthermore, after some controversy, both thick and thin filaments were confirmed to exist also in smooth muscle (*8*), suggesting the possibility of the same type of sliding mechanism as that believed to operate in skeletal muscle.

We have tried to locate a mechanism controlling the rate of superprecipitation which is common to both types of muscle tissue. Since, as described above, myosin S seems to lack the "S_1 group" and also has a higher activation free energy of ATP-splitting than myosin A (Table I), myosin A blocked with NEM at S_1 (referred to as $S_1^{\times}$-myosin A), which possesses a higher activation energy for the reaction, was used in a series of experiments as a third type of myosin.

The rate of superprecipitation and the ATPase activity were measured for three actomyosin samples which were made with rabbit skeletal muscle F-actin as a common component from myosin A (rabbit skeletal), $S_1^{\times}$-myosin A (rabbit skeletal) and myosin S (horse esophagus), respectively.

As can be seen in Fig. 3, it was found that the rate of superprecipitation on a logarithmic scale shows an excellent linear correlation with the loga-

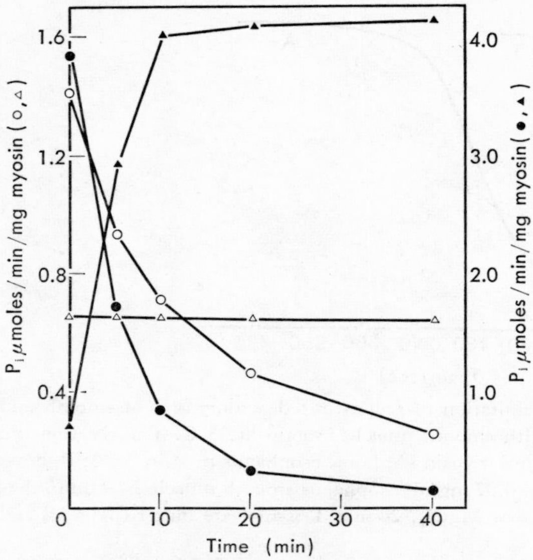

FIG. 2. Change in the ATPase activity of myosin S and skeletal myosin A (horse) caused by reaction with NEM (*2*). Myosins (open symbols, myosin S; closed ones, myosin A) were preincubated with NEM at 0°C in 0.5 M KCl and Tris buffer (pH 7.0). ATPase activity was measured at 37°C in a system containing 0.5 M KCl, 20 mM histidine buffer (pH 7.6), 1 mM ATP, and either 5 mM $CaCl_2$ (△, ▲) or 1 mM ethylenediaminetetraacetic acid (EDTA) (○, ●).

rithm of the ratio of actomyosin ATPase (A_{AM}) to myosin ATPase (A_M), though not with each ATPase alone (*13*). Linear plots were obtained not only in the presence of 1 mM or 10 mM $MgCl_2$, but also with 1 mM $MnCl_2$ (*9*).

Since the velocity of a reaction can be related to the free energy of activation, we obtain a simple equation from this result,

$$-E_{suppt} = \alpha(E_M - E_{AM}) + \beta .$$

E_{suppt}, E_M, and E_{AM} are free energies of activation of superprecipitation and ATP-splitting reactions by myosin and actomyosin, respectively, α and β being constants.

This equation shows that the decrease of the barrier level of superprecipitation is in linear correlation with the decrease of the activation energy of ATP hydrolysis by myosin, when actin is added to this system. This suggests

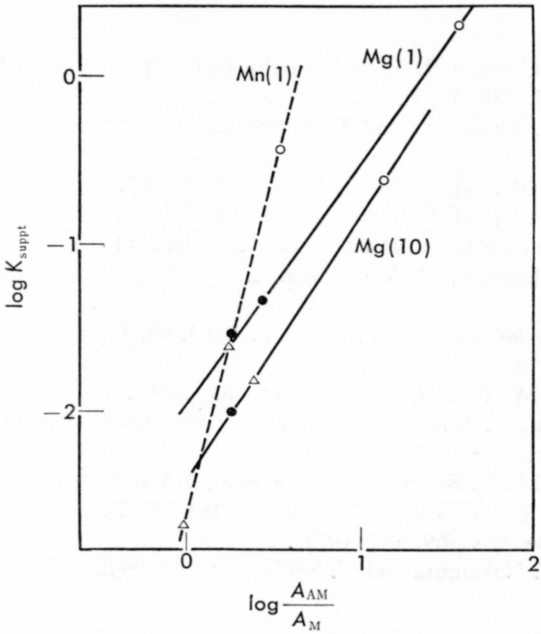

FIG. 3. Relation of the rate of superprecipitation to the degree of activation of myosin ATPase activity induced by F-actin. The reaction mixture contained 0.2 mg/ml myosin (○ myosin A, ● S_1-myosin A, △ myosin S) and 0.1 mg/ml rabbit skeletal F-actin in 0.028 M KCl at pH 6.5 and 25°C. Metal concentration is indicated as mM in parenthesis in the figure. ATPase activity and superprecipitation of the same actomyosin sample were measured under the same conditions described above. $S_1^{\times}$-myosin was prepared according to Sekine's method (*11*). The rate of superprecipitation, K_{suppt}, was defined as follows, according to Watanabe (*12*),

$$K_{suppt}=(E/E_o)/t_{1/2},$$

where E and E_o are the values of the maximum level (Δ O.D. at 660 nm) in each experiment and in an arbitrary standard, respectively, and $t_{1/2}$ is the time required to reach half of the maximum level.

that superprecipitation involves a cyclic dissociation-association process between actin and myosin, its rate being controlled by this process.

REFERENCES

1. M. Bárány, T. E. Conover, L. H. Schiselfeld, E. Gaetjens, and M. Goffart, *Eur. J. Biochem.*, **2**, 156 (1967).
2. M. Yamaguchi, Y. Miyazawa, and T. Sekine, *Biochim. Biophys. Acta*, **216**, 411 (1970).
3. D. M. Needham and J. M. Williams, *Biochem. J.*, **73**, 171 (1963).
4. D. M. Needham and J. M. Williams, *Biochem. J.*, **89**, 552 (1963).
5. T. Sekine and W. W. Kielley, *Biochim. Biophys. Acta*, **81**, 336 (1964).
6. A. Kotera, M. Yokoyama, M. Yamaguchi, and Y. Miyazawa, *Biopolymers*, **7**, 99 (1969).
7. M. Yamaguchi, *Seikagaku* (*J. Jap. Biochem. Soc.*), **43**, 185 (1971) (in Japanese).
8. R. E. Kelly and R. V. Rice, *J. Cell Biol.*, **42**, 683 (1969).
9. I. Ozawa, T. Fujii, N. Kaneko, and K. Maruyama, *Scient. Papers College Gener. Educ., Univ. Tokyo*, **20**, 93 (1970).
10. W. W. Kielley and L. B. Bradley, *J. Biol. Chem.*, **218**, 653 (1956).
11. T. Sekine and M. Yamaguchi, *J. Biochem.*, **54**, 196 (1963).
12. S. Watanabe, *J. Biochem.*, **69**, 387 (1971).
13. M. Yamaguchi, T. Nakamura and T. Sekine, unpublished.

The Role of Guanosine Triphosphate in the Polypeptide Elongation Reaction in *Escherichia coli*

Yoshito Kaziro

Institute of Medical Science, University of Tokyo, Tokyo, Japan

The role of GTP in the polypeptide elongation reaction has been investigated. Three polypeptide elongation factors, *i.e.*, EF-Tu, EF-Ts, and EF-G, involved in this reaction have been purified to homogeneity. A complex between EF-Tu and EF-Ts (Tu-Ts complex), catalyzes the binding of phenylalanyl-tRNA to ribosomes in the presence of GTP. In this reaction, (phenylalanyl-tRNA)-Tu-GTP complex is transferred to the A site of a poly U-ribosome complex having an N-acetylphenylalanyl-tRNA prebound to its P site (Complex I), resulting in the formation of N-acetyl-diphenylalanyl-tRNA. GMP-PCP, a nonhydrolyzable analog of GTP, can substitute for GTP in promoting the binding of phenylalanyl-tRNA to ribosomes. However, very little dipeptide is formed under these conditions. When Tu-GMP-PCP is removed from the ribosomal complex by ultracentrifugation through 10% sucrose solution, approximately 50% of the bound radioactivity was recovered as N-acetyl-diphenylalanine. It was concluded that the role of GTP hydrolysis in Tu-promoted binding reaction is to remove EF-Tu from ribosomes after the completion of the binding of aminoacyl-tRNA, in order to facilitate peptidyl transfer and also to initiate the cyclic reutilization of EF-Tu. In the EF-

This work was carried out in collaboration with Dr. K. Arai, Mrs. N. Inoue-Yokosawa, Dr. M. Kawakita, and Dr. H. Yokosawa. The following abbreviations are used: GDP, guanosine diphosphate ; GTP, guanosine triphosphate ; PP_i, pyrophosphate ; EF-Tu, polypeptide elongation factor Tu; EF-Ts, elongation factor Ts; EF-G, elongation factor G; Tu-Ts, a complex of EF-Tu and EF-Ts; IF-2, polypeptide initiation factor 2; RF, polypeptide release factor; and GMP-PCP, 5′-guanylyl methylenediphosphonate.

G-catalyzed translocation reaction, N-acetyldiphenylalanyl-tRNA is transferred from the P site to the A site of ribosomes with concomitant release of a deacylated tRNA from the P site. One mole of GTP is required for this reaction. It has been demonstrated that GMP-PCP can substitute for GTP in this reaction. Since it has also been demonstrated that EF-G is tightly bound to ribosomes with GMP-PCP (but not with GDP), the role of GTP cleavage in this reaction also appears to be the removal of EF-G from ribosomes in order to reinitiate a new cycle of reactions. A similar mechanism is discussed for two other GTP-requiring reactions in protein synthesis, and also for other mechanochemical reactions utilizing the energy of the phosphate bond of ATP.

It has been known for some time that GTP, besides ATP, is required for protein synthesis (*1*). Later, it became apparent that each process of protein synthesis, *i.e.*, initiation, elongation, and termination of polypeptide bonds, has a specific requirement for GTP, and that in each reaction GTP is split into GDP and P_i (see Refs. *2* and *3*). The role of GTP split in these reactions has not been elaborated from the viewpoint of energy transformation. However, as will be discussed below, these systems have certain advantages in the study of the molecular mechanism of mechanochemical reactions. In this paper, I should like first to describe our recent studies on the role of GTP in the process of polypeptide elongation, and then to discuss the general mechanism of GTP hydrolysis in the process of protein synthesis.

In protein synthesis, amino acids are first activated in the form of aminoacyl-tRNA according to the following equations:

$$\text{amino acid} + \text{ATP} + \text{enzyme} \rightleftarrows \text{(aminoacyl-AMP)-enzyme} + \text{PP}_i \quad (1)$$

$$\text{(aminoacyl-AMP)-enzyme} + \text{tRNA} \rightleftarrows \text{aminoacyl-tRNA} + \text{AMP} + \text{enzyme} \quad (2)$$

$$\text{PP}_i \rightarrow 2\text{P}_i \quad (3)$$

$$\text{Sum:}\quad \text{amino acid} + \text{ATP} + \text{tRNA} \rightarrow \text{aminoacyl-tRNA} + \text{AMP} + 2\text{P}_i \quad (4)$$

As shown in Reaction 4, the formation of 1 mole of aminoacyl-tRNA is achieved at the expense of 2 moles of phosphodiester bonds of ATP. One is consumed to form an enzyme-bound aminoacyl-adenylate in which the bond between the amino acid and AMP posesses an energy comparable to the P-O-P bond of ATP (Reaction 1). The other high-energy phosphate bond which is consumed in Reaction 3, is utilized to shift the equilibrium of the overall reaction (Reaction 4) towards the synthesis of aminoacyl-tRNA. In the presence of inorganic pyrophosphatase, the activation of amino acids becomes practically irreversible.

It has been established that 2 moles of GTP are still required for the elongation of 1 peptide bond starting from aminoacyl-tRNA as a substrate (Reaction 5) (*4, 5*). Since the ΔF of peptidyl-tRNA is about −7,000 cal/mole (*6*), it is already ample for the synthesis of a peptide bond containing about −3,000 cal/mole. Therefore, the energy from these 2 moles of GTP is probably utilized for a purpose other than the synthesis of covalent bonds.

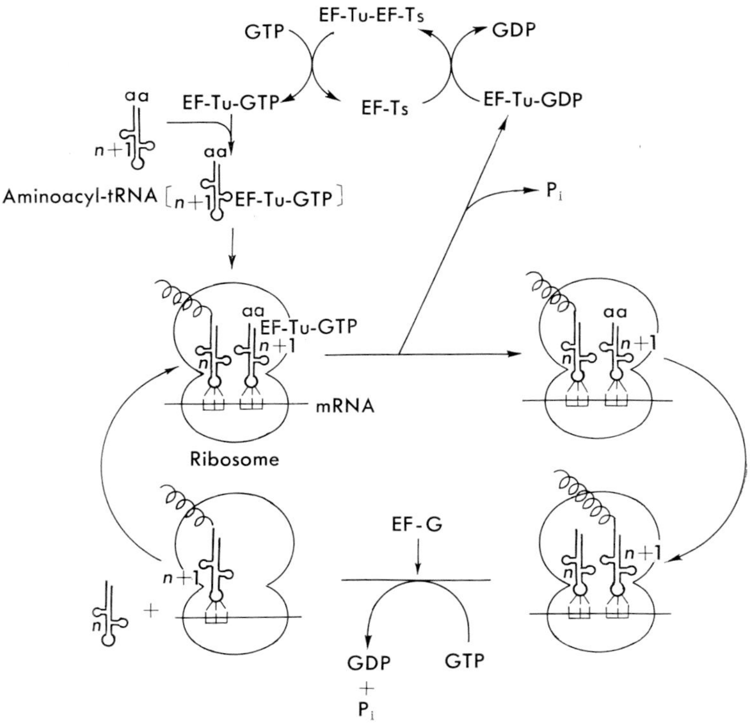

FIG. 1. Elongation of polypeptide bonds in *Escherichia coli*.

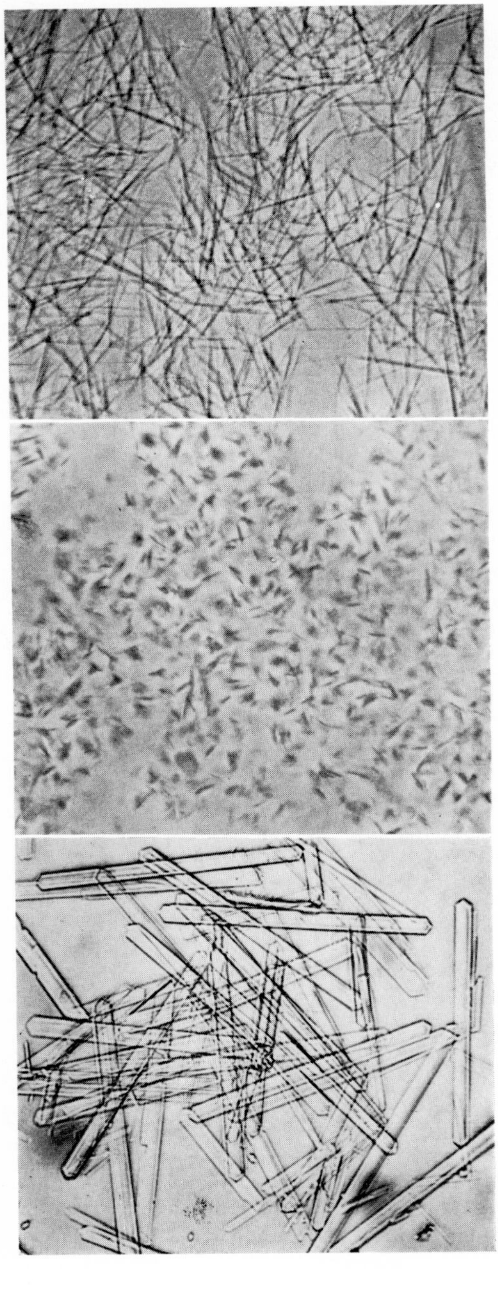

Tu-GDP Tu-Ts G

FIG. 2. Crystalline EF-Tu, EF-Tu-Ts, and EF-G (*35, 36*).

$$\text{(amino acid)}_n + \text{aminoacyl-tRNA} + 2\text{GTP} \rightarrow \text{(amino acid)}_{n+1} + \text{tRNA} + 2\text{GDP} + 2\text{P}_i \quad (5)$$

Figure 1 summarizes the current knowledge of polypeptide elongation reactions in *E. coli* (see Ref. *2*). The three protein factors, the elongation factors Tu (EF-Tu), Ts (EF-Ts), and G (EF-G), are involved in these reactions (*7*). The complex of EF-Tu and EF-Ts, Tu-Ts complex, reacts with GTP to yield Tu-GTP which immediately interacts with aminoacyl-tRNA to form a ternary complex, (aminoacyl-tRNA)-Tu-GTP (*4*, *8–16*). The complex is then transferred to the A site of a ribosome having a peptidyl-tRNA prebound to its P site, GTP is hydrolyzed, and Tu-GDP is released (*4*, *5*, *11*, *17–22*). Tu-GDP subsequently reacts with Ts to regenerate Tu-Ts complex with the displacement of GDP (*22–25*).

The next reaction is the transfer of the carboxyl group of peptidyl-tRNA to the adjacent amino group of newly bound aminoacyl-tRNA, and this reaction is catalyzed by an enzyme (or enzymes) in the 50S ribosomal subunit (*26–28*). The resulting ribosome possesses a deacylated tRNA in its P site, and a peptidyl-tRNA, having a chain one amino acid longer, in its A site. EF-G catalyzes the translocation of peptidyl-tRNA from the A site to the P site (*17*, *29*, *30*) with concomitant release of deacylated tRNA from the P site (*19*, *31*, *32*) and the movement of mRNA by three-nucleotide distances on the 30S ribosomal subunit (*33*, *34*). One mole of GTP is hydrolyzed in the translocation reaction catalyzed by EF-G.

The net result of the above sequences of reactions is the elongation of 1 peptide bond using 1 molecule of aminoacyl-tRNA and 2 molecules of GTP. One molecule of GTP is consumed for the entry of an aminoacyl-tRNA, while another molecule of GTP is utilized for the exit of deacylated-tRNA.

In order to obtain more insight into the role of GTP in the above 2 reactions, and to understand the molecular mechanism by which the energy derived from GTP is utilized for a process other than the synthesis of covalent bonds, it seemed essential to obtain the elongation factors in a highly purified form and in a substrate quantity. We have therefore attempted large-scale purification of the elongation factors from *E. coli*, and have succeeded in obtaining these factors in crystalline forms (Fig. 2). The details of the purification have been reported elsewhere (*35*, *36*).

Role of GTP in EF-Tu-Promoted Binding of Aminoacyl-tRNA to Ribosomes

In order to investigate the role of GTP in the EF-Tu-promoted binding reaction of aminoacyl-tRNA to ribosomes, we have studied the effect of GMP-PCP, a nonhydrolyzable analog of GTP, on this reaction. Haenni and Lucas-Lenard (*17*), Skoultchi *et al.* (*37*), and Shorey *et al.* (*38*) demonstrated that GMP-PCP could substitute for GTP in the binding reaction. However, under these conditions, no peptide bond formation between the newly bound aminoacyl-tRNA and the prebound peptidyl-tRNA could be observed (Fig. 3). Therefore, it is conceivable that the hydrolysis of GTP is required in some step prior to the formation of the new peptide bond.

There are two possible interpretations for the role of GTP hydrolysis in

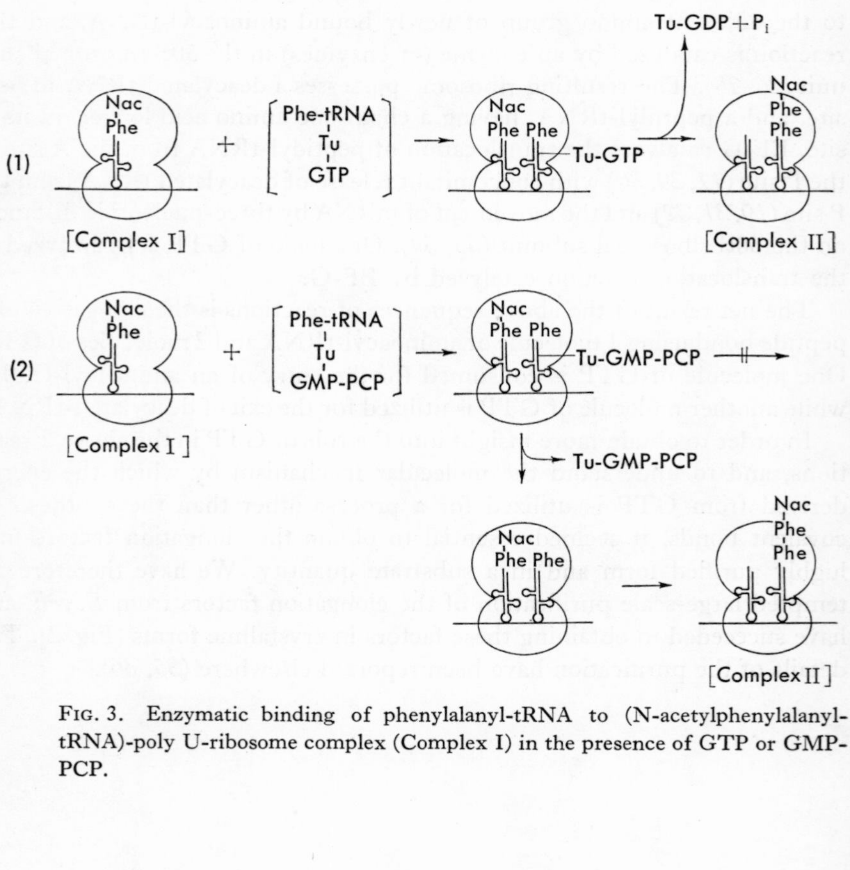

FIG. 3. Enzymatic binding of phenylalanyl-tRNA to (N-acetylphenylalanyl-tRNA)-poly U-ribosome complex (Complex I) in the presence of GTP or GMP-PCP.

the EF-Tu-promoted binding reaction of aminoacyl-tRNA to ribosomes. First, GTP hydrolysis may be required for the activation or accommodation of bound aminoacyl-tRNA to facilitate peptide bond formation (*33*). Second, GTP hydrolysis may be required for the removal of EF-Tu from ribosomes in the form of Tu-GDP. The presence of EF-Tu on ribosomes could be inhibitory to the subsequent peptidyl transfer (*38*). The results to be described in this paper indicate that the second alternate is correct. When complex formed in the presence of GMP-PCP was reisolated by centrifugation through 10% sucrose solution, Tu-GMP-PCP was released from the ribosomes and the peptidyl transfer reaction took place. Apparently, the presence of Tu-GMP-PCP on the ribosomes is the reason for the inability of the bound aminoacyl-tRNA to engage in the peptidyl transfer reaction (see Fig. 3).

Ribosomal complex having N-acetyl-[^{14}C]-phenylalanyl-tRNA at the P site (Complex I) was prepared by incubation of N-acetyl-[^{14}C]-phenylalanyl-tRNA with poly U and ribosomes. The complex was isolated by centrifugation through 10% sucrose solution and suspended in a buffer containing 20 mM Tris-HCl buffer, pH 7.5, 10 mM magnesium acetate, 20 mM NH_4Cl, and 5 mM 2-mercaptoethanol (buffer A). To A site of Complex I, [^{14}C]-phenylalanyl-tRNA was bound enzymatically in the presence of Tu-Ts and GTP or GMP-PCP. After incubation, the newly formed complex was isolated either by filtration through nitrocellulose membrane filters or by centrifugation through 10% sucrose solution, and the radioactive reaction products were analyzed by paper electrophoresis after alkaline hydrolysis of aminoacyl- or peptidyl-tRNA.

As can be seen from Fig. 4, when Complex I was incubated with Tu-Ts, [^{14}C]-phenylalanyl-tRNA and GTP, most of the radioactivity was recovered in the N-acetyldiphenylalanine peak after incubation for 10 min at 0°C (Fig. 4,a). On the other hand, in experiments in which GMP-PCP was substituted for GTP, only negligible dipeptide formation was observed even after incubation for 120 min at 0°C (Fig. 4,b). In an attempt to test the possibility that the inability to form the dipeptide might be due to the presence of Tu-GMP-PCP on the ribosomes, the complex was incubated for 5 min at 30°C and reisolated by centrifugation through buffer A containing 10% sucrose. This yielded a ribosome complex in which approximately 50% of the radioactivity was present as N-acetyldiphenylalanyl-tRNA (Fig. 4,c). These results suggest that the peptidyl transfer reaction could occur in the absence of GTP hydrolysis, provided EF-Tu was released from

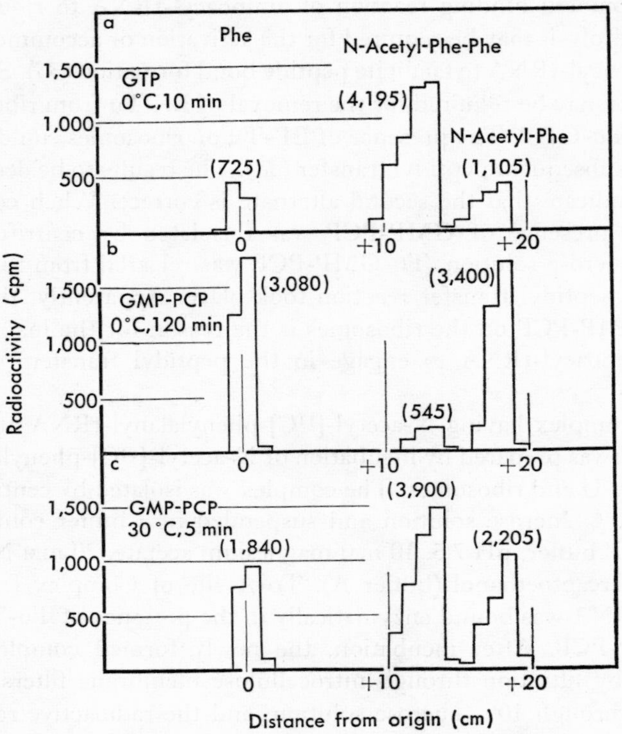

FIG. 4. Product analysis of the EF-Tu-promoted binding reaction. The (N-acetyl-[^{14}C]-phenylalanyl-tRNA)-poly U-ribosome complex (Complex I) was prepared by incubation of 690 pmoles of N-acetyl-[^{14}C]-phenylalanyl-tRNA (specific activity, 500) with 55 A_{260} units of ribosomes in the presence of poly U, and 13 mM Mg^{2+} and 160 mM NH_4^+. The complex isolated by centrifugation through 10% sucrose contained 10–12 pmoles of N-acetyl-[^{14}C]-phenylalanyl-tRNA per 1.0 A_{260} unit of ribosomes. To 0.75 A_{260} unit of Complex I, 40 units of Tu-Ts, 25 pmoles of [^{14}C]-phenylalanyl-tRNA (specific activity, 500) were added and incubated with 0.1 mM GTP or GMP-PCP at 0°C or 30°C for the time period indicated. a: Incubation was carried out for 10 min at 0°C with 0.1 mM GTP. b: For 120 min at 0°C with 0.1 mM GMP-PCP. c: For 5 min at 30°C with 0.1 mM GMP-PCP and the complex was reisolated by centrifugation for 180 min at 50,000 rpm through 10% sucrose in buffer A. The pellet was suspended in the same buffer. After alkaline hydrolysis of peptidyl- and aminoacyl-tRNA, product analysis was carried out by high-voltage paper electrophoresis, as described elsewhere. From Yokosawa *et al.* (*39*).

the ribosomes by ultracentrifugation through 10% sucrose solution.

The hypothesis that the presence of EF-Tu near the A site of ribosomes is inhibitory to the peptidyl transfer reaction is further strengthened by the following experiment in which [^{14}C]-phenylalanyl-tRNA was bound nonenzymatically to the A site of Complex I. When [^{14}C]-phenylalanyl-tRNA was bound to Complex I in the absence of EF-Tu, GTP or GMP-PCP using a higher concentration of Mg^{2+} ions, it was found that 30% and 65% of the radioactivity of N-acetyl-[^{14}C]-phenylalanyl-tRNA in Complex I was converted to N-acetyl-[^{14}C]-diphenylalanyl-tRNA following incubation for 60 min at 0°C and for 30 min at 30°C, respectively. Since neither GTP nor GMP-PCP was present during incubation, the peptidyl transfer appears to be an intrinsic property of ribosomes, which does not require cleavage of the phosphate bond of GTP.

It must be pointed out that these experiments have been carried out in the presence of substrate amounts of EF-Tu. When only a catalytic amount of EF-Tu was employed, GMP-PCP could not substitute for GTP, since the repeated utilization of EF-Tu was inhibited in the absence of GTP hydrolysis.

The role of GTP in EF-Tu-promoted binding of [^{14}C]-phenylalanyl-tRNA might be interpreted as follows. First, a unique conformation of EF-Tu complexed with GTP can select exclusively the aminoacylated form of tRNA in preference to its deacylated form. In contrast, EF-Tu-GDP interacts neither with aminoacyl-tRNA nor with deacylated tRNA. Second, the ternary complex, (aminoacyl-tRNA)-Tu-GTP complex, is transferred to a precise location on the ribosomes through the interaction of EF-Tu and 50S ribosomal proteins. The nonenzymatic binding of aminacyl-tRNA to ribosomes requires higher concentrations of Mg^{2+}, and the reaction is more sluggish. Third, after the transfer of aminoacyl-tRNA to the A site of ribosomes is completed, EF-Tu has to come off the ribosomes to initiate a new cycle of reactions. This could be accomplished by the hydrolysis of bound GTP to GDP. An additional advantage of the breakdown of GTP is presumably to shift the equilibrium irreversibly towards the binding of aminoacyl-tRNA to ribosomes. An experiment not included in this report demonstrated that phenylalanyl-tRNA bound to Complex I in the presence of EF-Tu and GMP-PCP is exchangeable with external phenylalanyl-tRNA while one bound with GTP is not.*

* M. Kawakita, K. Arai, and Y. Kaziro, unpublished.

Role of GTP in EF-G-Promoted Translocation Reaction

As in the case of EF-Tu, the role of GTP in EF-G-catalyzed translocation of (peptidyl-tRNA)-mRNA complex from the A site to the P site of ribosomes was investigated using GMP-PCP. Two possible mechanisms of the translocation reaction with respect to GTP hydrolysis are depicted in Fig. 5. In Model I, the hydrolysis of GTP precedes, while in Model II, it follows the translocation. We have attempted to distinguish between these two possibilities by use of GMP-PCP. If Model II is correct, the deacylated tRNA will be released from the ribosomes by addition of EF-G and GMP-PCP, whereas no tRNA release should occur in the absence of GTP hydrolysis according to Model I. The results represented in Table I indicate that Model II is correct and that translocation precedes the hydrolysis of GTP.

Complex II, prepared by incubating Complex I with EF-Tu and GTP, was isolated by ultracentrifugation and suspended in buffer A. Complex II contains deacylated tRNA in its P site, and N-acetyldiphenylalanyl-tRNA in its A site. After incubation of Complex II with GTP or GMP-PCP in the presence of a substrate amount of EF-G, the reaction mixture was filtered through a nitrocellulose membrane filter. The filtrate was collected in a test-tube and tRNA released from the ribosomes was determined by amino-acylation with [^{14}C]-phenylalanine in the presence of purified phenylalanyl-tRNA synthetase and ATP.

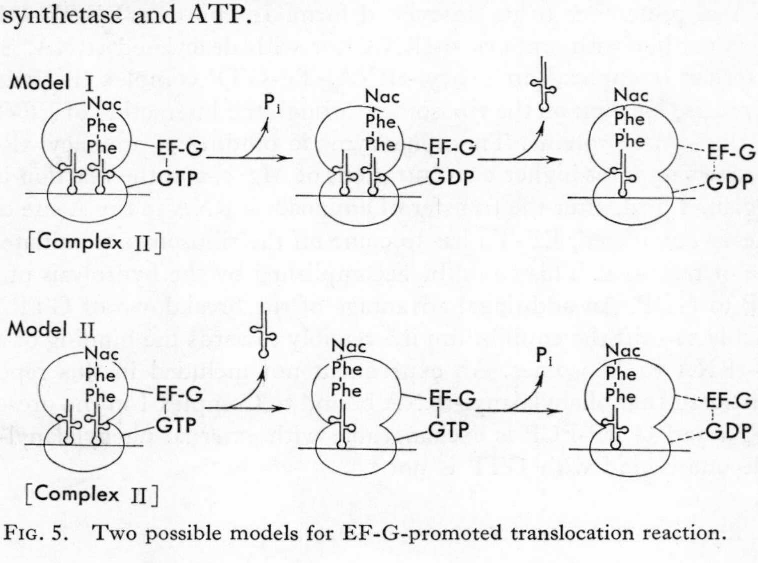

FIG. 5. Two possible models for EF-G-promoted translocation reaction.

TABLE I
Release of tRNA from Ribosomes

Additions	tRNA released (pmoles)
Complete system[a]	12.6
minus EF-G, GTP	0.8
minus EF-G	1.8
minus GTP	2.4
minus GTP, plus 0.1 mM GMP-PCP	7.5

[a] Complete system contained in 0.1 ml; 50 mM Tris-HCl, pH 7.5, 0.15 M NH_4Cl, 10 mM magnesium acetate, 10 mM 2-mercaptoethanol, tRNA-ribosome (poly U)-(N-acetyl $(Phe)_2$-tRNA) complex containing 2.0 OD_{260} units of ribosomes and 14 pmoles of N-acetyl-[^{14}C]-diphenylalanine, 0.1 mM GTP, and 350 milliunits of EF-G. The incubation was for 5 min at 30°C. The tRNA released was measured as described in the text. (N. Inoue-Yokosawa and Y. Kaziro, unpublished).

As shown in Table I, the release of tRNA is dependent on EF-G and GTP, and GMP-PCP can substitute for GTP to the extent of approximately 60%. From these results, it may be considered that, in the translocation reaction, EF-G binds to Complex II in the presence of GTP or GMP-PCP, and the conformation of the complex is altered so as to facilitate the shift of N-acetyl-[^{14}C]-diphenylalaninyl-tRNA from the A site to the P site of ribosomes. After the translocation reaction is completed, GTP is hydrolyzed to release EF-G from ribosomes and to reutilize it for a new cycle of reactions. We have previously demonstrated that EF-G binds tightly with ribosomes in the presence of GMP-PCP to form a stable ternary complex EF-G-ribosome-GMP-PCP (*40*). On the other hand, a similar complex formed with GTP or GDP, *i.e.*, EF-G-ribosome-GDP, easily dissociates unless stabilized by the addition of fusidic acid (*40*), an antibiotic which is known to inhibit reactions catalyzed by EF-G.

It must be noted that the above experiments were carried out with substrate amounts of EF-G. As in the case of the EF-Tu-promoted reaction, when a catalytic amount of EF-G was used, GMP-PCP cannot replace GTP, because the catalytic reutilization of EF-G is inhibited in the absence of GTP hydrolysis. As expected, fusidic acid did not inhibit the one-step translocation reaction catalyzed by substrate amounts of EF-G, for it inhibits dissociation of EF-G from the ternary complex after the reaction has been completed (data not shown in this paper).*

* N. Inoue-Yokosawa and Y. Kaziro, unpublished.

Thus, the roles of GTP in EF-Tu- and EF-G-promoted reactions are quite analogous in that the one-step reactions are accomplished by a conformation change induced in the protein molecule(s) by the association with GTP (and not with GDP), and the cleavage of the terminal phosphate is only required to release EF-Tu or EF-G from the ribosomes and to reinitiate a new cycle of reactions.

An analogous mechanism of GTP cleavage was also found in IF-2-promoted binding of formylmethionyl-$tRNA_F$ to ribosomes by Benne and Voorma (*41*), Lockwood and Maitra (*42*), and Dubnoff *et al.* (*43*). Formylmethionyl-$tRNA_F$ bound to ribosomes with IF-2 and GTP was able to react with puromycin whereas the same complex formed with GMP-PCP was not reactive with puromycin. However, after passing the complex through a Sephadex column, bound GMP-PCP and IF-2 were removed from the (formylmethionyl-tRNA)-ribosome complex and the formylmethionyl-tRNA became reactive to puromycin.

Therefore, in this case also, the hydrolysis of GTP is not necessary for the binding of formylmethionyl-$tRNA_F$ to the P site of ribosomes, and appears to be required for the removal of IF-2, the presence of which may probably be inhibitory to the subsequent puromycin (or peptidyl transfer) reaction. A similar mechanism might also operate with another GTP-utilizing reaction in protein synthesis, *i.e.*, the release factor-promoted termination of the peptide chain. It is conceivable that the four GTP-utilizing reactions in protein synthesis, catalyzed by IF-2, EF-Tu, EF-G, and RF, respectively, are probably mediated through similar and common reaction mechanism.

Finally, I should like to point out the analogy between the reaction mechanism of GTP hydrolysis in protein synthesis and that of other ATP-utilizing mechanochemical reactions such as muscle contraction, active transport, and oxidative phosphorylation. For example, in EF-Tu-promoted binding of aminoacyl-tRNA to ribosomes, EF-Tu first binds with GTP to form EF-Tu-GTP complex which, because of its unique conformation, can specifically bind with aminoacyl-tRNA to form a ternary complex, (aminoacyl-tRNA)-(EF-Tu)-GTP. After the transfer of the complex to ribosomes, aminoacyl-tRNA is released from EF-Tu by hydrolysis of GTP through the function of the 50S ribosomal subunit. Therefore, one can consider EF-Tu as a carrier protein which mediates an energy-dependent unidirectional transport of aminoacyl-tRNA. In the transport of K^+ across the biological membrane, Na^+,K^+-ATPase might play a role comparable to EF-Tu.

E-P or E-ATP complex may possess a unique conformation that can bind effectively with K^+. The resulting K^+-E-P (or K^+-E-ATP) complex may be transferred to the inner surface of membranes by interaction with some specific protein molecule followed by dephosphorylation and release of K^+. It would be of interest to search for a protein which interacts with the K^+-E-P (or K^+-E-ATP) complex and stimulates dephosphorylation at the inner surface of the cellular membrane.

REFERENCES

1. E. B. Keller and P. C. Zamecnik, *J. Biol. Chem.*, **221**, 45 (1956).
2. J. Lucas-Lenard and F. Lipmann, *Ann. Rev. Biochem.*, **40**, 409 (1971).
3. P. Lengyel, *Cold Spring Harbor Symp. Quant. Biol.*, **34**, 827 (1969).
4. R. L. Shorey, J. M. Ravel, C. W. Garner, and W. Shive, *J. Biol. Chem.*, **244**, 4555 (1969).
5. Y. Ono, A. Skoultchi, J. Waterson, and P. Lengyel, *Nature*, **223**, 697 (1969).
6. Y. Nishizuka and F. Lipmann, *Arch. Biochem. Biophys.*, **116**, 344 (1966).
7. J. Lucas-Lenard and F. Lipmann, *Proc. Natl. Acad. Sci. U.S.*, **55**, 1562 (1966).
8. J. M. Ravel, *Proc. Natl. Acad. Sci. U.S.*, **57**, 1811 (1967).
9. J. Gordon, *Proc. Natl. Acad. Sci. U.S.*, **58**, 1574 (1967).
10. J. Gordon, *Proc. Natl. Acad. Sci. U.S.*, **59**, 179 (1968).
11. J. Lucas-Lenard and A.-L. Haenni, *Proc. Natl. Acad. Sci. U.S.*, **59**, 554 (1968).
12. D. Cooper and J. Gordon, *Biochemistry*, **8**, 4289 (1969).
13. J. M. Ravel, R. L. Shorey, and W. Shive, *Biochem. Biophys. Res. Commun.*, **32**, 9 (1968).
14. J. M. Ravel, R. L. Shorey, S. Froehner, and W. Shive, *Arch. Biochem. Biophys.*, **125**, 514 (1968).
15. R. Ertel, B. Redfield, N. Brot, and H. Weissbach, *Arch. Biochem. Biophys.*, **128**, 331 (1968).
16. A. Skoultchi, Y. Ono, H. M. Moon, and P. Lengyel, *Proc. Natl. Acad. Sci. U.S.*, **60**, 675 (1968).
17. A.-L. Haenni and J. Lucas-Lenard, *Proc. Natl. Acad. Sci. U.S.*, **61**, 1363 (1968).
18. J. Gordon, *J. Biol. Chem.*, **244**, 5680 (1969).
19. J. Lucas-Lenard, P. Tao, and A.-L. Haenni, *Cold Spring Harbor Symp. Quant. Biol.*, **34**, 455 (1969).
20. Y. Ono, A. Skoultchi, J. Waterson, and P. Lengyel, *Nature*, **222**, 645 (1969).

21. A. Skoultchi, Y. Ono, J. Waterson, and P. Lengyel, *Biochemistry*, **9**, 508 (1970).
22. D. L. Miller and H. Weissbach, *Arch. Biochem. Biophys.*, **132**, 146 (1969).
23. D. L. Miller and H. Weissbach, *Biochem. Biophys. Res. Commun.*, **38**, 1016 (1970).
24. H. Weissbach, D. L. Miller, and J. Hachmann, *Arch. Biochem. Biophys.*, **137**, 262 (1970).
25. J. Waterson, G. Beaud, and P. Lengyel, *Nature*, **227**, 34 (1970).
26. R. E. Monro, *J. Mol. Biol.*, **26**, 147 (1967).
27. B. E. H. Maden, R. R. Traut, and R. E. Monro, *J. Mol. Biol.*, **35**, 333 (1968).
28. T. Staehelin, D. Maglott, and R. E. Monro, *Cold Spring Harbor Symp. Quant. Biol.*, **34**, 39 (1969).
29. L. Skogerson and K. Moldave, *Arch. Biochem. Biophys.*, **125**, 497 (1968).
30. R. W. Erbe, M. M. Nau, and P. Leder, *J. Mol. Biol.*, **39**, 441 (1969).
31. J. Lucas-Lenard and A.-L. Haenni, *Proc. Natl. Acad. Sci. U.S.*, **63**, 93 (1969).
32. H. Ishitsuka, Y. Kuriki, and A. Kaji, *J. Biol. Chem.*, **245**, 3346 (1970).
33. S. S. Thach and R. E. Thach, *Proc. Natl. Acad. Sci. U.S.*, **68**, 1791 (1971)
34. S. L. Gupta, J. Waterson, M. L. Sopori, S. M. Weissman, and P. Lengyel, *Biochemistry*, **10**, 4410 (1971).
35. Y. Kaziro, N. Inoue-Yokosawa, and M. Kawakita, *J. Biochem.*, **72**, 853 (1972).
36. K. Arai, M. Kawakita, and Y. Kaziro, *J. Biol. Chem.*, **247**, 7029 (1972).
37. A. Skoultchi, Y. Ono, J. Waterson, and P. Lengyel, *Biochemistry*, **9**, 508 (1970).
38. R. L. Shorey, J. M. Ravel, and W. Shive, *Arch. Biochem. Biophys.*, **146**, 110 (1971).
39. H. Yokosawa, N. Inoue-Yokosawa, K. Arai, M. Kawakita, and Y. Kaziro, *J. Biol. Chem.*, **248**, 375 (1973).
40. Y. Kuriki, N. Inoue, and Y. Kaziro, *Biochim. Biophys. Acta*, **224**, 487 (1970).
41. R. Benne and H. O. Voorma, *FEBS Letters*, **20**, 347 (1972).
42. A. H. Lockwood and U. Maitra, *Fed. Proc.*, **31**, 410 Absr. (1972).
43. J. S. Dubnoff, A. H. Lockwood, and U. Maitra, *J. Biol. Chem.*, **247**, 2884 (1972).

STRUCTURAL AND ENERGY STATES OF CHLOROPLASTS AND MITOCHONDRIA

Energy Coupling and Structural Organization of Membrane Particles in Mitochondria

Lester Packer

Department of Physiology-Anatomy, Univeristy of California, Berkeley, California, U.S.A.

Membrane particles of mitochondria seen by the technique of freeze-fracture electron microscopy have been studied and it has been found that: 1) The outer and inner mitochondrial membranes have an asymmetric organization. 2) Membrane particles are sensitive to metabolic states of mitochondria and treatments which effect lipoprotein structure in mitochondria, thus the membrane particles which are presumably the functional components in the hydrophobic core of the membranes are mobile. 3) The areas of the membrane hydrophobic core which are not interrupted by particles are presumed to be ordered lipid domains. Approximately 12% of the hydrophobic core of the outer membrane and 24% of the inner membrane are interrupted by particles. Studies with spin labeled mitochondrial membranes using hydrocarbon nitroxide spin labels have also been employed as a more sensitive probe of the fluid structure of the lipid domains in the membrane. Changes in the partitioning of this spin label are seen in different metabolic and structural states of mitochondria, *e.g.*, when mitochondria are expanded or contracted as a result of ion transport. A phase transition of 23–24 in the Arhenius plots of the rate of mitochondrial oscillation of ion transport, of mitochondrial respiration, of mitochondrial ATPase, and of the partitioning of lipid spin labels suggests that lipids play a central role in the coupling of energy in the mitochondria. The results of this study are consistent with a fluid mosaic model of the membrane structure in which the energy generated by either the oxido reduction or ATPase systems of the mitochondria are transferred to each other through changes in the particle network in which the reactions are initiated and subsequently transmitted to lipid domains.

To comprehend more fully the structural basis for the functions of mitochondrial membranes, information is required on the organization of membrane components.

Occurrence and distribution of membrane particles in mitochondrial inner and outer membranes observed by electron microscopy after freeze-etching has been under study in this laboratory following the discovery of such structures by Wrigglesworth *et al.* (*1*). Similar structures have also been observed by Ruska and Ruska (*2*), Chilcroft (*3*), and Hackenbrock (*4*). Further studies have now been undertaken to clarify intramembrane particle interactions. In particular, we will report on:

1) Membrane particles of mitochondria—the occurrence of these structures and the influence of metabolic states of mitochondria upon the organization and distribution of these particles.
2) Different environmental conditions, in particular lipid composition and some lipid depletion and spin label studies.
3) Relation of the protein and lipid components of the membrane to function.

Membrane Particles of Mitochondria

1. Distribution

An asymmetrical distribution of membrane particles in purified outer and inner mitochondrial membranes was reported by Melnick and Packer (*5*), who observed that mitochondrial membranes have a densely and a lightly particulated fracture face, as frequently found for other membranes observed by electron microscopy and freeze-cleavage (*cf.* Ref. *6*). The densely particulated fracture face (designated A face) and the lightly particulated fracture face (designated B face) show a particle ratio of about 2: 1 for inner membranes and 4: 1 for outer membranes. The A face is toward the cytoplasm in the case of the outer membrane and toward the matrix in the case of the inner membrane.

Figure 1 shows a composite electron micrograph showing respective A and B faces of the outer and inner membranes. These have been put together to reveal how we interpret the freeze-cleavage half-membranes to arise from the outer and inner membranes at present. The membrane particle

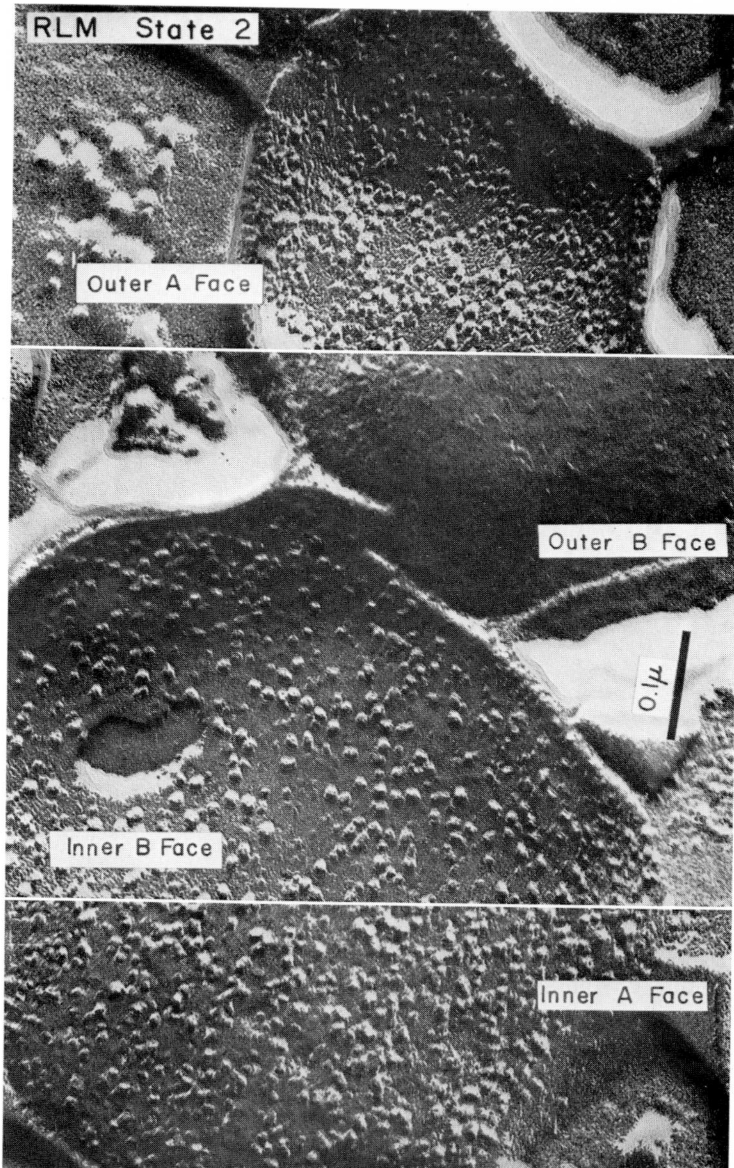

FIG. 1. Composite electron micrograph showing the four half-membranes that arise from the fracturing of outer and inner rat liver mitochondrial membranes.

organization of outer and inner membranes is so characteristic that by suitable examination of concave and convex fracture faces of mitochondria *in vivo*, we can generally easily identify the half-membrane under observation. These results correlate well with the isolated outer and inner membrane preparations studied by Melnick and Packer (*5*).

A diagramatic representation of the organization of outer and inner membranes is shown in Fig. 2, which summarizes the characteristic appearance of the fracture faces and the membrane particle densities found upon examination of these membranes in intact rat liver mitochondria.

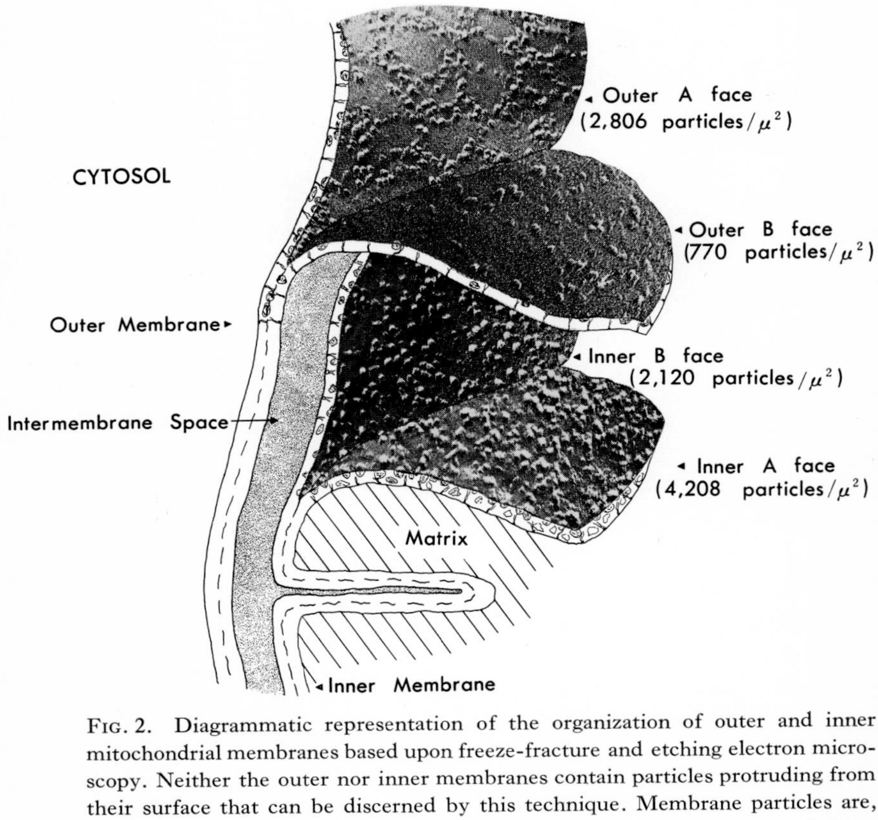

FIG. 2. Diagrammatic representation of the organization of outer and inner mitochondrial membranes based upon freeze-fracture and etching electron microscopy. Neither the outer nor inner membranes contain particles protruding from their surface that can be discerned by this technique. Membrane particles are, however, evident in the four half-membranes that arise as a result of freeze-cleavage of the outer and inner membranes. Each half-membrane has a characteristic particle density and arrangement. Description in text.

The implications of these results may be of considerable interest. (1) For example, orientation of membrane vesicles suggests a greater concentration of components facing the side of the membrane expected to contain compartmentalized reactions. This agrees with recent quantitative studies of staining density made by Muscatello and Pasquaili-Ronchetti (*7*), which also reveal structural asymmetry of the inner membrane. (2) These result suggest that upon fragmentation of outer-inner membranes, outer membranes should form "right-way-round" and inner membranes should form "inside-out" because the preponderance of protein-associated charged groups on the outside of vesicles is the thermodynamically stable configuration for vesicle formation (R. Melnick and P. John, unpublished results). In other words, the orientation of membrane vesicles bears the same relation as 1° structure of proteins has to 3° structure—the 1° structure of the membrane determines reorientation after fragmentation but has nothing to do with pinching off of cristae or folding.

2. Metabolic states

In our earlier study of particle distribution, we noted that under certain circumstances concave and convex fracture faces of mitochondrial membranes showed a peculiar clustering of the particles. These particle clusters appeared to be almost hexagonal or circular in configuration. Thus, concentrations of particles occurred together with large areas of smooth fracture faces from which particles were excluded. These results suggested that a special organization may exist for membrane components and that this may reflect the organization of electron transport complexes. It was therefore of interest to determine the influence of metabolic states upon the distribution of particles.

To perform this experiment, mitochondria were placed in the five different metabolic states as defined by Chance and Williams (*8*). Neither the A or B faces of outer or inner membranes revealed significant differences. To pursue this question further, we decided to employ mitochondria that were synchronized morphologically by taking advantage of oscillatory state conditions. Deamer *et al.* (*9*) had previously shown that mitochondria swollen in the oscillatory state, prior to damping of the oscillations, can be converted from a contracted to an expanded gross morphology and configuration. The results of this experiment are shown in Table I.

Mitochondria prior to swelling and at the peak of the first oscillation, when the inner membranes would be expected to be maximally expanded,

TABLE I
Analysis of Particle Distribution in Outer Membrane A Face of Rat Liver Mitochondria

Metabolic state	Area analyzed μ^2	Total number particles[a]	Particles per μ^2	Particle network diameter (μ)[b]
Contracted	0.054	225.3	4136	0.0293
Swollen	0.050	167.6	3404	0.0327

[a] Data are averaged from >5 fracture faces. [b] Average of 50 networks.

were quickly frozen in liquid nitrogen. Analysis of the half-membrane fracture faces showed that only the outer A face manifests a significantly changed particle density. The inner membrane, which conventional electron microscopy shows to undergo the greatest configurational changes, showed no apparent change in particle distribution.

The diameter of the particle network of the outer A face of rat liver mitochondria was also analyzed. This was found to change from a diameter of 0.0293 μ in contracted, as compared to 0.0327 μ in expanded mitochondria. (Data were averaged from more than five fracture faces and represent an average of at least 50 particle networks.) Hence, swelling of the inner membrane system may expand the outer membrane, causing a change in the clustering and distribution of membrane particles. We have previously noticed that expansion of the inner membrane compartment under these conditions does increase the diameter of intact mitochondria about 10%. Brinkman and Packer (unpublished observations) have also found a decrease in the number of intact outer membranes of rat liver mitochondria after oscillations.

It must be concluded from these studies that changes in metabolic state do not give rise to changes in the size and/or distribution of membrane particles large enough to be detected by the freeze-fracture technique. The resolution of this technique is estimated to be about 20 Å, as a consequence of the necessity for thin-platinum shadowing. Thus, we would not detect changes in the distribution and organization of particles less than 20 Å, as might be expected for normal metabolic state changes.

Some Lipid Depletion and Spin Label Studies

More drastic changes in the structure of membranes do lead to changes in the number and distribution of membrane particles. We have already reported (*10*) that high sucrose concentrations and low pH cause aggregation of membrane particles in mitochondria. The environmental changes induced by such treatments obviously fall outside of the physiological range. It was therefore of interest to test more physiological changes. In particular, Williams *et al.* (*11*) have examined the effects of changing the essential unsaturated fatty acid composition of membranes by dietary means.

A) *Unsaturated fatty acid:* The fatty acid composition of mitochondrial membranes was altered by growing rats on diets completely deficient in essential unsaturated fatty acids. Analysis of mitochondrial membrane preparations showed differences in the distribution of particles in the A and B faces of the half-membranes. The value of the ratio of particles in the A face to particles in the B face was higher in the membranes from linolenic acid-fed animals, whose mitochondria also show a higher degree of energy coupling, as compared with the fatty acid-deficient preparations. These two types of mitochondria have remarkably different capacities for energy coupling, as observed by the period of the oscillatory state of ion transport. The period of the oscillations is considerably lengthened in mitochondria from unsaturated fatty acid-deficient rats. The proportion of total unsaturated fatty acids in mitochondrial membranes under these conditions does not change, but the unsaturated fatty acids contain fewer double bonds; that is, their unsaturation index changes. These experiments appear to provide direct evidence that changing the fluidity of the membrane is reflected by changes in the pattern of distribution of particles in the half-membranes and in energy coupling.

B) *Lipid extraction:* In collaboration with S. Fleischer, C. Mehard, and W. Zehler, we have studied lipid-depleted heart submitochondrial vesicles by freeze-fracture electron microscopy, or the freeze-fracture technique. Electron microscopic examinations of extracted preparations of known lipid composition showed that as the lipids were gradually removed, the smooth fracture faces disappeared, and eventually only clusters of particles were deposited. Very extensive lipid depletion caused these clusters of particles to become dispersed.

It is clear from both the unsaturated fatty acid studies and the lipid depletion studies that lipid plays a crucial role in establishing the proper conditions for the dispersal of components in the membrane. In view of this, it was important to test whether or not structural changes were occurring in the lipids under conditions of mitochondrial function.

Previous studies by Brinkman and Packer (*12*), and Raison and Lyons (*13*) have established that mitochondrial respiration undergoes a change in activation energy at around 23°C. We (*14*) have now observed that the oscillatory state of mitochondrial swelling also displays a discontinuous

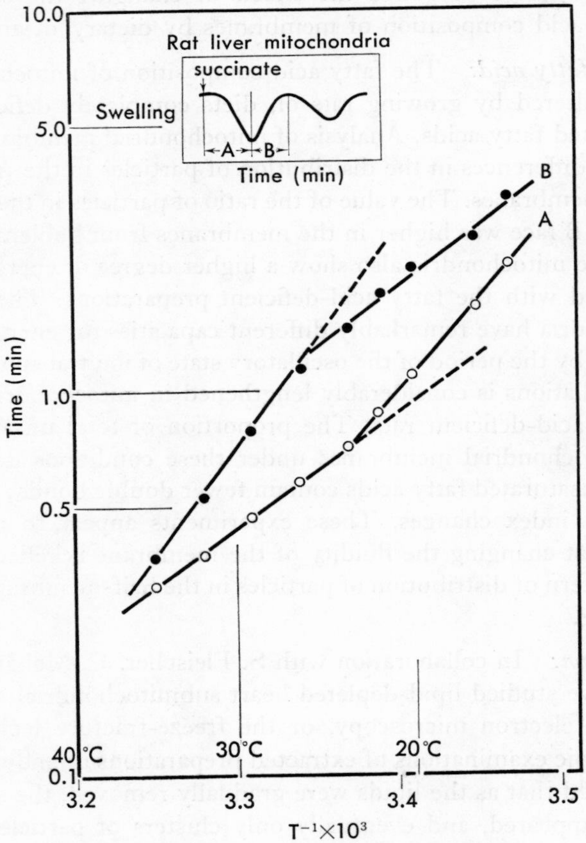

FIG. 3. Temperature dependence of mitochondrial oscillations. To a solution containing 100 mM sucrose and 1.0 mM ethylenediaminetetraacetic acid (EDTA) (pH 7.8) were added mitochondria (1.0–2.0 mg protein per ml) and sodium phosphate (0.5 M, 0.5 ml, pH 7.8). The oscillatory state was initiated by the addition of sodium succinate (0.02 ml, 1.0 M, pH 7.85). The total volume was 8.0 ml. Light scattering was measured at 90°C. The system was thermoregulated at the temperature indicated. Insert: An oscillation showing the time intervals determined.

temperature response with a change in activation energy occurring at around 23°C (Fig. 3). These results suggest that mitochondrial energy coupling is sensitive to the physical state of membrane lipid. The involvement of lipid in the oscillations is also indicated by recent studies showing that changes in oscillation period accompany changes in unsaturated fatty acid composition (Williams *et al.*).

C) *Lipid spin label:* We next examined the organization of lipids in relation to mitochondrial metabolic state. Rat liver mitochondria were put in an

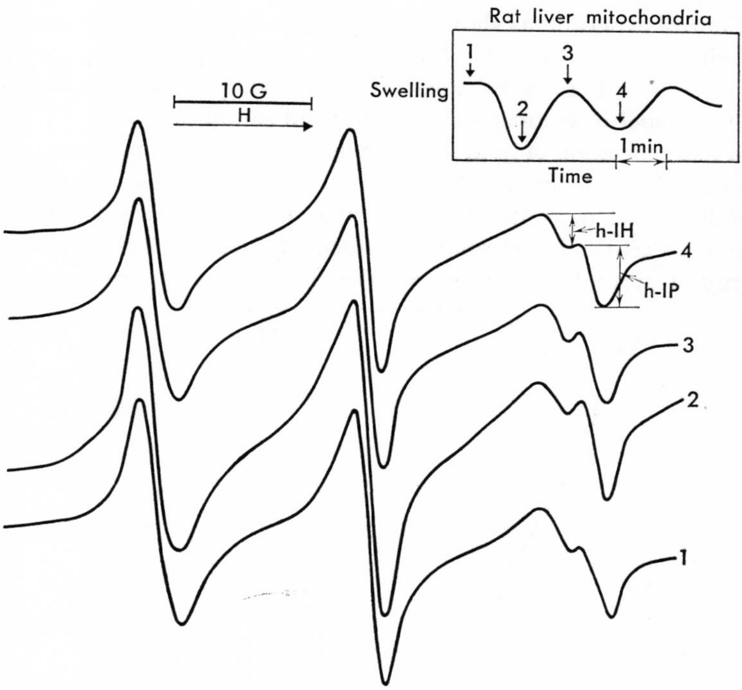

FIG. 4. EPR spectra of 6N11 in glutaraldehyde-fixed mitochondria. Mitochondria were placed in the oscillatory state as described in Fig. 1. Glutaraldehyde (0.5%) was added at the various points indicated in the insert. The preparations were placed on ice for 15–30 min prior to centrifugation at 10,000 rpm for 10 min (Spinco No. 40 rotor). The fixed mitochondria were washed with a medium containing 100 mM sucrose, 1.0 mM EDTA (pH 7.8) and centrifuged as before. The samples were resuspended in the same solution at 10–20 mg protein per ml. 6N11 was added at a bulk concentration of 4×10^{-5} M. EPR spectra were recorded at 25±1.0°C on a JEOLCO X-band Electron Spin Resonance Spectrometer.

oscillatory state by adding permeant ions and oxidizable substrate. At various times during the oscillation, glutaraldehyde was added to fix the samples. Fixed samples were washed and resuspended in oscillation medium (100 mM sucrose, 1 mM EDTA, pH 7.8). The hydrocarbon nitroxide spin label 6N11 was added and EPR spectra recorded (Fig. 4). Swollen mitochondria show a 35% decrease in the ratio of high field line components (h_{-1H}/h_{-1P}), an empirical index of partitioning of the spin label between polar and hydrophobic domains. These observations indicate a reorganization of lipid structure coincident with mitochondrial swelling.

Control experiments show that the concentration of fixative used in these experiments (50 mM glutaraldehyde for 15 min at 4°C in a mitochondrial suspension of 1–2 mg/ml) does not significantly affect the partitioning of the spin label in comparison with unfixed samples.

A reorganization of lipid structure is also suggested by studies of the temperature dependence of 6N11 partitioning. This was examined in mitochondria fixed in contracted and expanded configuration and in unfixed material. Partitioning in unfixed mitochondria shows a change in activation energy at around 23°C (Fig. 5). We suggest that as the temperature in-

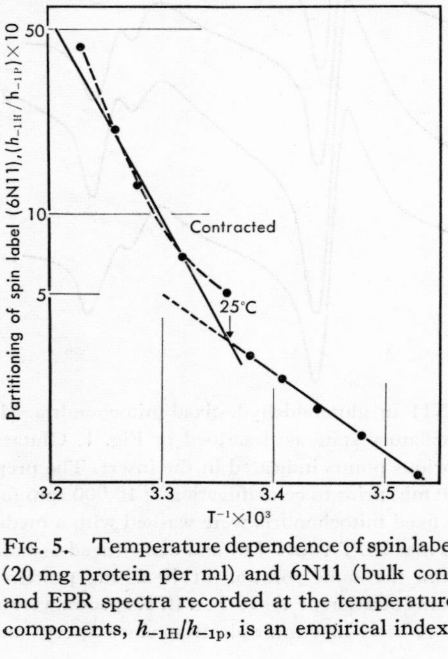

FIG. 5. Temperature dependence of spin label partitioning. Unfixed mitochondria (20 mg protein per ml) and 6N11 (bulk concentration=4×10^{-5} M) were mixed and EPR spectra recorded at the temperatures indicated. The ratio of third line components, h_{-1H}/h_{-1p}, is an empirical index of spin label partitioning.

creases, more spin label is solubilized within the hydrophobic domain of membrane lipids. Furthermore, at all temperatures examined, the two third-line components are adequately resolved, indicating that temperature increases do not appear to induce a breakdown in the sharp boundary between polar and hydrophobic domains. Partitioning, therefore, is useful in detecting a phase change in lipid alkyl chains. Partitioning in expanded mitochondria does not show a break in an Arrhenius plot; a break was retained in contracted mitochondria (Fig. 6). These results show that mitochondrial swelling is accompanied by changes in the physical state of the lipid. Previous studies have shown that swollen mitochondria display decreased energy coupling as well as reduced efficiency in maintaining proton gradients under these conditions (*15*). Comparison of the temperature dependence of parti-

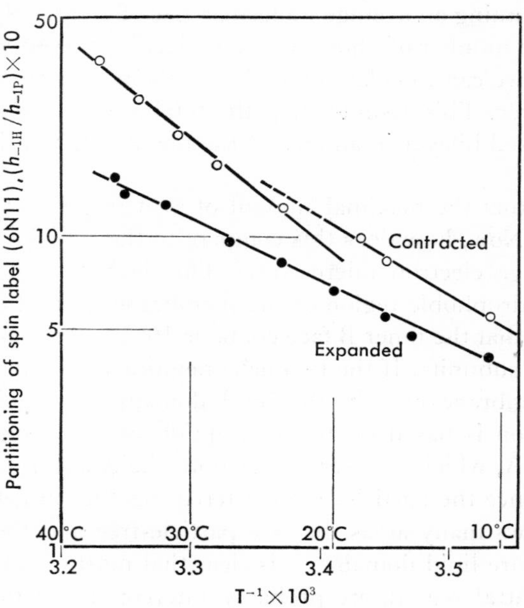

FIG. 6. Temperature dependence of spin label partitioning in glutaraldehyde-fixed mitochondria. Samples were fixed in contracted and expanded states during the oscillatory state with glutaraldehyde as described in Fig. 2. The contracted state was obtained by one addition of fixative prior to succinate. The expanded state was obtained by adding glutaraldehyde at the point of maximum light scattering decrease. For EPR measurements, one sample contained 10–20 mg protein per ml and 4×10^{-5} M 6N11.

tioning in fixed and unfixed mitochondria indicates also that fixation of the proteins partially restricts thermal disordering of the lipid alkyl chains (*cf.* Figs. 5 and 6).

Relation of the Protein and Lipid Components of the Membrane to Function

Now what about the consequences of the organization of the inner mitochondrial membrane in relation to energy coupling? From a number of different viewpoints it seems clear that the organization of the electron transport system in mitochondria and of the ATP synthetase-ATPase system must be lateral within the plane of the membrane and separate from one another. Some pertinent considerations are as follows: Measurements of the phospholipid and protein composition indicate 0.29 mmoles/mg of inner membrane protein. Assuming an average molecular size of 800 for the lipid component and a known number of phospholipid molecules of 0.46 μmoles/mg membrane protein, we can calculate that 37% of the membrane can be covered by a lipid bilayer. This assumes that the total area of the lipid is present as a tightly packed bilayer in an area of the membrane that is 80 Å thick.

It follows from this that the maximal amount of protein subunit which can be present is 63%. Now, how does this compare to the actual data observed by freeze-cleavage electron microscopy? This technique observes the 50–55 Å central, hydrophobic region of the membrane. Actual experiments (Table II) reveal that the inner B face contains 16% and the inner A face 31% as membrane subunits. If the two half-membranes are averaged, about 24% of the membrane area in the lipid domain is occupied by particles. This calculation is based on the assumption of an average diameter of particles of 49 Å, which is observed for both the A and B faces of the inner membrane. Since the particles are clustered together in networks and the membrane shows many areas that are particle-free and therefore presumably represent pure lipid domains, it is clear that most of the particles, at least in the central region, are probably interconnected to other particles in clusters ranging from 2 to 6, with a predominant clustering occurring on the half-membrane that faces the mitochondrial matrix. Hence the membrane is probably composed of a mosaic with the electron transport chains located laterally in the membrane; each of these may be associated with perhaps only a small number of specific input components as the dehydrogenases.

TABLE II
Percent Area of RLM Membrane Covered by Lipid and Protein

	Calculated data		Actual data	
	Lipid bilayer (%)	Protein subunit (%)	Protein subunit (%)	
Outer membrane	49	51	Outer A face 16.2 Outer B face 5.68	10.94
Inner membrane	37.2	62.8	Inner B face 16 Inner A face 31.7	23.85

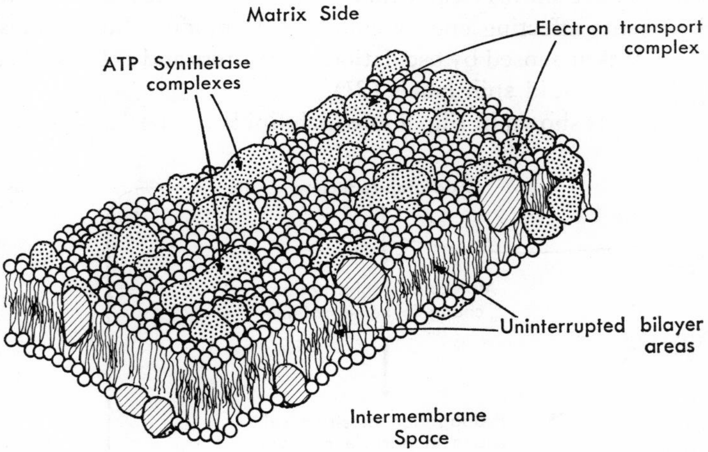

FIG. 7. Diagrammatic representation of lateral organization of electron transport and ATP synthetase complex of the inner mitochondrial membrane.

The huge molecular size of the 9 subunit ATPase system of 355,000 appears incompatible with direct association with the respiratory chain. ATPase seems too large to interact directly with three sites of the respiratory chain. Also much circumstantial evidence that ATPase can exist in other membranes in association with an electron transport system exists. This is shown diagrammatically in Fig. 7. How then can energy coupling occur between ATPase-ATP synthetase and the electron transport chain?

This problem is relevant to recent hypotheses advanced in Singer's (*16*) and McConnell's (*17*, *18*) laboratories of rapid lateral translational mobility of lipid molecules in functional membranes. The question of the extent to which lateral translational movement of lipid molecules can occur in various membranes, and the speed with which they occur, requires further study. In phospholipid multilayers, the speed of lateral translational movement has been found to be of the order of 10^{-8} cm^2/sec (*19*). However, in experiments with bacterial membranes, Morrison and Morowitz (*20*) showed that spin labels in these membranes are immobilized and do not equilibrate. This suggests that in mitochondrial inner membranes there may be lipid domains within which lateral translational movement of lipid molecules may be restricted to small areas as a result of the intrusion of protein into hydrophobic regions.

Our studies have shown clearly that various metabolic states of mitochondria and reagents affecting energy coupling do modify the organization of lipid components as sensed by oscillation parameters and a changed environment of nitroxide lipid spin labels (*21*).

A hypothesis is shown in Fig. 8 in which lipids can be considered to con-

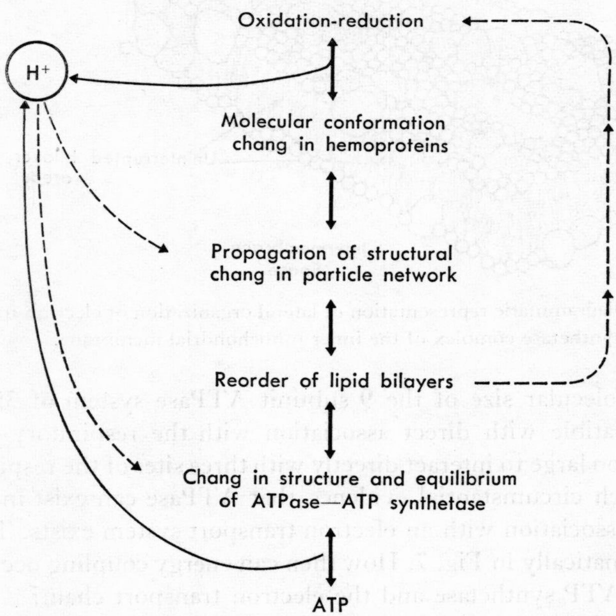

FIG. 8. Hypothesis for the role of lipid structure in the propagation of structural changes between the ATPase-ATP synthetase complex and the electron transport complexes.

serve and transfer energy between the oxido-reduction and/or ATP systems. Changes in lipid order would affect energy coupling by modulating the structure of the ATPase and electron transport components, and membrane permeability. In support of this, Raison *et al.* (*22*) have now found that the oligomycin-sensitive ATPase also shows a phase transition around 23°C, which is abolished by oligomycin. This is similar to the temperature dependence of mitochondrial respiration found earlier by Raison and Lyons (*13*) and Brinkman and Packer (*12*).

Lipid bilayers act not only as an insulation and barrier against the free diffusion of charged ions, but also affect the dispersal of the catalytic protein components. Conformational changes arising at the level of molecular structure in response to oxidation-reduction changes can be translated to the membrane as a whole in much the same way as are movements that occur when one perturbs the structure of a waterbed. The structural change initiated at the level of the hemoprotein is amplified by the particle network and translated laterally through the lipid bilayer. The flexibility of the lipid bilayer allows for propagation of the structural change. Such changes would be expected to affect markedly the redox potential of hemoproteins and ATPase-ATP synthetase because of changes in the lipid environment.

Acknowledgments

The experimental work presented in this article represents contributions by several of my colleagues; Dr. Ronald L. Melnick, Lou Worthington, and Susan Tinsley for the freeze-cleavage and etching studies of outer and inner mitochondrial membranes; by Drs. Charles W. Mehard, W. Zahler, and Sidney Fleischer for the lipid extraction studies; by Dr. Mary Ann Williams and Richard C. Stancliff and Alec D. Keith for the unsaturated fatty acid-deficient mitochondrial studies; and Drs. Harold Tinberg, Rolf Melhorn, and Alec D. Keith for the lipid spin label studies.

This research was supported by grants from the United States Public Health Service (AM 06438) and the National Science Foundation (GB 20951).

REFERENCES

1. J. M. Wrigglesworth, L. Packer, and D. Branton, *Biochim. Biophys. Acta*, **205**, 125–135 (1970).

2. C. Ruska and H. Ruska, *Z. Zellforsch.*, **97.**, 298 (1969).
3. J. P. Chilcroft, Ph. D., Dissertation, University of New Zealand, Aukland (1972).
4. C. R. Hackenbrock, *J. Cell Biol.*, in press.
5. R. Melnick and L. Packer, *Biochim. Biophys. Acta*, **253**, 503–508 (1971).
6. D. Branton and D. Deamer, "Membrane Structure," ed. by M. Alfert *et al.*, Springer-Verlag, Wien-New York, pp. 1–70 (1972).
7. U. Muscatello and I. Pasquaili-Ronchetti, private communication.
8. B. Chance and G. R. Williams, *J. Biol Chem.*, **217**, 383 (1955).
9. D. Deamer, K. Utsumi, and L. Packer, *Arch. Biochem. Biophys.*, **121**, 641–651 (1967).
10. L. Packer, *J. Bioenergetics*, in press.
11. M. A. Williams, R. C. Stancliff, L. Packer, and A. D. Keith, *Biochim. Biophys. Acta*, in press.
12. K. Brinkman and L. Packer, *J. Bioenergetics*, **1**, 523–526 (1970).
13. J. K. Raison and J. M. Lyons, *Plant Physiol.*, **45**, 382–385 (1970).
14. H. Tinberg, A. Keith, and L. Packer, unpublished results.
15. K. Utsumi and L. Packer, *Arch. Biochem. Biophys.*, **121**, 633–640 (1967).
16. S. S. Singer and G. L. Nicolson, *Science*, **175**, 720–731 (1972).
17. P. Devaux and H. M. McConnell, *J. Am. Chem. Soc.*, in press.
18. R. D. Kornberg and H. M. McConnell, *Proc. Natl. Acad. Sci. U.S.*, **68**, 2564–2568 (1971).
19. R. D. Kornberg and H. M. McConnell, *Biochemistry*, **10**, 1111–1120 (1971).
20. D. C. Morrison and H. J. Morowitz, *J. Mol. Biol.*, **49**, 441 (1970).
21. G. Zimmer, A. D. Keith, and L. Packer, *Arch. Biochem. Biophys.*, in press.
22. J. K. Raison, J. M. Lyons, and W. W. Thomson, *Arch. Biochem. Biophys.*, **142**, 83–90 (1971).

Thermal Denaturation of Thylakoids and Inactivation of Photophosphorylation in Isolated Spinach Chloroplasts

Yasuo Mukohata

Department of Biology, Faculty of Science, Osaka University, Osaka, Japan

Chloroplasts isolated from spinach in suspension were shaken with a small amount of lipid solvent, such as *n*-alkane or ethyl *n*-alkanoate, and assayed for physical and biological activities. The light scattering response was found to be much more sensitive to the shaking treatment than other activities, such as the light-induced pH shift, ferricyanide-Hill reaction and photophosphorylation. Electron micrography revealed that the thylakoids of alkane-treated chloroplasts seemed to be softened, probably by penetration of alkane molecules. The relationship between the lipid architecture in the thylakoid membranes, which is thought to be essential to maintain the spatial positioning of the components in the photosynthetic system, and biological activities, especially the coupling of electron transport with phosphorylation, was studied by means of light scattering as a measure of the intactness of the lipid architecture. Chloroplasts transiently warmed above 30°C showed uncoupling with decrease of the light scattering activity. When chloroplasts were warmed transiently in the presence of amphipathic molecules, such as alcohols, the uncoupling temperature of warming was lowered. There were systematic relationships between the effectiveness of alcohols in lowering the uncoupling temperature and their molecular characteristics. The cooperative effect of these molecules with heat was discussed in connection with denaturation of the lipid architecture in thylakoids and photosynthetic activities governed by it.

Reversible increase in the intensity of light scattered from isolated spinach chloroplasts was first observed by Packer (*1*) in 1962 and called the light scattering response (increment; ΔLS). ΔLS is known to be electron transport-dependent and has been extensively studied in connection with photophosphorylation (*1–3*) or the so-called high-energy state (*4*), light-induced proton uptake measured as the pH shift (pH increment; ΔpH) of suspensions (*5–7*), changes in gross structure of chloroplasts (*3, 8, 9*), configuration and conformation of thylakoids (*6, 7, 10, 11*) and lipophilic structure in lamellae (*12*). At present, except for some conflicting opinions, ΔLS is thought to reflect changes in the internal structure of chloroplast lamellae (*5–7, 13–15*).

Results showing a close relationship of ΔLS to lipophilic structure in lamellae revealed that lipophilic structure was also important in maintaining the biological activities of chloroplasts (*12*), as briefly described below.

Effects of Lipid Solvents on Thylakoids

On a suspension of chloroplasts isolated from spinach leaves with 0.35 M NaCl in a small test tube, one of a series of liquid *n*-alkanes, from hexane to dodecane, was layered to the extent of 5 v/v %. The 2 layers were vigorously mixed by a flash mixer for several sec. Then the aftereffects of shaking, *i.e.* the effect of wetting chloroplasts with alkane molecules, on the physical and biological activities of chloroplasts were investigated (*12*).

In Fig. 1, the activities of alkane-treated chloroplasts relative to the initial activities are plotted against shaking time. Activities to produce ΔLS and ΔpH and of ferricyanide(Fecy)-Hill reaction and photophosphorylation gradually decrease with prolongation of shaking time. In particular, ΔLS is more sensitive to the shaking treatment with *n*-alkanes. In general, activities are much more depressed by shorter *n*-alkanes, as partly shown in Fig. 1 regarding ΔLS activity.

Comparative work by electron micrography on alkane-treated and -untreated chloroplasts shows (Fig. 2) that the difference in the dimensions of thylakoid sacs is significant. The sacs of the nonane-treated specimen (Fig. 2,B) seem to be swollen, probably due to softening of the thylakoid membranes. Of course, these pictures are the final results of fixation by glutaraldehyde and OsO_4, embedding in epoxy resin after dehydration and staining by uranyl acetate and lead citrate, but they strongly suggest that the

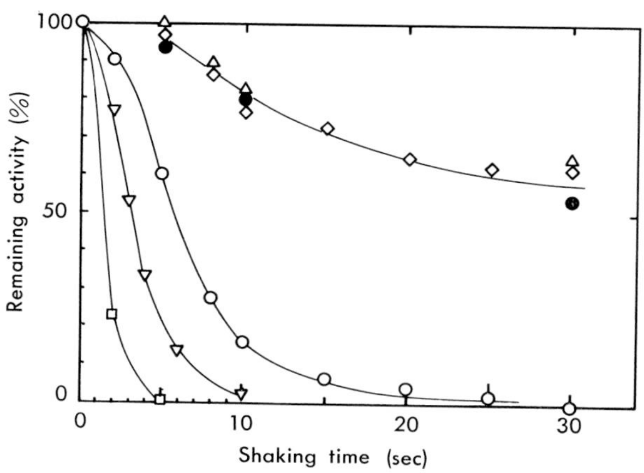

FIG. 1. Shaking-time dependence of activities for Fecy-Hill reaction (◇, potassium ferricyanide was used), phosphorylation (●, noncyclic), ΔpH (△) and ΔLS (○) in isolated spinach chloroplasts shaken with *n*-dodecane. The ΔLS activities of the *n*-decane-treated (▽) and the *n*-heptane-treated (□) chloroplasts are also shown. ΔpH and ΔLS were measured with alkane-treated chloroplasts equivalent to 3.5 μg chlorophyll (*16*)/ml in a reaction medium containing 0.1 M NaCl, 1 mM $MgCl_2$ and 10 μM PMS at 23±1°C at pH 6.16±0.02 under red light illumination of 3,000 lux. Activities of the Fecy-Hill reaction and phosphorylation were measured with chloroplasts at 17.5 μg chlorophyll/ml in a medium containing 0.1 M NaCl, 1 mM $MgCl_2$, 400 μM Fecy, and 30 mM Tricine (Tris(hydroxymethyl) methylglycine) buffer (pH 7.9) at 25±0.2°C under 5×10^4 lux white light (*12*).

thylakoid membranes are susceptible to denaturation by *n*-alkanes. Similar pictures were obtained from specimens of chloroplasts shaken with other *n*-alkanes, differing only in the degree of swelling of the sacs. Shorter *n*-alkanes made the sacs more swollen. This corresponds to the relationship of the chain length of *n*-alkanes to the degree of inactivation of biological activities of the chloroplasts.

The effects of 5-sec shaking with ethyl esters of straight-chain saturated fatty acids, within a series from ethyl acetate to ethyl dodecanoate, (2.5 v/v %) are somewhat different from those with *n*-alkanes. Activities for ΔpH and phosphorylation are diminished in parallel with decrease of ΔLS activity, as shown in Fig. 3. Fecy-Hill activity, on the other hand, is highly enhanced by shorter esters. Since inhibition of phosphorylation together with stimulation of electron transport is conventionally called uncoupling, lipid solvents, such as these shorter esters, can act as uncouplers when they come in contact with

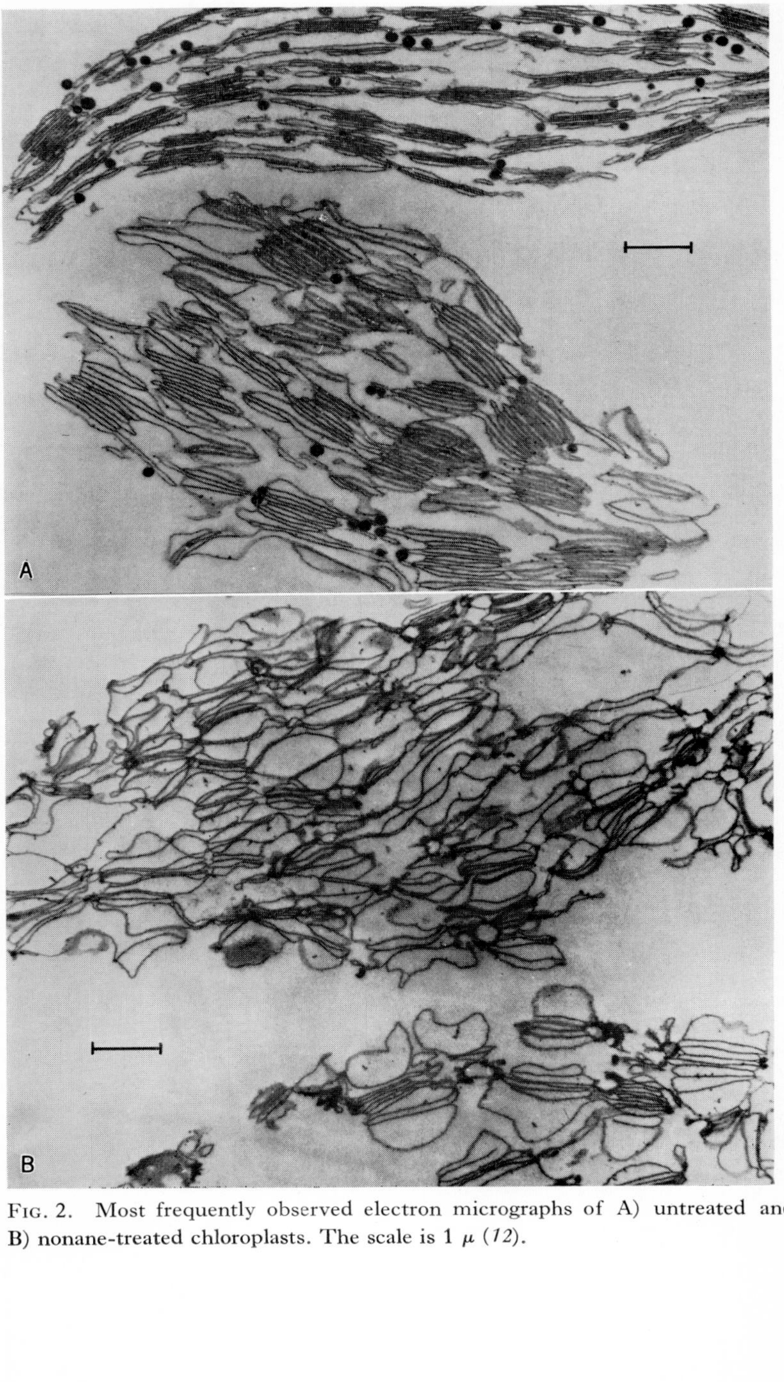

FIG. 2. Most frequently observed electron micrographs of A) untreated and B) nonane-treated chloroplasts. The scale is 1 μ (*12*).

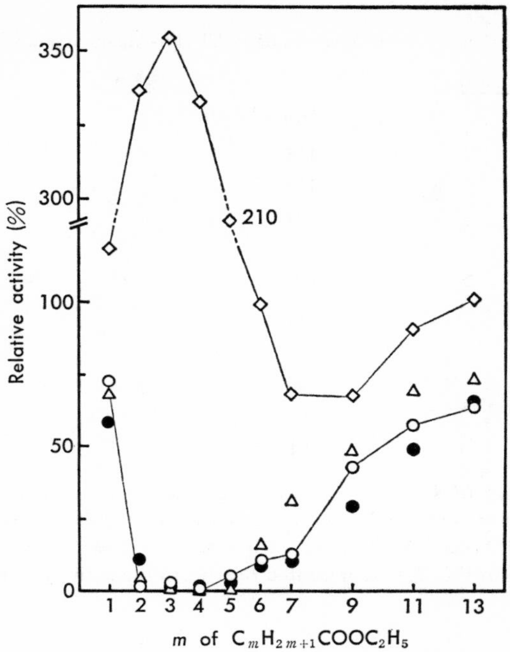

FIG. 3. Chain-length dependence of activities for the Fecy-Hill reaction, phosphorylation (P.P.), ΔpH, and ΔLS in isolated spinach chloroplasts shaken with ethyl esters of straight-chain fatty acids, represented relative to the corresponding activity of untreated material. Assay conditions were the same as in Fig. 1. ◇ Fecy; ○ ΔLS; △ ΔpH; ● P.P.

chloroplasts due to some process such as shaking. Extraction (by 2.5% of ester, 5-sec shaking) of some component(s) important to retain the activities under examination may be ruled out as a major cause of inactivation, since components related to the Fecy-Hill reaction seem to be retained (*cf.* Ref. *12*).

The effects of isomers of these esters are summarized in Table I. The degree of stimulation of the Fecy-Hill reaction is almost identical among isomers of the same chain length and is different among isomers of different chain lengths. There is no marked effect of the position of the ester linkage in a straight-chain molecule on stimulating the reaction.

It seems that dispersed droplets of longer esters (and perhaps, of *n*-alkanes) collide with chloroplasts, enwrap them and cause them to lose their

TABLE I
Effects of Shaking Treatment[a] with Some Isomers of *n*-Alkyl *n*-Alkanoates on Fecy-Hill Activity[b]

Isomers	Stimulation (%)[c]
Ethyl butanoate	115
Propyl propanoate	118
Butyl acetate	99
Ethyl hexanoate	123
Butyl butanoate	121
Hexyl acetate	120
Ethyl octanoate	98
Butyl hexanoate	99
Hexyl butanoate	96
Octyl acetate	91

These figures are the averages of 4 series of measurements and differ among different chloroplast preparations, but are comparable with each other as regards figures obtained with the same preparation. [a] With 2.5 v/v % of each ester, for 5 sec. [b] Experimental procedures were the same as in Fig. 3. [c] Represented relative to the activity of untreated chloroplasts.

activities. Thus, the effects of longer esters probably depend on their emulsifiability into aqueous suspensions under identical shaking conditions. Shorter esters, more or less soluble in water, interact molecularly with chloroplasts and bring about uncoupling. The degree of uncoupling will be related to the partition of these esters between aqueous medium and chloroplasts. That is, the effective concentration of lipid solvent (molecules and/or droplets) seems to determine the mode and extent of the effects of each solvent. Lipophilic interaction between chloroplasts and hydrocarbon chains is further implied by later results.

In any model proposed for the thylakoid membrane, lipids play a role to maintain spatial relationships among many components of thylakoids, such as chlorophylls, quinones, cytochromes and other proteins such as ATPase and carboxydismutase. It is postulated that: those components belonging to the so-called electron transport chain are inlaid or mosaiced in the lipid architecture in thylakoids; the sequence of the components in the chain is designated by the highest probability of occurrence of oxido-reduction, which depends on both redox potential and spatial positioning of each com-

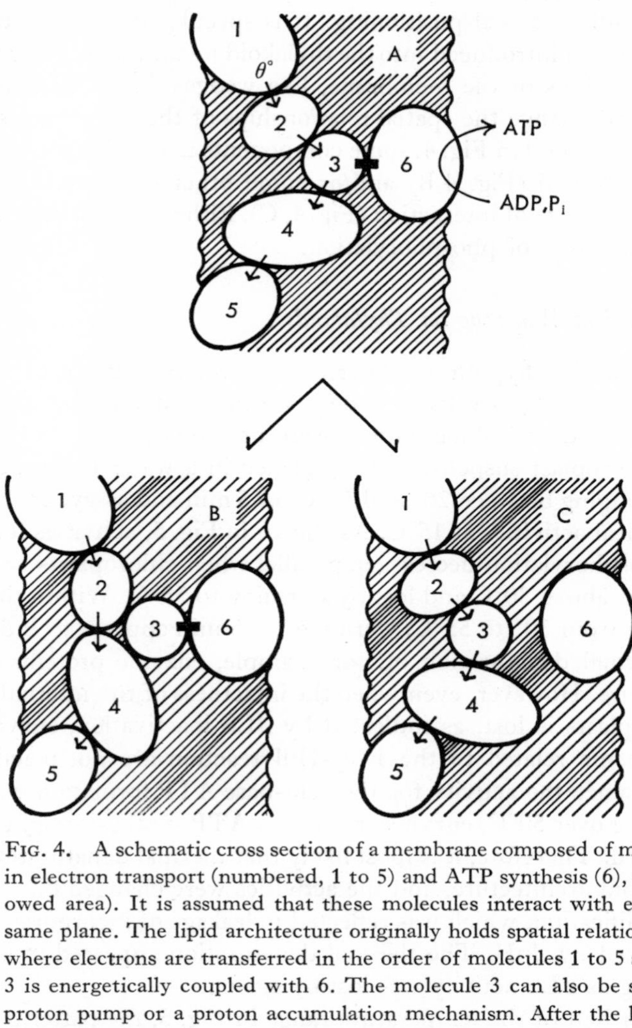

FIG. 4. A schematic cross section of a membrane composed of molecules engaged in electron transport (numbered, 1 to 5) and ATP synthesis (6), and lipids (shadowed area). It is assumed that these molecules interact with each other in the same plane. The lipid architecture originally holds spatial relationships as in (A), where electrons are transferred in the order of molecules 1 to 5 and the molecule 3 is energetically coupled with 6. The molecule 3 can also be supposed to be a proton pump or a proton accumulation mechanism. After the lipid architecture is modified, electron transport from 2 to 4 becomes more efficient and consequently 3 is skipped even though the coupling mechanism of 3 with 6 remains intact (B), or coupling between 3 and 6 is destroyed even though the electron transport system is intact (C). Phosphorylation inhibition will result from mechanisms between these 2 extreme examples.

ponent; coupling of electron transport with phosphorylation also depends on the mode of 3-dimensional positioning. It is speculated that molecules of lipid solvents, when introduced into the thylakoid membranes, penetrate into lipophilic portions of the membrane, like wedges driven into the lipid architecture, and change the spatial relationships of the components. Consequently, as illustrated in Fig. 4, some components in the electron transport chain will be skipped (Fig. 4,B) and/or some other components will be separated from the site of interaction (Fig. 4, C). Either one of these changes results in inactivation of phosphorylation.

Effects of Transient Warming on Thylakoids

Based on the working hypothesis described above, the effects of thermal agitation, instead of lipid-solvent wedges, on the lipid architecture, which would be of a liquid-crystalline nature, were investigated.

Spinach chloroplast suspension was warmed in a water bath at a given temperature ranging between 20° and 55°C for 5 min and assayed for physical and biological activities at 15°C. As shown in Fig. 5, activities for ΔLS, ΔpH, and phosphorylation decrease in parallel when chloroplasts are transiently warmed above 30°C and lost by warming to 55°C. Within the temperature range from 30° to 55°C, warming for 5 min appears to induce irreversible thermal denaturation of, for example, enzyme proteins or the lipid architecture. However, even when the intactness of the internal structure of chloroplasts is lost, as indicated by ΔLS inactivation, enzymes or redox components related to the Fecy-Hill reaction are not inactivated. Furthermore, since the activity for the acid-base ATP formation was only lost on warming over 50°C, enzymes related to ATP synthesis might not be fully inactivated. Therefore, it is most likely that thermal denaturation took place in the lipid architecture, and the activities were changed due to such structural modification, which was reflected in leakage of incorporated protons, as analyzed by ΔpH. The thermal denaturation appeared to be irreversible as long as ΔLS activity was lost by ageing.

In the lower half of the temperature range in Fig. 5, the patterns of the activity curves are typical of uncoupling and are similar to those of chloroplasts shaken with shorter esters. In the higher half, the whole system tends to be thermally inactivated, possibly due to denaturation of the oxygen evolution site (*e.g.*, Mn release (*17*)) and the stimulation maximum of the Fecy-Hill reaction occurs nearly at the temperature where the phosphoryla-

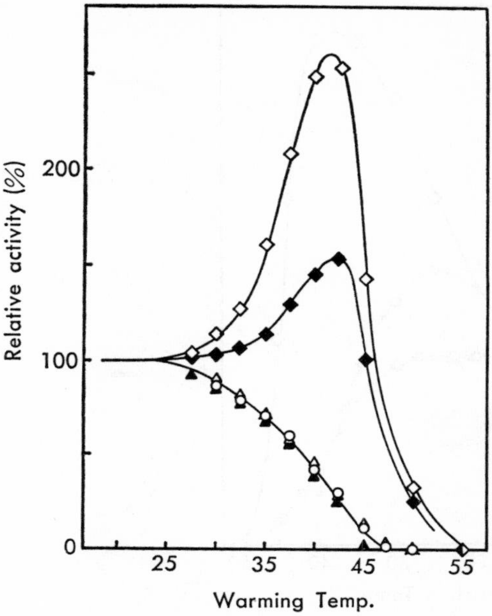

FIG. 5. Activities for Fecy-Hill reaction, phosphorylation (P.P.), ΔpH, and ΔLS in chloroplasts warmed for 5 min at temperatures given in the abscissa, represented relative to those of untreated chloroplasts. Experimental conditions were the same as in Fig. 1, except for the reaction temperature (15°C). The reaction medium for Fecy reduction at pH 6.0 was buffered by 30 mM phosphate. ○ ΔLS (pH 6); △ ΔpH (pH 6); ▲ P.P. (pH 8); ◇ Fecy (pH 6); ◆ Fecy (pH 8).

tion activity falls away to zero. Generally, the percentage enhancement of Fecy-Hill activity at the stimulation maximum, relative to the activity of untreated material, is found to be larger with lowering the reaction pH within the range from 8.8 to 6.3, while the absolute rate of Fecy reduction is higher at higher reaction pH's.

In Fig. 6, the activities of untreated chloroplasts measured at various reaction temperatures are plotted relative to those obtained at 15°C. With rise of reaction temperature, ΔLS and ΔpH decrease. The patterns are almost identical with those obtained from chloroplasts warmed transiently at the corresponding temperatures (Fig. 5). ΔLS mediated by Fecy tends to follow the profile of Fecy-Hill activity, which is enhanced by raising the reaction temperature and then depressed at temperatures higher than 40°C. Dif-

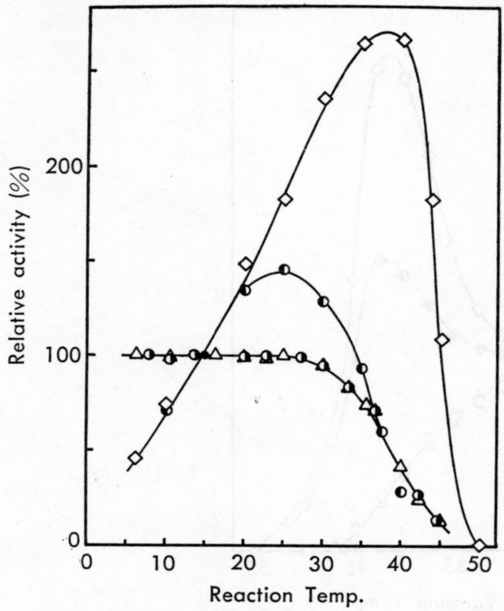

FIG. 6. Temperature dependence of activities for Fecy-Hill reaction, ΔpH, and ΔLS mediated by Fecy or phenazine methosulfate (PMS). Experimental conditions were the same as in Figs. 1 and 5. Replotted from Fig. 4 in (*15*) by setting the activities at 15°C to be 100%, with permission. ◑ ΔLS (PMS); ◐ ΔLS (Fecy); △ ΔpH; ◇ Fecy.

ferences in these results (Figs. 5 and 6) are readily understood on the basis of the following considerations: Reaction temperature is 15°C in the former and the temperature of warming in the latter; below 30°C all reaction systems examined were intact and above 30°C the activities change according to the heat sensitivity of individual systems; if chloroplasts once experienced a temperature higher than 30°C their thermal history is retained by a heat-sensitive mechanism, possibly in the lipid architecture.

Fecy-Hill activities of transiently warmed chloroplasts (0.5 M sucrose preparation, for phosphorylation experiments) in the presence and absence of ADP and/or orthophosphate are plotted against warming temperature in Fig. 7, A. It is apparent that up to the temperature of the stimulation maximum, activities under different conditions can fit 2 curves, one of which is for the complete system and the other for all other incomplete systems. In

Fig. 7,B, phosphorylation activities are plotted relative to the activity of untreated material. The results again indicate that the stimulation maximum can be a marker of phosphorylation inactivation or uncoupling.

When the difference between those 2 curves in Fig. 7,A is approximately presented in Fig. 7,B (the solid line) on the same percentage scale, the plots and the line are almost coincident. The same correlation has been found in

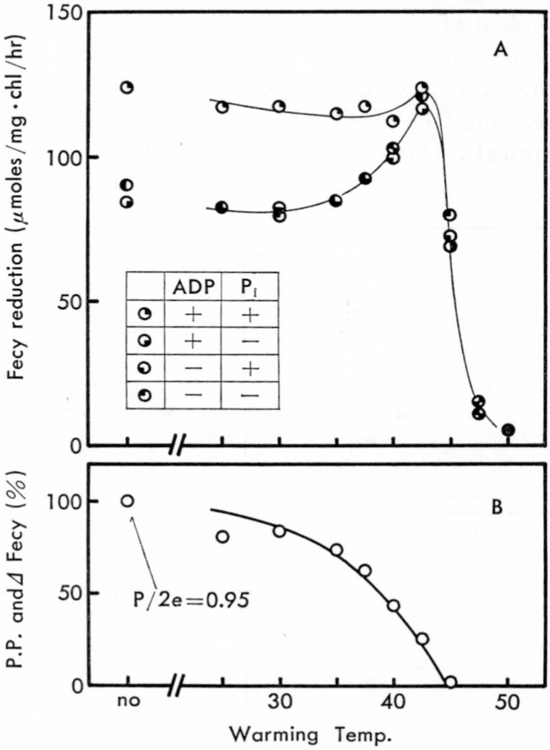

FIG. 7. A: Plots of the rate of Fecy reduction in the presence and absence of ADP and/or orthophosphate in chloroplasts warmed transiently at the temperature given in the abscissa. B: Plots of the relative activity for phosphorylation (○) and the difference between 2 curves in (A) (—). Chloroplasts were warmed in a suspending (=isolation) medium containing 0.5 M sucrose, 5 mM $MgCl_2$, and 10 mM Tricine buffer, pH 7.8 for 5 min. The complete reaction medium contained 0.1 M sucrose, 5 mM $MgCl_2$, 10 mM Tricine buffer at pH 8.3, 500 μM Fecy, 1 mM ADP, and 1 mM orthophosphate. Experiments were carried out at 15°C under 5×10^4 lux white light.

every result obtained, even under different experimental conditions (pH, presence of alcohol, *etc.*). Therefore, if the difference in Fig. 7,A is tentatively taken to be the amount of Fecy reduced under coupling with phosphorylation, phosphorylation and coupled electron transport are inactivated in parallel with rise of warming temperature, with a P/2*e* ratio maintained constant. Then there will be a speculation that there are 2 electron transport systems, one of which is tightly coupled (never uncoupled; heat inactivated) and the other independent of phosphorylation (basal; heat stimulating).

In Fig. 8, phosphorylation activities mediated by PMS or Fecy are plotted in an absolute (Fig. 8,A) and a relative (Fig. 8,B) scale against warming temperature. Since the profiles of inactivation of 2 phosphorylation systems are identical within experimental error (Fig. 8,B), the loca-

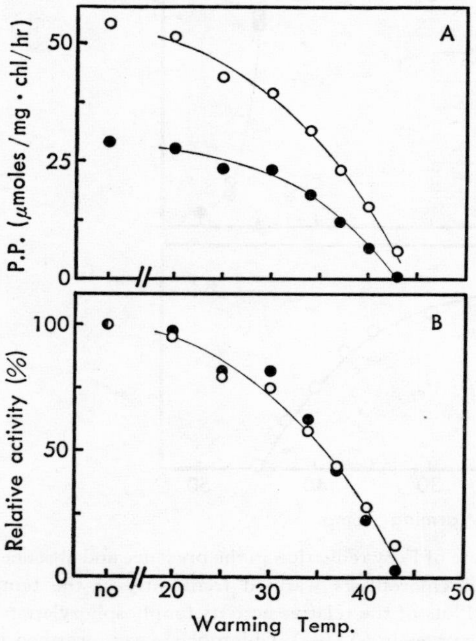

FIG. 8. A: Activities for phosphorylation mediated by PMS or Fecy in transiently warmed chloroplasts. Experimental conditions were the same as in Fig. 7, except for illumination with red actinic light instead of white. Fecy in the reaction medium was replaced by 10 μM PMS for the assay of cyclic phosphorylation. B: Replots from (A) in a relative scale. ○ P.P. (PMS); ● P.P. (Fecy).

tion (or mechanism) damaged by warming treatment is common to both systems. As described above, neither the electron transport system nor the enzyme(s) for ATP synthesis is thought to be responsible for the observed phosphorylation inactivation. A membrane structure sensitive to heat, most likely the lipid architecture, is expected to be modified by warming treatment.

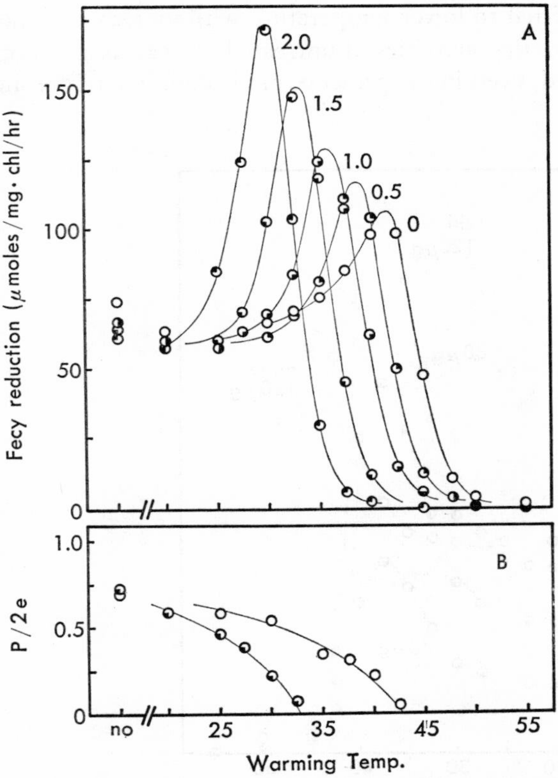

FIG. 9. A: Effects of *n*-butyl alcohol with warming treatment of chloroplasts at various temperatures on their activity for Fecy reduction under the conditions of phosphorylation. B: Plots of the phosphorylation efficiency (P/2*e*) in chloroplasts warmed in the absence of and in 1.5% *n*-butyl alcohol. Chloroplasts (120 μg chlorophyll) were warmed for 5 min at a given temperature in 1 ml of a suspending medium involving additional *n*-butyl alcohol at an indicated final concentration. After the warming treatment, the suspension was diluted with 5 ml of the same reaction medium as in Fig. 7. Succeeding experiments were carried out as in Fig. 7.

Effects of Alcohols on Thylakoids under Transient Warming

Thus, the effects of hydrocarbon segments on the lipid architecture under warming treatment were repeatedly studied. When chloroplasts are warmed in the presence of varied amounts of *n*-butyl alcohol, at various temperatures for 5 min, and assayed for Fecy reduction (under the conditions of phosphorylation) at 15°C after adding an aliquot of reaction medium, the stimulation maximum is shifted to lower temperature with increase of amount, as shown in Fig. 9. Note that activities of untreated (no warming) chloroplasts are almost unchanged, even in the presence of alcohol. Therefore, heat and

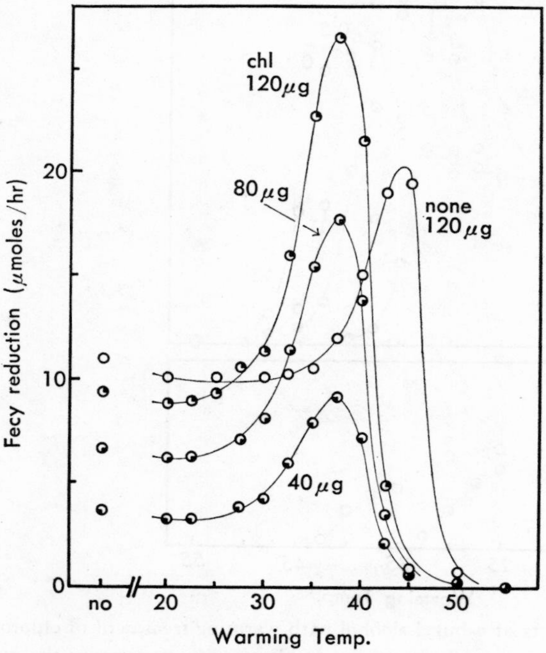

FIG. 10. Effects of chloroplast concentration with warming treatment in 40 mM *n*-amyl alcohol on Fecy reduction. Experimental procedures were similar to those in Fig. 9, except for the amounts of chloroplasts which were indicated as weight of total chlorophyll (*16*) in 1 ml of suspension under warming treatment. Butyl alcohol in the suspending medium in Fig. 9. was replaced by 40 mM *n*-amyl alcohol.

n-butyl alcohol appears to be cooperative in uncoupling, possibly in modifying the lipid architecture.

Extensive survey on the combined effects of alcohols and thermal energy revealed that the effectiveness in lowering the uncoupling temperature is, in general, much greater with increase of the hydrocarbon chain length of *n*-alcohols, and smaller with increase in the number of hydroxyl groups in an alcohol.

As shown in Fig. 10, since reduction of the quantity of chloroplasts upon warming in a fixed volume of 40 mM *n*-amyl alcohol results in proportional

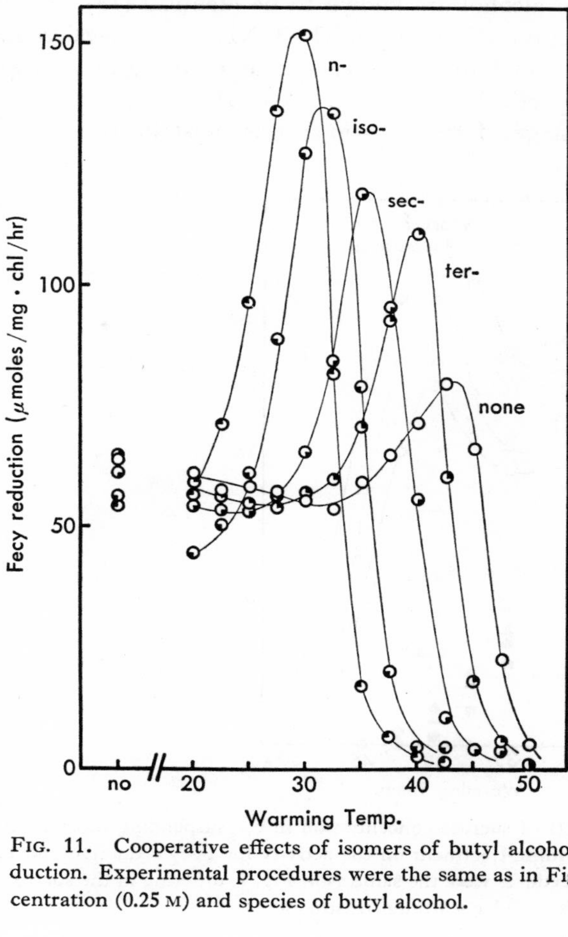

FIG. 11. Cooperative effects of isomers of butyl alcohol with heat on Fecy reduction. Experimental procedures were the same as in Fig. 9, except for the concentration (0.25 M) and species of butyl alcohol.

decrease of the amounts of Fecy reduced, without shift of the stimulation maximum, the effectual factor assigned to cooperation of *n*-alcohol is not the quantitative ratio of alcohol to chloroplast (*cf.* Fig. 9) but the concentration of alcohol upon warming chloroplasts. Then physicochemical properties such as the surface tension of aqueous alcohol seem to be cooperative with heat.

However, as shown in Fig. 11, where the effects of 4 structural isomers of butyl alcohol are presented, there are clear distinctions in their effects, while the surface tension, for example, of aqueous solutions of *n*-butyl alcohol and isobutyl alcohol are known to be identical (*18*). Besides, the effectiveness of *n*-butyramide (*n*-propyl-$CONH_2$) and *n*-valeramide (*n*-butyl-$CONH_2$) are found to be rather similar to those of *n*-propyl and *n*-butyl alcohol, respectively.

At the present stage of investigation, it is most likely that hydroxyl

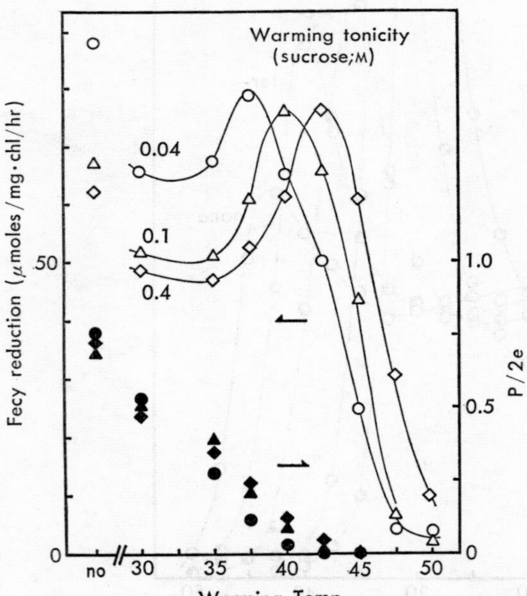

FIG. 12. Effects of sucrose concentration in the suspending medium of chloroplasts with warming treatment on the activity for Fecy reduction and P/2*e*. Experimental procedures were the same as in Fig. 7, differing in the sucrose content in the suspension under warming treatment.

groups in alcohols and amide groups in amides are necessary only for dissolving these molecules into aqueous media and the hydrocarbon chain is essential to cooperation with heat, as long as the results with lipid solvents (shorter esters, especially) described earlier are taken into account. It is plausible that introduction of these lipophilic segments of amphipathic molecules into thylakoids results in loosening of the original lipid architecture. Consequently, heat works more effectively on such a loosened construction and alters biological activities after the indispensable spatial relationship is lost. The degree of loosening, and thus the extent of the shift of the stimulation maximum will depend greatly on the amphipathic properties or penetratability of these molecules, and on their quantity.

This speculation is fairly well supported by the results which are shown in Figs. 12 and 13. When the sucrose concentration in the suspending

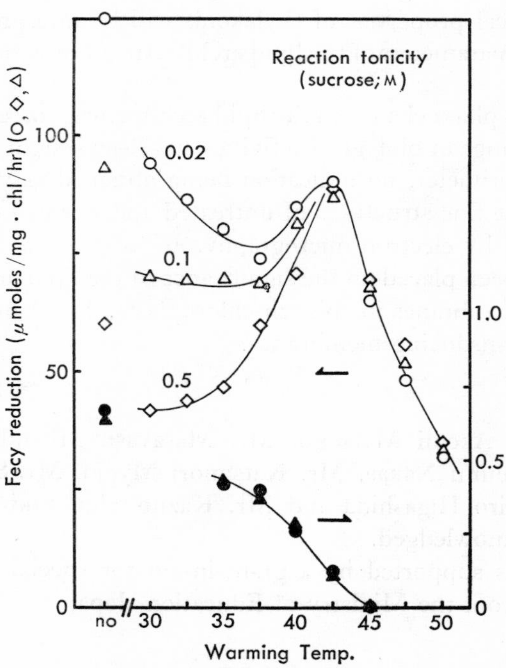

FIG. 13. Effects of sucrose concentration in the reaction medium on the activity for Fecy reduction and P/2*e* in transiently warmed chloroplasts. Experimental procedures were the same as in Fig. 7, differing in the sucrose content in the reaction medium.

medium (ordinarily 0.5 M) is reduced during warming, the photosynthetic activities are lost at lower warming temperatures (Fig. 12), although the results are not so clear (shoulders on the curves) as those of alcohols. When chloroplasts are warmed in 0.4 M sucrose and tested for their activities in reaction media at varied sucrose concentrations, only one stimulation maximum is observable (Fig. 13). Therefore, in these cases, lowering of the uncoupling temperature can be ascribed to the cooperative effect of heat and tonicity. It is plausible that swollen thylakoids in a hypotonic medium, or more loosened construction of the lipid architecture, are more vulnerable to heat than a normal construction.

Results obtained with chloroplasts warmed at low temperatures in the presence of alcohols suggest that heat activation of lipases and/or release of unsaturated fatty acids cannot be the major cause of phosphorylation inactivation (*19*, *20*) due to warming treatment. The effects of temperature on the physical and biological properties of thylakoids will be interpreted in terms of thermodynamic parameters of the lipid architecture interacting with functional proteins.

Thermal evidence of a phase change in the lipid architecture, corresponding to the irreversible change of biological activities, has been sought using a differential scanning calorimeter, no indication being obtained as yet. Besides, no difference in the fine structures of untreated and warmed chloroplasts has been observed by electron micrography.

So far, emphasis has been placed on the significance of the lipid architecture in photosynthetic membranes in spinach chloroplasts. The same must be true of any energy-transducing membranes.

Acknowledgments

The assistance of Miss Akemi Matsuno, Mr. Masayasu Mitsudo, Mr. Takao Yagi, Mr. Tatsubumi Nakae, Mr. Katsunori Myojo, Mr. Shigeru Kakumoto, Mr. Mitsuhiro Higashida and Mr. Kazuo Shinozaki in this research is gratefully acknowledged.

Part of this work was supported by a grant-in-aid for special project research on biophysics from the Ministry of Education, Japan.

REFERENCES

1. L. Packer, *Biochim. Biophys. Acta*, **75**, 12–22 (1963).
2. L. Packer, *Biochem. Biophys. Res. Commun.*, **9**, 355–360 (1962).

3. R. A. Dilley and L. P. Vernon, *Biochemistry*, **3**, 817–824 (1964).
4. G. Hind and A. T. Jagendorf, *J. Biol. Chem.*, **240**, 3195–3209 (1965).
5. Y. Mukohata, M. Mitsudo, and T. Isemura, *Ann. Rep. Biol. Works, Fac. Sci. Osaka Univ.*, **14**, 107–119 (1966).
6. D. W. Deamer, A. R. Crofts, and L. Packer, *Biochim. Biophys. Acta*, **131**, 81–96 (1967).
7. S. Murakami and L. Packer, *J. Cell Biol.*, **47**, 332–351 (1970).
8. M. Itoh, S. Izawa, and K. Shibata, *Biochim. Biophys. Acta*, **66**, 319–327 (1963).
9. S. Izawa and N. E. Good, *Plant Physiol.*, **41**, 533–543 (1966).
10. S. Murakami and L. Packer, *Biochim. Biophys. Acta*, **180**, 420–423 (1969).
11. J. E. Sundquist and R. H. Burris, *Biochim. Biophys. Acta*, **223**, 115–121 (1970).
12. Y. Mukohata, M. Mitsudo, T. Nakae, and K. Myojo, *Plant Cell Physiol.*, **12**, 859–868 (1971).
13. Y. Mukohata, *Ann. Rep. Biol. Works, Fac. Sci. Osaka Univ.*, **14**, 121–134 (1966).
14. Y. Mukohata, *in* "Comparative Biochemistry and Biophysics of Photosynthesis," ed. by K. Shibata *et al.*, University of Tokyo Press, Tokyo, 89–96 (1968).
15. Y. Mukohata, M. Mitsudo, S. Kakumoto, and M. Higashida, *Plant Cell Physiol.*, **12**, 869–880 (1970).
16. D. I. Arnon, *Plant Physiol.*, **24**, 1–15 (1949).
17. M. Kimimura and S. Katoh, *Plant Cell Physiol.*, **13**, 287–296 (1972).
18. H. N. Dunning and E. R. Washburn, *J. Phys. Chem.*, **56**, 235–237 (1952).
19. R. McCarty and A. T. Jagendorf, *Plant Physiol.*, **40**, 725–735 (1965).
20. Y. G. Molotokovsky and I. M. Zheskova, *Biochim. Biophys. Acta*, **112**, 170–172 (1966).

Systems for Hydrolysis of ATP and Pyrophosphate in Chromatophores from *Rhodospirillum rubrum*

Takekazu Horio, Jinpei Yamashita, Katsuzo Nishikawa,*1 Tomisaburo Kakuno, Kazuo Hosoi, Junnosuke Suzuki,*2 and Setsuko Yoshimura*3

Division of Enzymology, Institute for Protein Research, Osaka University, Osaka, Japan

The formation and hydrolysis of ATP and pyrophosphate (PP_i) were studied with chromatophores from *Rhodospirillum rubrum.* The possibility was suggested that the photosynthetic ATP and PP_i formation are catalyzed by two different enzyme systems. The system for ATP formation may be composed of latent nucleoside triphosphate-nucleoside diphosphate (NTP-NDP) kinase, bound ADP, ubiquinone-10 and another redox-component, whereas the system for PP_i formation, of latent pyrophosphatase and ubiquinone-10. The dark ATP hydrolysis is mostly, and the dark PP_i hydrolysis is partly, catalyzed by the backward reactions of the photosynthetic ATP and PP_i formation, respectively. The remainder of the dark PP_i hydrolysis is catalyzed by the pyrophosphatase that is active before sonication. The possibility was also presented that protons associated and dissociated by the electron transfer energize the latent NTP-NDP kinase and pyrophosphatase, and the energized states are responsible for ATP and PP_i formation, respectively.

Rhodospirillum rubrum, a photosynthetic bacterium, if grown in the light, forms a few hundred chromatophores in each cell. Electron microscope pictures indicate that the chromatophore is shaped like a sphere of 600 Å

*1 Present address: The Department of Biology, University of California, San Diego, La Jolla, California 92037, U.S.A. *2 Present address: Tokyo Research Laboratories, Kowa Co., Ltd., Higashi-murayama-shi, Tokyo 189, Japan. *3 Present address: Enzyme Laboratories, Oriental Yeast Co., Ltd., Suita-shi, Osaka 564, Japan.

average diameter with a membrane, which is 55 Å thick (*1*). The dry weight of such a chromatophore is 7.8×10^{-17} g, of which the protein weighs 2.4×10^{-17} g (*2*). If all molecules of the protein had a molecular weight of 25,000, typical of a structural protein (*3*), the chromatophore would contain 580 molecules of protein. It has been found that most structural protein molecules possess non-heme iron.

In the presence of ADP and P_i, chromatophores are able to synthesize both ATP and pyrophosphate (PP_i) in the light (*6*). Earlier, Baltscheffsky *et al.* (*4*, *5*) found that PP_i is formed when chromatophores are illuminated in the presence of P_i without addition of ADP, and that the inhibitor of the photosynthetic ATP formation, oligomycin, does not inhibit the photosynthetic PP_i formation.

The present paper deals with studies on the formation and hydrolysis of ATP and PP_i by chromatophores, with attempts to elucidate the relationship between the photosynthetic ATP and PP_i formation.

Formation and Hydrolysis of ATP and PP_i

Although the photosynthetic ATP formation proceeds until nearly the whole amount of either ADP or P_i present in the reaction medium is exhausted, the photosynthetic PP_i formation reaches a steady state when approximately 1/100 of the amount of P_i has been incorporated into PP_i. Under optimum conditions for activity, the PP_i formation is one-tenth as low in initial rate as the ATP formation. When the light is turned off in the steady state, both ATP and PP_i formed in the preceding light period are hydrolyzed. The initial rate of the ATP hydrolysis is less than 1/10 that of the ATP formation (*6*), whereas that of the PP_i hydrolysis is the same as that of the PP_i formation. Doubtless, at a steady state in the light, the rate of photosynthetic PP_i formation keeps a balance with the rate of PP_i hydrolysis; thus, when [^{32}P]-P_i is further added as a P_i tracer in the steady state, [^{32}P]-PP_i is formed at the same initial rate as that of the reaction in the light.

Mg^{2+} is required for the photosynthetic ATP and PP_i formation. The cation can be replaced by Mn^{2+} in the ATP formation, but not in the PP_i formation (Table I). If Mn^{2+} is further added in the presence of Mg^{2+}, the Mg^{2+}-activated ATP formation is depressed slightly, whereas the Mg^{2+}-activated PP_i formation is suppressed completely. On the other hand, the dark ATP and PP_i hydrolysis also require Mg^{2+}. Mn^{2+} substitutes Mg^{2+} completely in the former activity, but only slightly in the latter activity.

TABLE I
Relative Rates of Photosynthetic PP_i and ATP Formation, and Dark PP_i and ATP Hydrolysis in Presence of Various Reagents

Additions (mM)		Photosynthetic formation (%)		Dark hydrolyses (%)	
$MgCl_2$	Other additions	PP_i	ATP	PP_i	ATP
6.7	——	(100)	(100)	(100)	(100)
0	$MnCl_2$ (6.7)	0	73	20	109
6.7	$MnCl_2$ (6.7)	2	77	42	93
6.7	ADP (2.0)[a]	54	(100)[a]	103	—
6.7	ATP (2.0)[a]	12	100	96	(100)[a]
20	ATP (2.0)[a]	92	98	—	—
6.7	Antimycin A (1 μg/ml)	8	7	100	100
6.7	Oligomycin (3.3 μg/ml)	129	15	93	12
6.7	ADP (2.0)+oligomycin (3.3 μg/ml)	119	—	—	—
6.7	DCPI (0.067)	5	6	152	330
6.7	DCPI (0.67)	0	0	168	30
6.7	DCPI (0.67)+DNP (4.0)	—	—	—	30
6.7	DNP (4.0)	3	12	166	320
6.7	DCPI (0.067)+ascorbate (0.33)	100	99	—	93
6.7	DCPI (0.067)+ascorbate (0.33)+DNP (4.0)	—	—	—	260
6.7	DCPI (0.067)+ascorbate (6.7)	92	43	104	246

The initial reaction rates (mmoles/mole bacteriochlorophyll·min) are 280 for PP_i formation, 4,900 for ATP formation, 750 for PP_i hydrolysis and 320 for ATP hydrolysis, in the presence of 6.7 mM $MgCl_2$ with no addition. [a] Same conditions as for no addition.

TABLE II
Effect of Sonication on Photosynthetic PP_i and ATP Formation, Dark PP_i and ATP Hy-

Sonication (min)	Photosynthetic formation (%)		Dark hydrolysis (%)	
			PP_i	
	PP_i	ATP	—	+DNP
0	(100)	(100)	(100)	170
10	23	18	140	175
30	0	0	165	170

DNP, 4 mM. The Dark hydrolyses and the dark exchanges were measured without addition III). [a] Remarkably low in well-washed chromatophores; other activities are not influenced

The cytoplasmic fluid contains nucleoside triphosphate-nucleoside diphosphate (NTP-NDP) kinase and pyrophosphatase. It may be worth noting that the former enzyme is activated either by Mg^{2+} or by Mn^{2+} ("Mg^{2+}": "Mn^{2+}"=10: 7), and the latter enzyme only by Mg^{2+}, but not by Mn^{2+}.

Effect of Sonication and UV Illumination on Formation and Hydrolysis of ATP and PP_i

When chromatophores are sonicated, the photosynthetic ATP and PP_i formation decrease in parallel with each other. The dark ATP hydrolysis decreases in parallel with the photosynthetic ATP and PP_i formation, whereas the dark PP_i hydrolysis increases in rate to 160–200% (Table II). The percentage value varies from one chromatophore preparation to another. All the activity for PP_i hydrolysis by chromatophores is tightly bound to the membrane before and after sonication, although the enzyme, pyrophosphatase (pyrophosphate phosphohydrolase, EC 3.6.11), is present in a water-soluble state also in the cytoplasmic fluid. On the other hand, well-washed "intact" chromatophores barely show the activity for ATP-ADP exchange, and on sonication, a significant amount of NTP-NDP kinase (previously called ADP-ATP exchange enzyme) is solubilized and simultaneously becomes active (*8*), indicating that the enzyme is latent in the bound state. The cytoplasmic fluid also contains the enzyme. It has been shown that the bound enzyme has the same properties as the cytoplasmic enzyme (*8*).

All attempts so far made in our laboratory, which involve treatments with various proteolytic enzymes, phospholipases and detergents, have failed to solubilize the enzyme ATPase. The cytoplasmic fluid, if concentrated, does

drolysis, and Dark PP_i-P_i, ATP-P_i and ATP-ADP Exchanges

Dark hydrolysis (%)		Dark exchange (%)		
ATP				
—	+DNP	PP_i-P_i	ATP-P_i	ATP-ADP
(100)	360	(100)	(100)	(100)[a]
15	70	14	11	1,000
0	0	0	0	1,200

of DCPI, at appropriate concentrations of which they were stimulated (see Tables I and
by repeated washings with appropriate buffer solutions.

not catalyze ATP hydrolysis. It seems reasonable to conclude that chromatophores do not possess ATPase, unlike mitochondria or chloroplasts. Perhaps the ATP-hydrolyzing activity in chromatophores is catalyzed by an enzyme system that is composed of latent NTP-NDP kinase, bound ADP (*10*) and oxidation-reduction components (*7*). It is currently being attempted to modify NTP-NDP kinase into ATPase.

When chromatophores are illuminated with *UV* light, the activities for NADH-cytochrome c_2 reduction and succinate-cytochrome c_2 reduction are impaired slowly. Although the photosynthetic ATP and PP_i formation are also impaired, the latter formation is significantly more resistant to illumination than the former formation (*9*). It is conceivable that the slow decrease of the PP_i formation is mainly due to the impairment of the electron transport system, and that the rapid decrease of the ATP formation is mostly due to the impairment of the energy conservation system.

Effect of ATP and PP_i on Photosynthetic ATP and PP_i Formation

The photosynthetic ATP formation is approximately 30% inhibited by PP_i. The initial rate of photosynthetic PP_i formation is approximately 50% inhibited by low concentrations (0.2–2.0 mM) of ADP, at which the rate of photosynthetic ATP formation is maximum. However, it is not influenced by ATP. The ADP inhibition of the PP_i formation is not observed when the ATP formation has been inhibited by the addition of oligomycin. These results suggest that the electron transport system is common for the ATP and PP_i formation, and that both somehow compete with each other for the energy produced by the electron transfer.

TABLE III
Various Reactions Catalyzed by Chromatophores

Photosynthetic and dark reaction	Relative maximum rate[a] (%)	DCPI-sensitive rate (%)	Remarks
1) $ADP + 3P_i \xrightarrow{h\nu} ADP\text{-}P + P\text{-}P_i$			
ADP-P formation	340	100 ↓	i) Partially depressed by addition of PP_i ii) Uncoupled by DNP iii) All inactivated by sonication
$P\text{-}P_i$ formation	24	100 ↓	i) Partially depressed by addition of ADP ii) Uncoupled by DNP iii) All inactivated by sonication
2) $2P_i \xrightarrow{h\nu} P\text{-}P_i$	48	100 ↓	i) The same as those for $P\text{-}P_i$ formation in Reaction 1
3) $ATP + H_2O \longrightarrow ADP + P_i$	270	90↑↓	i) All inactivated by sonication ii) Stimulated by DNP
4) $ATP + P_i^* \longrightarrow ADP\text{-}P^* + P_i$	100	100 ↓	i) All inactivated by sonication ii) Uncoupled by DNP
5) $ATP + ADP^* \longrightarrow ADP^*\text{-}P + ADP$	19	—	i) Relative rate increased to 225% on sonication
6) $PP_i + H_2O \longrightarrow 2P_i$	320	50 ↑	i) Activated by sonication ii) Stimulated by DNP
7) $PP_i + P_i \longrightarrow P\text{-}P_i + P_i$	12	100 ↓	i) All inactivated by sonication ii) Uncoupled by DNP

↓ and ↑ represent depression and stimulation, respectively. ↑↓ represents stimulation followed by depression. For example, the rate of ATP hydrolysis (Reaction 3) is stimulated by low concentrations of DCPI, reaches a maximum (270%) at 67 μM, and approximately 90% of the maximum rate is depressed at 0.67 mM. The values for relative maximum rates vary from one chromatophore preparation to another to some extent. [a] Measured under appropriate conditions.

Partial Reactions for Photosynthetic ATP and PP_i Formation

Chromatophores are able to catalyze ATP-P_i exchange and PP_i-P_i exchange in darkness; the maximum rates of both exchange reactions are 1/3–1/4 as high as those of photosynthetic ATP and PP_i formation, respectively. However, neither the dark ATP formation from PP_i and ADP nor the dark PP_i formation from ATP and P_i takes place to a meaningful extent (Table III). Since all the activities for dark ATP-P_i and PP_i-P_i exchanges are depressed by sonication and uncoupled by 2,4-dinitrophenol (DNP) in the same manner as those for photosynthetic ATP and PP_i formation, it is probable that all of the activities for ATP-P_i and PP_i-P_i exchange are due to the partial reactions for the photosynthetic ATP and PP_i formation, respectively.

Effect of 2,6-Dichlorophenol Indophenol and DNP on Formation and Hydrolysis of ATP and PP_i

When 2,6-dichlorophenol indophenol (DCPI), a redox-dye ($E_{m,7}=+0.217$ V), is added to the reaction medium, the photosynthetic ATP and PP_i formation, and the dark ATP-P_i and PP_i-P_i exchanges decrease almost in parallel with one another with increasing concentrations of the dye; their rates become negligible at 0.067 mM (Table I). On the contrary, the dark ATP and PP_i hydrolysis increase with increasing concentrations of the dye. Compared with the cases without addition of DCPI, the rates of ATP and PP_i hydrolysis in the presence of 0.067 mM DCPI are 330% and 152%, respectively. These values vary to some extent from one chromatophore preparation to another; with some preparations, the rate of PP_i hydrolysis increases twice as high. At higher concentrations of DCPI, the rate of ATP hydrolysis decreases, whereas that of PP_i hydrolysis is barely influenced. At 0.67 mM, the rate of ATP hydrolysis is approximately one-tenth that at 0.067 mM. When ascorbate, a reducing reagent, is further added to reaction medium containing 0.067 mM DCPI, the activities for photosynthetic ATP and PP_i formation are restored, provided that the concentrations of ascorbate are appropriate. At 0.33 mM, the restorations are nearly complete. At 6.7 mM, the rate of ATP formation is again depressed to 1/2, whereas that of PP_i formation is barely influenced. At 67 mM, the latter is depressed to 1/2.

Effect of Light on PP_i Hydrolysis

The rate of photosynthetic PP_i formation is maximum at pH 8, and negligible at pH 6. The dark PP_i hydrolysis has a much broader pH-activity curve; the rate is highest at pH 6, and at pH 8 it is nearly as high. In the light, the PP_i hydrolysis is reduced to approximately 70% of the rate in darkness throughout the range of pH at which the PP_i hydrolysis is active. This suggests that the depression in the light is caused by a change in the oxidation-reduction state of the electron transport system.

In fact, the DCPI- and DNP-sensitive part of the dark PP_i-hydrolyzing activity is eliminated when quinones are extracted from chromatophores, and the activity is restored when ubiquinone-10 is added in a similar manner to that of the dark ATP-hydrolyzing activity (*11*).

Effect of DNP on Formation and Hydrolysis of ATP and PP_i

The uncoupler of the photosynthetic ATP and PP_i formation, DNP, stimulates not only the dark ATP hydrolysis but also the dark PP_i hydrolysis (Table I). The maximum rates of ATP and PP_i hydrolysis in the presence of DNP are the same as those in the presence of DCPI, respectively. In the case of the PP_i hydrolysis, it is almost the same as the maximum rate attained by sonication of chromatophores (Table II). The rate of ATP hydrolysis in the presence of DNP, however, decreases in parallel with that in the absence of the uncoupler.

Discussion

The distribution of oxidation-reduction components in the chromatophore membrane is thought to be as shown schematically in Fig. 1 (*2*). There is evidence that most of the structural protein is non-heme iron protein ($E_{m,7}=+0.28$ V, $n=1$) (Higuti *et al.*; Erabi *et al.*, to be published). If the molecules are distributed uniformly on the membrane, the distance is 38 Å between 2 adjacent bacteriochlorophyll molecules, and 47 Å between 2 adjacent quinone molecules. If the quinone and non-heme iron are mixed together and distributed uniformly, the distance between 2 adjacent molecules of quinone and non-heme iron is 37 Å. These distances are applicable to the resonance transfer of excited energy (*12*), whereas they appear to be too great to bring about electron transfer, especially among the quinones

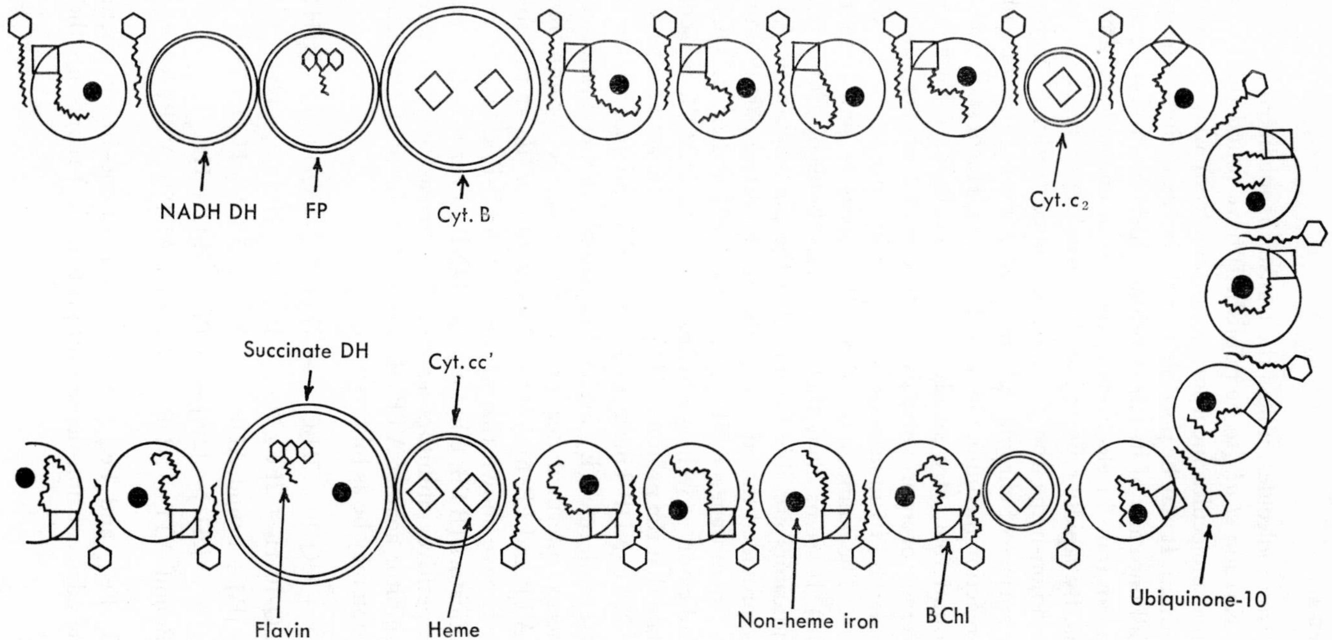

FIG. 1. Distribution of oxidation-reduction components in chromatophore membrane. The figure shows part of a section of chromatophore membrane, which has been bent at will. BChl, bacteriochlorophyll; DH, dehydrogenase; Cyt, cytochrome; FP, flavoprotein.

and non-heme iron molecules, unless the transfer is mediated by other factors. Recently, Kakuno *et al.* (to be published) have found that on reduction, the quinone molecules incorporate protons from water molecules in the reaction medium. It may be possible that the benzene ring of the quinone molecule can fall onto any of its surrounding non-heme iron protein molecules, and then can transfer electrons to the iron atoms and simultaneously liberate protons. By rapidly repeated oxidation-reduction reactions between the quinones and non-heme iron proteins, the concentration of protons may increase on the surface of chromatophore vesicles, raising the concentration of OH^- in the reaction medium. On the other hand, Hosoi *et al.* (to be published) have found that, at a steady state in the light, 1 chromatophore liberates an average of approximately 10 protons, which are adsorbed by bromthymol blue (BTB) molecules bound to the surface of the chromatophore vesicles. The number of adsorbed protons is similar to that of bound ADP (*10*). It seems likely that appropriate kinds of proton removers such as pH indicators are able to uncouple the photosynthetic ATP and PP_i formations, provided that the reagents are at least in part bound to the chromatophore membrane. DNP is a typical pH indicator. DCPI also can be a pH indicator, as well as a hydrogen acceptor. At the present time, we may speculate that the protons ($[H^+]$) present in the active sites of nearly all the bound NTP-NDP kinase molecules (E_1) and half of the bound pyrophosphatase molecules (E_2) in the chromatophore membrane are in so-called high-energy intermediates or states; thus, the proton removers stimulate the ATP and PP_i hydrolysis. Yoshimura, *et al.*, and Hosoi, *et al.* (to be published) have shown that in the dark ATP-P_i exchange reaction, several pH indicators (BTB, ethyl orange, DNP, alizarin yellow, *etc.*) and DCPI are competitive inhibitors against P_i, whereas only the latter is a competitive inhibitor against ATP. Reactions for the photosynthetic ATP and PP_i formation may be as follows:

$$2Fe_{(1)}^{2+} + \text{UQ-10} + 2H_2O \longrightarrow 2Fe_{(1)}^{3+} + \text{UQ-10}\cdot H_2 + 2OH^-$$

$$\text{UQ-10}\cdot H_2 + 2Fe_{(2)}^{3+} \longrightarrow \text{UQ-10} + 2Fe_{(2)}^{2+} + [2H^+]$$

$$[2H^+] + [P_i] + \text{bound ADP} \longrightarrow \text{bound ATP} + H_2O + H^+$$

$$[2H^+] + [2P_i] \longrightarrow PP_i + H_2O + H^+$$

$$\text{Bound ATP} + \text{ADP} \longrightarrow \text{bound ADP} + \text{ATP}\,,$$

where UQ-10, $Fe_{(1)}$ and $Fe_{(2)}$ represent ubiquinone-10, non-heme iron protein and the adjacent non-heme iron protein. [ ] represents the active site of bound NTP-NDP kinase or bound pyrophosphatase, which has latent activity against the NTP or the PP_i added to the reaction medium. Reactions for uncoupling in the presence of the pH indicators such as DNP may be as follows:

$$[H^+] + \text{bound DNP} \longrightarrow [H^+] + \text{bound DNP}\cdot H^+$$
$$\text{Bound DNP}\cdot H^+ + \text{buffer} \longrightarrow \text{bound DNP} + \text{buffer } H^+ .$$

REFERENCES

1. T. Oda and T. Horio, *Exp. Cell Res.*, **34**, 414 (1964).
2. T. Kakuno, R. G. Bartsch, K. Nishikawa, and T. Horio, *J. Biochem.*, **70**, 79 (1971).
3. R. C. Criddle, R. M. Bock, D. E. Green, and H. Tisdale, *Biochemistry*, **1**, 827 (1962).
4. H. Balscheffsky and L. von Stedingk, *Biochem. Biophys. Res. Commun.*, **22**, 722 (1966).
5. H. Baltscheffsky, L. von Stedingk, H. W. Heldt, and M. Klingenberg, *Science*, **153**, 1120 (1966).
6. K. Nishikawa, K. Hosoi, J. Suzuki, S. Yoshimura, and T. Horio, *J. Biochem.*, **73**, 537 (1973).
7. Y. Horiuti, K. Nishikawa, and T. Horio, *J. Biochem.*, **64**, 577 (1968).
8. N. Yamamoto, Y. Horiuti, K. Nishikawa, and T. Horio, *J. Biochem.*, **72**, 599 (1972).
9. T. Horio, K. Nishikawa, Y. Horiuti, and T. Kakuno, *in* "Comparative Biochemistry and Biophysics of Photosynthesis," ed. by K. Shibata *et al.*, University of Tokyo Press, Tokyo, p. 408 (1968).
10. N. Yamamoto, S. Yoshimura, T. Higuti, K. Nishikawa, and T. Horio, *J. Biochem.*, **72**, 1397 (1972).
11. N. Yamamoto, H. Hatakeyama, K. Nishikawa, and T. Horio, *J. Biochem.*, **67**, 587 (1970).
12. R. K. Clayton, *in* "Molecular Physics in Photosynthesis," Blaisdell Publishing Co., New York (1965).

Thiophosphate as a Probe of Phosphorylation Reactions and Permeability

Lester Packer, Joaquim Tavares de Sousa, Kozo Utsumi, and Gregory R. Schonbaum

Department of Physiology-Anatomy, University of California, Berkeley, California, U.S.A.

Monothiophosphate was used as a probe in substrate level phosphorylation, mitochondrial oxidative phosphorylation, and membrane permeability. Thiophosphate did not support ATP synthesis either in mitochondria or model enzyme reactions, conservation of oxidative energy occurring through formation of ATP(γ)S. Thiophosphoryl transfer was slower than phosphoryl (group) transfer. Such behavior was compatible with an addition-elimination mechanism. Thiophosphate accumulated in mitochondria by an energy-dependent process and substituted for phosphate in the catalysis of metabolite transport. Relative to phosphate, a slower and less extensive permeation was observed.

The energy derived from the mitochondrial redox reactions is coupled to the production of adenosine triphosphate, providing that ADP and orthophosphate are present. The reaction may involve an intermediate formation of a phosphorylated derivative, X$\sim$P, but, however plausible, this proposal is not sufficiently substantiated to warrant general acceptance. Indeed, the strongest evidence in its favor are observations of Cross *et al.* (*1*) showing

Abbreviations used: TMPD, N,N,N′,N′-tetramethyl-*p*-phenylene-diamine dihydrochloride; EDTA, ethylenediaminetetraacetic acid; MES, 2, N (morpholino) ethane sulfonic acid; TES, N-Tris (hydroxymethyl) methyl-2-aminoethane sulfonic acid; PEI polyethyleneimine; ATP(γ)S, thiophosphoryl ADP; DTT, dithiothreitol; RLM, rat liver mitochondria; SP, thiophosphate; GAPDH, glyceraldehyde-3-phosphate dehydrogenase; PGK, phosphoglycerate kinase.

that incorporation of ^{32}P into rat liver mitochondria is greater in the presence of aurovertin than that noted using oligomycin. Significantly, the presumed ^{32}P-labeled intermediate is labile and is rapidly discharged on treatment with uncouplers. Thus, the above procedure does not seem promising for the identification of ^{32}P-derivative. Seeking a method which would lead to the formation of an intermediate of enhanced stability, we have therefore directed our attention to the possible use of monothiophosphate as a probe of phosphorylation mechanisms.

The rationale of this approach is that relative to the organophosphate derivatives the corresponding organothiophosphate compounds should be more stable providing that the rate-limiting step of their solvolysis involves a nucleophilic attack at the phosphorus atom.

The use of thiophosphate also offers other advantages. Thus, monothiophosphate shares with orthophosphate similar acid-base properties; but it is readily assayed, either directly—exploiting its distinctive absorption spectra—or by using the typical -SH reagents. In addition, thiophosphate, labeled at the phosphorus or sulfur, is readily available.

Materials

Mitochondria were prepared as described previously (*2*), except that the isolation medium contained 0.25 M sucrose and 1 mM EDTA.

Thiophosphate (Ventron Corp., Na_3PO_3S) was recrystallized from water and dried. The resulting anhydrous powder was assayed by ferricyanide oxidation at pH 7.0 and found to be 98% Na_3SPO_3. Elemental analysis found: P 16.69%, S 17.00%; expected: P 17.21%, S 17.81%.

Glyceraldehyde-3-phosphate dehydrogenase from rabbit muscle, phosphoglycerate kinase, type IV from yeast, and DL-glyceraldehyde-3-phosphate were obtained from Sigma Chemical Co. All other chemicals were of the highest available purity.

Methods

Glyceraldehyde-3-phosphate dehydrogenase was activated with dithiothreitol and assayed essentially as described by Velick (*3*).

The concentration of D-isomer in DL-glyceraldehyde-3-phosphate was determined enzymatically in the presence of arsenate (*3*).

Mitochondrial protein was determined by the Biuret method using bovine serum albumin as a standard.

Oxygen uptake by mitochondrial suspensions was determined polarographically with a Clark-type electrode, at 25°C.

Mitochondrial ATP-synthesis was estimated by measuring absorbance at 260 nm after separation of ATP by column chromatography according to Pressman (*4*). ATP-P_i exchange was investigated using ^{32}P-labeled orthophosphate, the separation of the nucleotides being obtained by the method of Hagihara and Lardy (*5*).

Membrane permeability studies were performed in a Brice-Phoenix light-scattering photometer. Particle volume changes occurring as a consequence of solute permeation were monitored indirectly by changes in the intensity of the scattering light at 90°. Reaction conditions are given in the appropriate figure legends.

Ion exchange thin-layer chromatography of the products of both model and mitochondrial reactions on MN-Polygram CEL 300 PEI/UV (Machery and Nagel, Duren, Germany) was done, eluting with 0.75 M KH_2PO_4 adjusted to pH 3.4 with concentrated HCl, as described by Goody and Eckstein (*6*). Sulfur compounds were detected by spraying the plates with azide-iodine (*7*). ATP(γ)S was oxidized with H_2O_2 (*6*), and the resulting disulfide identified by its chromatographic mobility (see above).

Spectrophotometric measurements were carried out using a Cary Model 14 Spectrophotometer.

Results

1. Model reaction

The feasibility of using thiophosphate as an analog of orthophosphate was qualitatively assessed in the glyceraldehyde-3-phosphate dehydrogenase-phosphoglycerate kinase system.

This substrate level oxidative phosphorylation may be expressed in terms of three partial reactions. The first step involves oxidation of D-glyceraldehyde-3-phosphate, giving an enzyme-bound thiol ester (*8*). In the next stage, free glyceraldehyde-3-phosphate dehydrogenase is released through phosphorolysis of the thiol ester with the concurrent formation of 1,3-diphosphoglycerate. The final reaction, yielding ATP, is mediated by phosphoglycerate kinase, and necessarily occurs only in the presence of ADP.

Our results, shown in Figs. 1 and 2, establish that an analogous scheme obtains using thiophosphate; but the rate of approach to equilibria and the extent of glyceraldehyde-3-phosphate dehydrogenase and phosphoglycerate kinase catalyzed reactions are lower than the corresponding reactions with orthophosphate.

The results summarized in Table I indicate that ATP is only a minor component of the reaction products. The major product appears to be ATP (γ)S

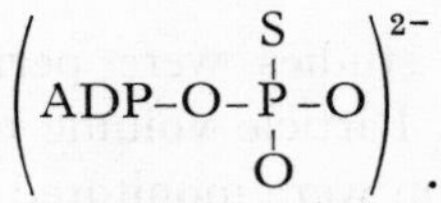

The latter was identified through thin-layer chromatography, using as a standard synthetic ATP(γ)S(*6*). Moreover, both the synthetic and "natural" compounds give—on mild oxidation with hydrogen peroxide—derivatives which have identical chromatographic mobilities. Such oxidation involves formation of a disulfide which should be reducible to ATP (γ)S. This was confirmed using mercaptoethanol as the reducing agent.

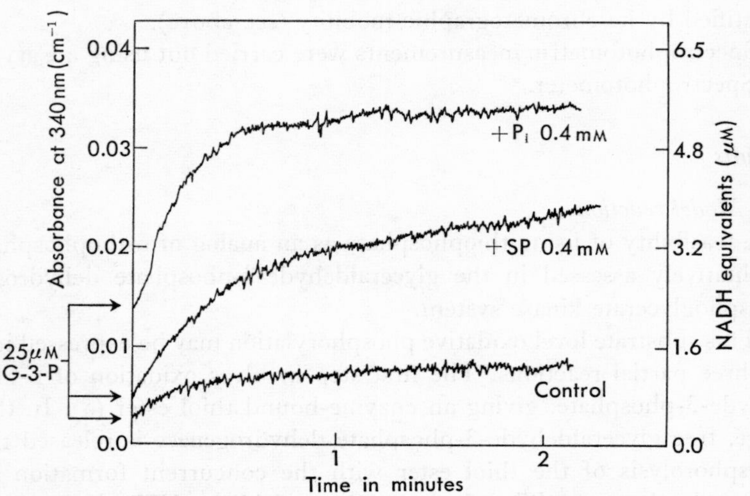

FIG. 1. Kinetics of NAD$^+$ reduction in the D-glyceraldehyde-3-phosphate dehydrogenase reaction. (On arsenolysis, obtained 24 μM NADH; expected for 100% conversion of D-glyceraldehyde-3-phosphate, 25 μM.)

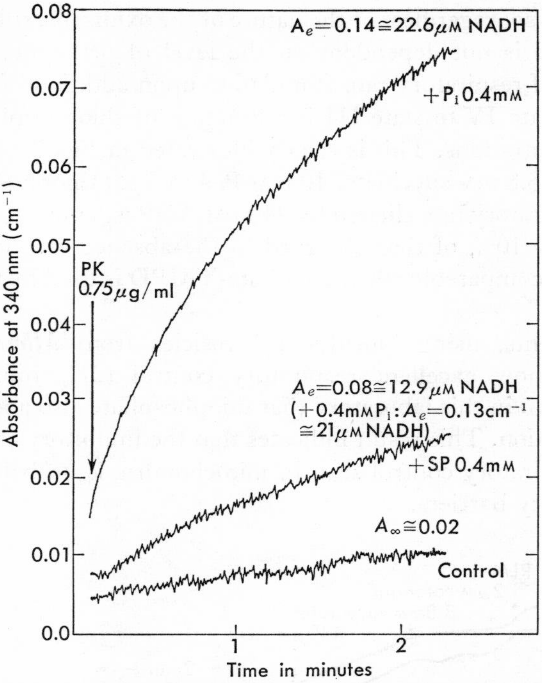

FIG. 2. Kinetics of phosphoryl and thiophosphoryl transfer in the coupled glyceraldehyde-3-phosphate dehydrogenase-phosphoglycerate kinase catalyzed reaction. The reaction was initiated with 0.022 μM phosphoglycerate kinase.

TABLE I
Effect of Phosphate and Thiophosphate on ATP Synthesis by the GAPDH-PGK Coupled System

	NADH nmoles ml^{-1}	ATP[a] nmoles ml^{-1}	(ATP : NADH) × 100
0.5 mM P_i	100	84	84
0.5 mM SP	80	9	11

The composition of the reaction mixture was: 0.1 M TES buffer, pH 7.6, 5 mM $MgCl_2$, 2.1 mM NAD, 0.5 mM ADP, 0.225 mM D-glyceraldehyde-3-phosphoric acid, 71 mM glyceraldehyde-3-phosphate dehydrogenase, and 0.22 μM phosphoglycerate kinase, at 25°C.
[a] Estimated through hexokinase-glucose-6-phosphate dehydrogenase assay.

2. *Mitochondrial oxidative phosphorylation*

In state IV, the respiratory rate of rat liver mitochondria is unaffected by thiophosphate. This is so, regardless of the nature of the oxidizable substrate (Figs. 3A and 3B) and is not dependent on the level of orthophosphate. However, the extent of respiratory rate stimulation upon addition of ADP, *i.e.*, transition from state IV to state III is a function of thiophosphate to orthophosphate concentrations. This is clearly illustrated in Fig. 3 showing that in the presence of 8.8 mM succinate, 10 mM P_i and 3 μM thiophosphate, an increase in the respiratory rate elicited by 440 μM ADP is, at most, equivalent to approximately 10% of that observed in the absence of thiophosphate. The results are comparable when ascorbate-TMPD is used instead of succinate (Fig. 3B).

In other experiments, using "inside-out" vesicles from *Micrococcus denitrificans*, which show excellent respiratory control (*21*), John and Schonbaum have shown in this laboratory that thiophosphate also abolishes the state IV-III transition. This result indicates that the inhibitory effect of thiophosphate on respiratory control seen in mitochondria is probably not due to any permeability barriers.

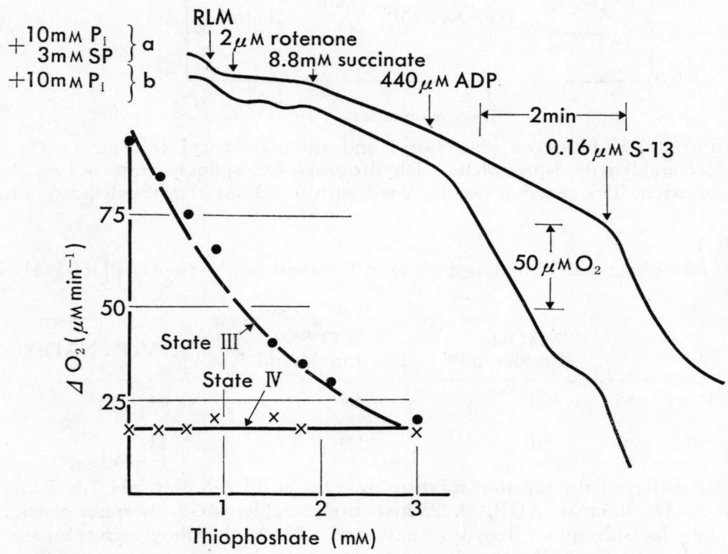

FIG. 3A. Effect of thiophosphate on the ADP-induced state IV-III transition accompanying succinate oxidation. Incubation medium contained: 2 mg protein/ml, 0.2 M mannitol, 0.05 M sucrose, 20 mM morpholinopropane sulfonic acid (MOPS) buffer at pH 7.4 and 25°C, and the other additions as shown.

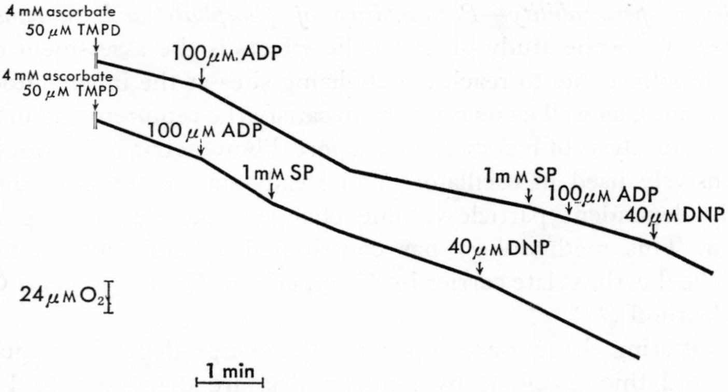

FIG. 3B. Effect of thiophosphate on the ADP-induced state IV-III transition accompanying ascorbate-TMPD oxidation. The incubation medium contained: 200 mM sucrose, 5 mM Tris-HCl pH 7.4, 10 mM KCl, 3 mM Mg^{2+}, 1.2 mM NaP_i, 0.6 mg/ml mitochondria at 25°C.

TABLE II
Effect of Thiophosphate on the ATP-P_i Exchange Reaction by Rat Liver Mitochondria

Thiophosphate (mM)	ATP-$^{32}P_i$ Exchange (μmoles/mg protein/10 min)
0.00	0.66
0.07	0.64
0.13	0.58
0.26	0.45
0.39	0.22
0.66	0.12

Mitochondria (6 mg protein) were incubated in a medium (5 ml) containing 0.2 M sucrose, 10 mM KCl, 5 mM $MgCl_2$, 5 mM Tris-HCl pH 7.4, 0.1 mM EDTA, 3 mM Na^+-ATP, 3 mM sodium phosphate ($^{32}P_i$, 0.1 μCi), and SP as indicated. Incubation: 10 min at 25°C.

Apparently thiophosphate competes for the phosphate binding site. And, if a phosphorylated intermediate is formed, it does not react sufficiently rapidly with ADP to effect a change in the respiratory rate. In this context, note that the uncoupler-stimulated respiration is not inhibited by thiophosphate. In accord with the above observations is the previously noted inhibition of ATP-P_i exchange (*9*)—which was confirmed in an independent series of experiments (Table II).

3. Membrane permeability—Permeability of phosphate and thiophosphate
Complementary to the study of metabolic effects is the assessment of the ability of thiophosphate to reach metabolizing sites at the inner mitochondrial membranes, as well as its capacity to satisfy the requirements in phosphate for the catalysis of metabolite transport. Photometric techniques have been extensively used in oscillatory and steady-state systems to monitor osmotically dependent particle-volume changes occurring during solute permeation. This methodology has contributed valuable information in studies of the dicarboxylate carrier by Chappell and Grofts (*10*) and Chappell and Haarhoff (*11*).

Light-scattering changes accompanying the energy-dependent uptake of phosphate and thiophosphate by mitochondria are compared in Fig. 4. A low concentration of succinate was provided as an energy source. As seen

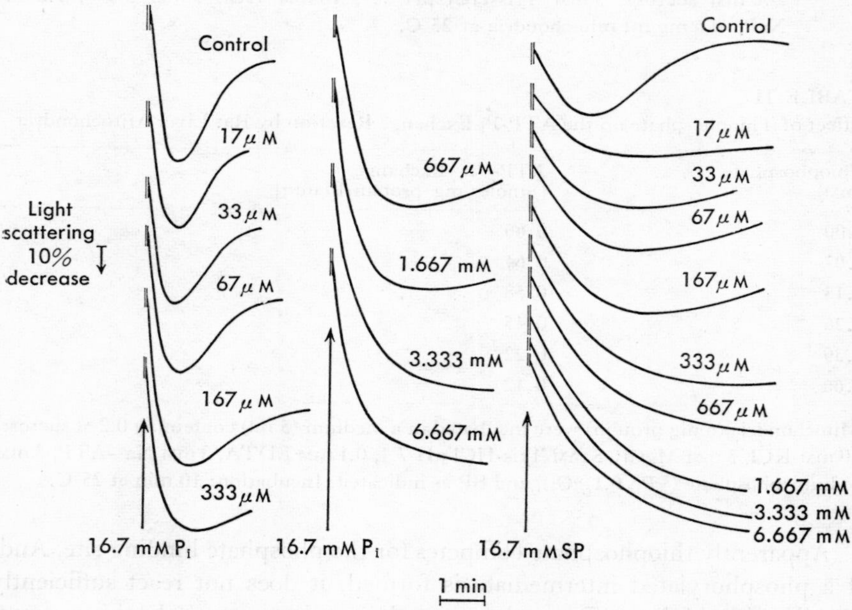

FIG. 4. Uptake of phosphate and thiophosphate by mitochondria. The incubation medium contained: 57 mM sucrose, 5 mM Tris-HCl pH 7.8, 33 μM EDTA, 3.3 mM sodium succinate, 1 mg protein/ml of mitochondria, and Mg^{2+} (as SO_4^{2-}) as indicated, at 25°C.

in the "control" traces, the permeation of thiophosphate is markedly slower and less extensive than that of phosphate.

At pH 7.8 and in the absence of bivalent cations, ion transport follows an oscillatory pattern. As previously reported (*12*), in the presence of bivalent cations, phosphate transport occurs, but the oscillatory movements of these ions are damped. This was confirmed for both anions by experiments at varying Mg^{2+} concentrations (Fig. 4, lower tracings). Moreover, Mg^{2+} is shown to increase the extent of swelling in both cases.

4. Phosphate- and thiophosphate-catalyzed metabolite transport

Thiophosphate is shown in Fig. 5 to substitute for phosphate in the catalysis of succinate uptake by rat liver mitochondria. Addition of Mg^{2+} leads to a more extensive uptake of succinate and abolishes the oscillatory kinetics. In this experiment, energy is provided by succinate oxidation. Further studies, shown in Fig. 6, show that the energy required for succinate transport can

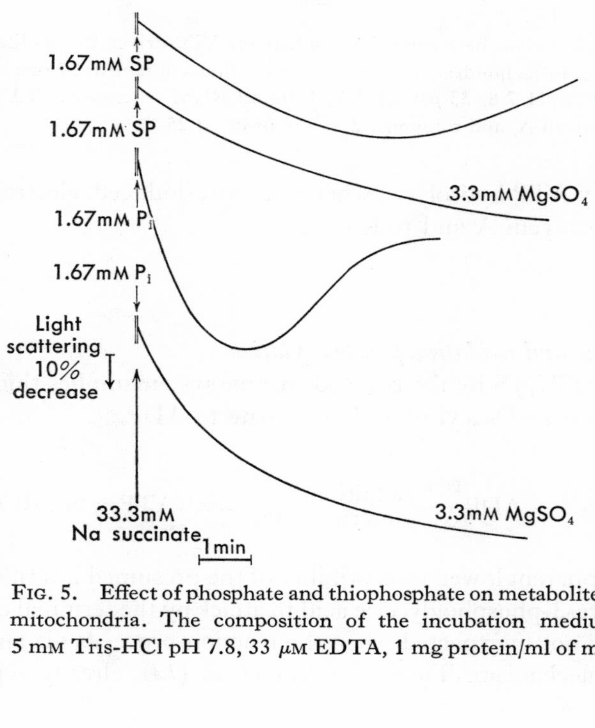

FIG. 5. Effect of phosphate and thiophosphate on metabolite transport by rat liver mitochondria. The composition of the incubation medium: 57 mM sucrose, 5 mM Tris-HCl pH 7.8, 33 μM EDTA, 1 mg protein/ml of mitochondria, at 25°C.

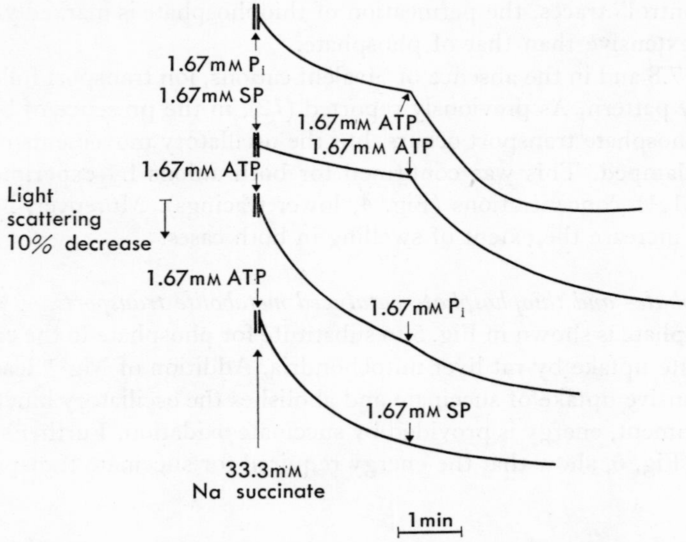

FIG. 6. Effect of phosphate and thiophosphate on ATP-driven metabolite transport by rat liver mitochondria. The incubation medium contained: 57 mM sucrose, 5 mM Tris-HCl pH 7.8, 33 μM EDTA, 1 mg/ml RLM (sucrose-EDTA prep.), 0.83 μM antimycin A, and rotenone, 2 μg/mg prot., at 25°C.

also be provided by ATP hydrolysis, when succinate-induced, electron flow is inhibited by antimycin A and rotenone.

Discussion

1. Model reactions and oxidative phosphorylation

The synthesis of ATP (γ) S by the coupled enzyme system implies thiophosphoryl transfer from an O-acyl phosphorothioate to ADP.

$$R-\overset{O}{\overset{\|}{C}}-O- \vdots\, PSO_3^- - + ADP \xrightarrow[Mg^{2+}]{PGK} \left(\begin{matrix}\text{addition}\\ \text{intermediate}\end{matrix}\right) \rightleftharpoons ATP\ (\gamma)S + RCO_2^-$$

Note that the apparent lower susceptibility of the presumed intermediate, 1-phosphorothioate-3-phosphoglyceric acid to attack by the terminal oxygen of ADP (Fig. 2) is not unexpected when the reaction proceeds via an addition-elimination mechanism. Thus, Ketelaar *et al.* (*13*), already reported several years ago that in the alkaline hydrolysis of phosphate and thiophosphate triesters, the ratio of the characteristic second-order rate constants $K_{p=s}/K_{p=0}$ is approximately 0.02: 0.1. Similarly, the work of Breslow and Katz (*14*) on *p*-nitrophenyl phosphorothioate and of Mushack and Coleman (*15*) on arylazoaryl phosphorothioates indicates the same type of kinetic constraint. These observations can be explained by the lesser electronegativity of sulfur as compared with oxygen, thereby rendering less favorable the ADP attack at phosphorus.

Thiophosphate is an ambident reagent and, *a priori*, we might expect it to react either *via* sulfur or oxygen atoms. Indeed, this is the case, but the nature of the dominant derivative is contingent on the properties of the substrate. Thus with alkyl halides, or epoxides, thiophosphate gives the corresponding S-alkyl compounds; while in reactions with acylating reagents, the product is predominantly an O-acyl derivative (*16*, *17*). These examples are an excellent illustration of Pearson's thesis (*18*, *19*) that in general, "soft" bases (RS^-) tend to react with "soft" acids (RCH_2^+); and conversely that "hard" bases (RO^-) tend to react with "hard" acids (RCO^+). The results presented in Tables II and III fully accord with these expectations.

Clearly, these results do not define the rates of ATP (γ) S formation, but indicate only that, relative to the kinetics of ATP generation, these are seemingly lower both in the glyceraldehyde-3-phosphate dehydrogenase-phosphoglycerate kinase and the mitochondrial systems. Whether this is simply due to a difference in the absolute rates of ATP (γ)S and ATP syntheses, or in addition, reflects a significant breakdown of ATP (γ)S to ADP and a phosphorylated precursor intermediate remains uncertain. This being the case, it is not surprising that only trace amounts of ATP (γ)S are detectable,

TABLE III
Effect of Thiophosphate on ATP Formation by Rat Liver Mitochondria

	ATP (nmoles)
−Phosphate	68
+Phosphate	3,400
+Phosphate + thiophosphate	540
+Thiophosphate	108

Mitochondria (3 mg/ml), 0.2 M sucrose, 10 mM KCl, 5 mM $MgCl_2$, 1 mM EDTA, 5 mM sodium succinate, 3 mM Na^+-ADP; as indicated 3 mM sodium phosphate and 1.3 mM sodium thiophosphate in 5 ml of reaction medium. Incubation: 10 min at 25°C.

even on prolonged incubation, using heavy suspensions of rat liver mitochondria.*

2. *Membrane permeability*

Current views on mitochondrial permeability, arising mainly from the work of Chappell *et al.* (*10*, *11*) and of Mitchell and Moyle (*20*), consider the phosphate carrier as a phosphoric acid-proton symporter, hydrodehydration reactions occurring at both sides of the membrane. The phosphoric acid-carrier complex is represented as

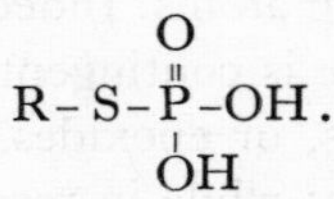

Our results indicate that thiophosphate shares with phosphate an energy-dependent mechanism of transport, and three limiting configurations can be assigned to the thiophosphoric acid-carrier complex:

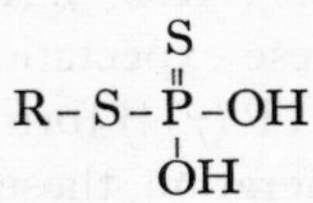

(1a)

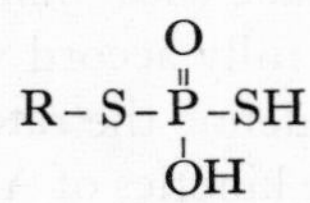

(1b)

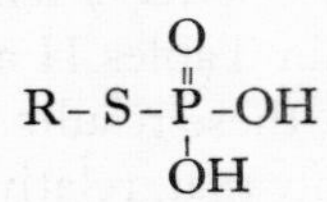

(2)

Transport as in (1b) and (2) is not likely in view of the low concentration of the thiolo form expected in a polar medium. Furthermore, mechanism (2) implies desulfuration and accumulation of orthophosphate by mitochondria. This mechanism is not favored in view of the absence of respiratory inhibition (Fig. 4) which should occur in H_2S liberation, and by the synthesis of ATP (γ)S.

Transport as in (1a) agrees with the observed slower kinetics, the lesser electronegativity of the sulfur rendering the phosphorus atom in the carrier complex less susceptible to OH^- attack.

Acknowledgments

We are indebted to Drs. D. Wilson, J. B. Jackson, and C. P. Lee, of the

* ATP (γ)S was determined after 30-min incubation of 5 mM thiophosphate with 5 mM ADP, in the presence of 0.35 M sucrose, 1 mM EDTA, 5 mM succinate, 10 mM KCl, 5 mM TES, and 5 mM $MgCl_2$ at pH 7.8, using rat liver mitochondria at 70 mg/ml of mitochondrial protein. The reaction mixture was continuously aerated.

University of Pennsylvania, for their cooperation in the initial survey of thiophosphate reactivities. The authors would like to acknowledge the assistance of Dr. Wendy Wu for some of the experiments involving ATP (γ)S, and to Dr. Philip John for the experiments with "inside-out" bacterial vesicles.

This research was supported by the United States Public Health Service (AM 6438), the Medical Research Council of Canada (MT 1270), and the Calouste Gulbenkian Foundation of Portugal.

REFERENCES

1. R. L. Cross, B. A. Cross, and J. H. Wang, *Biochem. Biophys. Res. Commun.*, **40**, 1155 (1970).
2. R. C. Stancliff, M. A. Williams, K. Utsumi, and L. Packer, *Arch. Biochem. Biophys.*, **131**, 629 (1969).
3. S. F. Velick, *in* "Methods in Enzymology," ed. by S. P. Colowick and N. O. Kaplan, Academic Press, New York, Vol. I, p. 401 (1955).
4. B. C. Pressman, *J. Biol. Chem.*, **232**, 967 (1958).
5. B. Hagihara and H. A. Lardy, *J. Biol. Chem.*, **235**, 889 (1960).
6. R. S. Goody and F. Eckstein, *J. Am. Chem. Soc.*, **93**, 6252 (1971).
7. R. M. C. Dawson, D. C. Elliott, W. H. Elliot, and K. M. Jones, *in* "Data for Biochemical Research," Clarendon Press, Oxford, 2nd ed., p. 531(1969).
8. C. S. Furfine and S. F. Velick, *J. Biol. Chem.*, **240**, 844 (1965).
9. T. R. Sato, J. F. Thomson, and W. F. Danforth, *Arch. Biochem. Biophys.*, **101**, 32 (1963).
10. J. B. Chappell and A. R. Crofts *in* "Regulation of Metabolic Processes in Mitochondria," ed. by J. N. Tager, S. Papa, E. Quagliarielo, and E. C. Slater, Elsevier, Amsterdam, Vol. 7, p. 293 (1966).
11. J. B. Chappell and K. N. Haarhoff, *in* " Biochemistry of Mitochondria," ed. by E. C. Slater, Z. Kaniuga and L. Wojtczak, Academic Press, London, p. 75 (1967).
12. L. Packer, K. Utsumi, and M. G. Mustafa, *Arch. Biochem. Biophys.*, **117**, 381 (1966).
13. J. A. A. Ketelaar, H. R. Gersmann, and K. Koopmans, *Rec. Trav. Chim.*, **71**, 1253 (1952).
14. R. Breslow, I. Katz, *J. Am. Chem. Soc.*, **90**, 7376 (1968).
15. P. Mushak and J. E. Coleman, *Biochemistry*, **11**, 201 (1972).
16. J. I. G. Cadogan, *J. Chem. Soc.*, 3067 (1961).
17. M. I. Kabachnik, T. A. Mastryukova, N. P. Radionova, and E. M. Popov, *Zhur. Obshchei Khim.*, **26**, 120 (1956).

18. R. G. Pearson, *J. Am. Chem. Soc.*, **85**, 3533 (1963).
19. R. G. Pearson, *Science*, **151**, 172 (1966).
20. P. Mitchell and J. Moyle, *Eur. J. Biochem.*, **9**, 149 (1969).
21. P. John and W. A. Hamilton, *FEBS Letters*, **10**, 246 (1970).

Mechanism of Ca^{2+} Transport Inhibition by Ruthenium Red and the Action of a Water-soluble Fraction of Mitochondria

Kozo Utsumi and Takuzo Oda

Department of Biochemistry, Cancer Institute, Okayama University Medical School, Okayama, Japan

Ca^{2+} and certain other bivalent cations are accumulated in mitochondria depending on an energy-linked process such as electron transport or ATP hydrolysis. Recently, it was found that $LaCl_3$ and ruthenium red inhibit mitochondrial Ca^{2+} transport specifically, and that a water-soluble factor extracted by hypotonic shock of mitochondria has a high binding affinity for Ca^{2+}. On the basis of these findings, it appeared likely that the water-soluble factor may be involved in the inhibition of Ca^{2+} transport by ruthenium red. We then studied the mechanism of inhibition of Ca^{2+} transport by ruthenium red and the effect of the water-soluble factor on this inhibition.At least a part of the specific inhibition by ruthenium red of active Ca^{2+} accumulation by mitochondria results from the inhibition of high-affinity Ca^{2+} binding. The inhibitory action of ruthenium red is decreased by the water-soluble fraction obtained from mitochondria by hypotonic shock. One of the fractions of the water-soluble fraction separated on columns of Sephadex G-200 bound ruthenium red and blocked the inhibitory activity of ruthenium red on mitochondrial Ca^{2+} transport. It is not clear whether this fraction, which contained protein, sialic acid, hexosamine and phospholipid, is identical with the water-soluble or -insoluble Ca^{2+} binding factor of Lehninger. CPC, a reagent reacting with acid polysaccharides, inhibited not only Ca^{2+} transport but also other energy transfer reactions of mitochondria.

Abbreviations used: DNP, dinitrophenol; EDTA, ethylenediaminetetraacetic acid; SDS, sodium dodecyl sulfate; TBA, 5, 5-diethyl-2-thiobarbituric acid; RR, ruthenium red; CPC, cetylpyridinium chloride; TMPD, tetramethyl-*p*-phenylenediamine.

As reported by many investigators, Ca^{2+} transport in mitochondria is a convenient system for studying the mechanism of active transport in biological membranes (*1*, *2*). Ca^{2+} and certain other bivalent cations are actively accumulated in mitochondria when coupled to an energy-linked process such as electron transport or ATP hydrolysis. This process of bivalent cation accumulation may proceed by the following 3 steps: 1) Binding of ions on the surface of the inner mitochondrial membrane. 2) Translocation of the bound ion by an energy-linked phenomenon across the membrane. 3) Release of the bound ion from the membrane into the matrix of the mitochondrion.

Thus, knowledge of the chemical nature or structure of the Ca^{2+} binding site may help to elucidate the mechanism of ion translocation across the mitochondrial membrane. In related studies, Reynafarje and Lehninger (*3*) have reported a high-affinity binding of Ca^{2+} to the mitochondrial membrane which is independent of respiration. However, the affinity of Ca^{2+} for the mitochondrial membrane is dependent on the inner membrane structure and the binding is inhibited by typical uncouplers (*e.g.*, DNP) at concentrations which uncouple energy transfer reactions.

Recently, Mela (*4*), Chance *et al.* (*5*) and Moore (*6*) found that polycations such as $LaCl_3$ and ruthenium red specifically inhibit the transport of Ca^{2+} into mitochondria. These inhibitors are thought to react with acid polysaccharides. Lehninger (*7*) also reported that a water-soluble factor extracted from mitochondria by hypotonic shock has a high affinity for Ca^{2+}. This high affinity for binding Ca^{2+} may be related to the energy-dependent accumulation of Ca^{2+} by mitochondria. On the basis of these findings it appeared likely that the water-soluble factor may be involved in the inhibition of Ca^{2+} transport by ruthenium red. In this communication the authors report on the mechanism of inhibition of Ca^{2+} transport by ruthenium red and the effect of the water-soluble fraction on this inhibition.

Rat liver mitochondria were isolated by a modification of the method of Hogeboom and Schneider using a medium containing 0.25 M sucrose, 3 mM Tris-HCl buffer (pH 7.4) in the presence or absence of 0.1–0.5 mM EDTA (*8*). Mitochondrial Ca^{2+} transport was measured by the combined use of oxygen and glass electrodes with a fluorometer or 90° light scattering apparatus (*9*). Ca^{2+} binding to the mitochondrial membrane was measured by the method of Reynafarje and Lehninger (*3*). The water-soluble fraction was extracted from rat liver mitochondria by a modification of the method of

Lehninger (*7*) and was fractionated on a columns of Sephadex G-200 or by 0.1% SDS disc gel electrophoresis. The various kinds of sugars were identified by means of the carbazole reaction (*10*) for uronic acid, direct Ehrlich (*11*) or TBA (*12*) reaction for sialic acid and the method of Rondle and Morgan (*13*) for hexosamine.

Ca^{2+} Transport in Rat Liver Mitochondria and the Effect of Ruthenium Red

It is well known that oxygen consumption is reversibly stimulated by addition of Ca^{2+} to a suspension of mitochondria in the presence of respiratory substrates. Studies on the stoichiometry of Ca^{2+} transport in the presence of different kinds of respiratory substrates showed that the Ca^{2+}/O ratios with NAD-linked substrate, succinate and ascorbate-TMPD were about 5, 3.4, and 1.7, respectively (*14, 15*). The Ca^{2+}/H^+ ratio has been measured and a value of 1 obtained by many investigators (*16*) though Azzone *et al.* (*17*) have reported a ratio of 2 under certain conditions. Figure 1 shows that Ca^{2+} transport is accompanied by the reversible oxidation and reduction of pyridine nucleotides and by changes in light scattering and it appears that Ca^{2+} transport is dependent on the formation of an H^+ gradient across the inner mitochondrial membrane, and therefore that energy is required for the maintenance of both the H^+ gradient and of the accumulated Ca^{2+} in mitochondria. Thus, when the energy supply was inhibited by addition of a respiratory inhibitor or uncoupler of oxidative phosphorylation the accumulated Ca^{2+} and the H^+ gradient were both discharged.

As reported by Mela (*4*), Chance *et al.* (*5*) and Moore (*6*) the changes in various parameters induced by Ca^{2+} transport in mitochondria were inhibited specifically by a small amount of ruthenium red or lanthanum chloride (Fig. 1). However, the inhibition of energy transfer reactions in mitochondria by ruthenium red was observed only in the case of Ca^{2+} transport and not with other energy transfer reactions such as ATP formation, reversed electron transport and other ion translocations such as the valinomycin-induced transport of K^+.

The results reported show a very important inconsistency between the concentration of ruthenium red for the inhibition of Ca^{2+}-induced respiration and that required for the inhibition of Ca^{2+} transport. About 0.03–0.1 nmoles of ruthenium red/mg of mitochondrial protein completely inhibited Ca^{2+}-stimulated respiration but 5 nmoles of ruthenium red/mg of

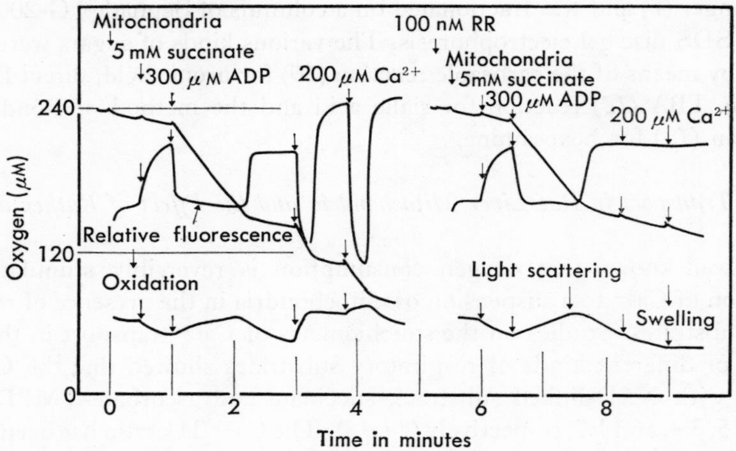

FIG. 1. Effect of ruthenium red on changes in various parameters of mitochondrial energy transfer reactions induced by ADP and Ca^{2+}. Rat liver mitochondria (1 mg protein/ml) were incubated in a medium containing 0.2 M sucrose, 5 mM phosphate buffer (pH 7.4), 3 mM $MgCl_2$, and 10 mM KCl at 25°C. Total volume of the incubation mixture was 2 ml. Oxygen consumption, oxidation, and reduction of pyridine nucleotides and 90° light scattering at 600 nm were recorded simultaneously. Other additions were as described in the figure. In the presence of 100 nM ruthenium red only Ca^{2+}-induced reversible changes in various parameters were inhibited.

mitochondrial protein was required for the inhibition of Ca^{2+} transport (*16*) (Fig. 2). The concentration of ruthenium red required for the inhibition of Ca^{2+}-stimulated respiration is almost half of the cytochrome *b* content of rat liver mitochondria, which is about 0.1 nmoles/mg protein of mitochondria. These data indicate that ruthenium red acts in 2 different ways.

Effect of CPC on the Energy Transfer Reactions of Mitochondria

Moore (*6*) and Martinez-Palomo *et al.* (*18*) believe that ruthenium red and lanthanum chloride react with acid polysaccharides and on this basis, it was predicted that an acid polysaccharide might be the site of Ca^{2+} transport in mitochondria. CPC is known to react with acid polysaccharides (*19*). As shown in Fig. 3, CPC increases State 4 respiration and decreases respiratory control and reversed electron transport activities. Thus, CPC at high con-

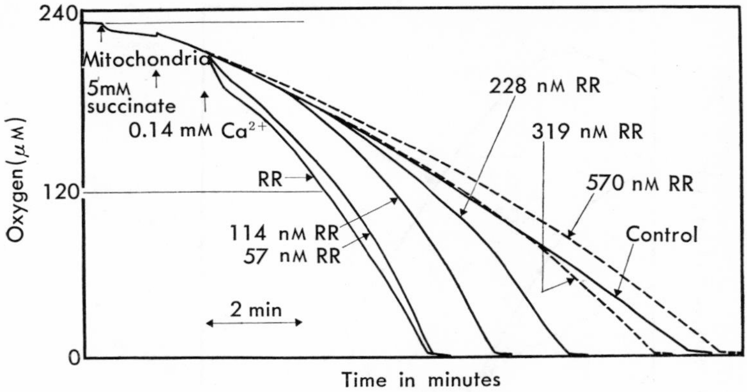

FIG. 2. Effect of various concentrations of ruthenium red on the Ca^{2+}-induced respiration of mitochondria. Incubation conditions were as in Fig. 1 except that mitochondrial protein was 2 mg/ml and the total incubation mixture was 3.5 ml.

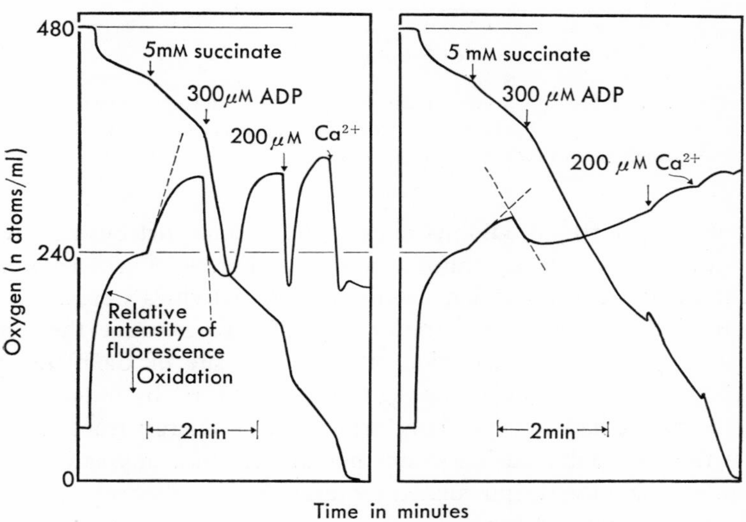

FIG. 3. Effect of CPC on ATP formation and Ca^{2+} transport in mitochondria. Incubation conditions were as in Fig. 1. 20 μM CPC decreased respiratory control and the reversible oxidation and reduction of pyridine nucleotides induced by ADP and Ca^{2+} in mitochondria (right figure).

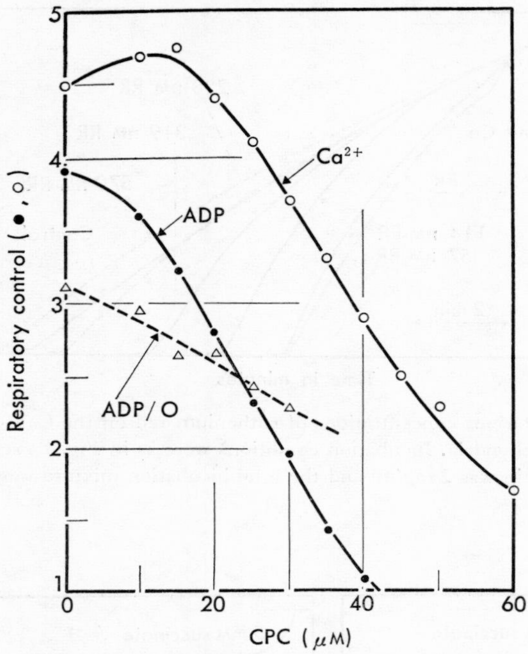

FIG. 4. Effect of various concentrations of CPC on respiratory control and on the ADP/O ratio of mitochondria. All of these reactions were inhibited by the increased concentration of CPC. Incubation conditions were as in Fig. 5.

centrations (20–50 μM) inhibits not only Ca^{2+} transport but also ATP formation and other energy transfer reactions (Fig. 4). Furthermore, Mg^{2+}-ATPase and DNP-dependent latent ATPase activities were both slightly stimulated by a high concentration of CPC. To examine the hypothesis that acid glycoproteins function at the active site of Ca^{2+} transport we studied the effect of sialidase on the transport of Ca^{2+} by mitochondria, but this treatment had no effect on mitochondrial energy transfer reactions. From the data obtained it was not possible to draw any satisfactory conclusions concerning the postulated requirement of polysaccharides for Ca^{2+} transport in mitochondria.

Effect of Ruthenium Red on the Binding of $^{45}Ca^{2+}$ to Mitochondria

As reported by Reynafarje and Lehninger (*3*), mammalian mitochondria can bind Ca^{2+} independently of respiration. There are 2 kinds of binding: One is the so-called high-affinity Ca^{2+} binding and the other is the low-affinity Ca^{2+} binding. The former is located in the inner membrane and the latter in both outer and inner mitochondrial membranes. The low-affinity binding site may apparently be represented as a nonspecific binding of Ca^{2+} to mitochondrial phospholipid, but the high-affinity binding site is considered to be the site of the Ca^{2+} carrier molecule. Therefore, as expected, ruthenium red showed a strong inhibition of binding affinity for Ca^{2+} (Fig. 5). However the concentration required for complete inhibition was very high (5 μM). This concentration of ruthenium red was about 10 times higher than that required for the complete inhibition of Ca^{2+}-induced oxygen consumption. According to the report of Vasington *et al.* (*20*), results similar to ours were obtained, but the concentration of ruthenium red required for complete inhibition of the high-affinity binding of Ca^{2+} was relatively high in comparison with our data.

The molecular mechanism of high-affinity binding of Ca^{2+} could not be

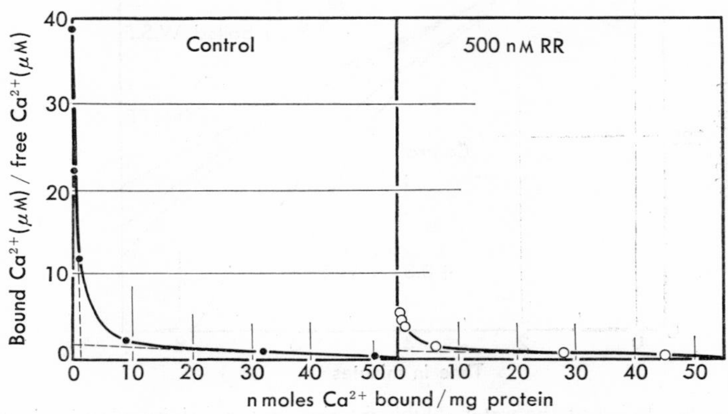

FIG. 5. Effect of ruthenium red on the binding of Ca^{2+} to the mitochondrial membrane. The reaction mixture contained 0.2 M sucrose, 10 mM KCl, 2 mM Tris-HCl buffer (pH 7.4), 0.25 μM rotenone, 0.25 μM antimycin A, mitochondria (1 mg protein/ml), and the total incubation mixture was 2 ml. The incubation was carried out at 0°C for 1 min. Ca^{2+} was measured isotopically.

elucidated, but the activity was found to be quite unstable. Uncouplers, aging or damage to the mitochondrial membrane structure decreased the binding activity. From these findings it was considered that the high-affinity binding of Ca^{2+} is closely related to the active accumulation of Ca^{2+} by a carrier or other similar substances. In opposition to this idea, some of the data do not support this hypothesis. For example, mitochondria can still accumulate Ca^{2+} actively after loss of high-affinity Ca^{2+} binding activity. Thus, with the available data it is not possible to decide whether or not the high-affinity binding of Ca^{2+} is directly related to the mechanism of active accumulation of Ca^{2+} in mitochondria (*21*).

Effect of a Water-soluble Fraction on Ruthenium Red-inhibited, Ca^{2+}-dependent Respiration of Mitochondria

According to Lehninger (7), the water-soluble Ca^{2+} binding factor obtained from mitochondria by hypotonic shock also has a high-affinity Ca^{2+} binding

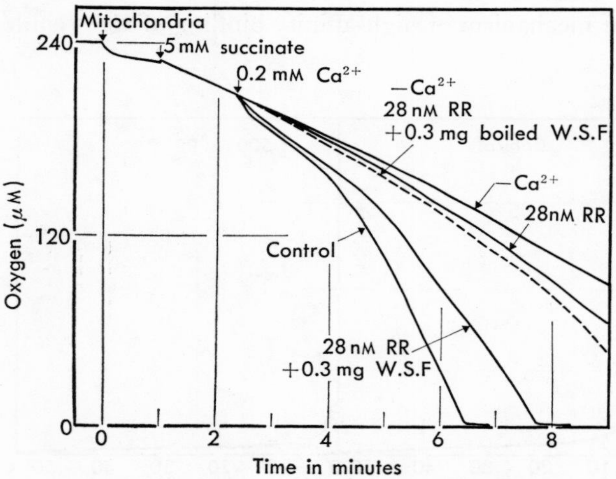

FIG. 6. Effect of the water-soluble fraction on the activity of ruthenium red on Ca^{2+} transport in mitochondria. Rat liver mitochondria (2 mg protein/ml) were incubated in a medium containing 0.2 M sucrose, and 5 mM phosphate buffer (pH 7.4) at 25°C. Total volume of the incubation mixture was 3.5 ml. The water-soluble fraction inhibited the action of ruthenium red and the boiled fraction lost all inhibitory activity. W.S.F.: Water soluble fraction.

capacity. Therefore, if the factor participates in the inhibitory action of ruthenium red, the inhibition should be decreased by addition of the factor to the incubation mixture for active mitochondrial Ca^{2+} transport. Figure 6 shows that the inhibitory action of ruthenium red on the reversible stimulation of respiration by Ca^{2+} is reduced by addition of the water-soluble fraction of mitochondria. This fraction is inactivated by heat treatment.

Character of the Water-soluble Fraction from Mitochondria

The water-soluble fraction was fractionated into 3 peaks on columns of Sephadex G-200 as shown in Fig. 7. The main component was located in the second peak of the fraction. The first peak eluted at the void volume and the second peak between hemoglobin (M.W. 68,000) and dextran blue (M.W. 2,000,000). The first peak fraction did not move when subjected to acrylamide gel electrophoresis but by using the SDS disc electrophoresis tech-

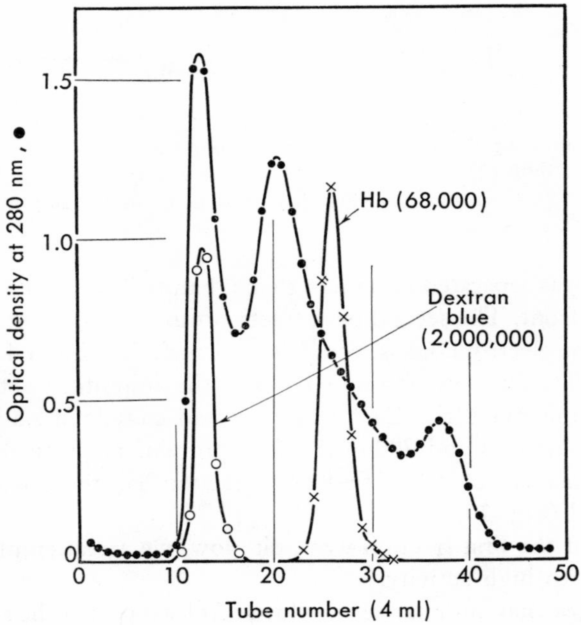

FIG. 7. Fractionation of the water-soluble fraction on columns of Sephadex G-200. The fractions were eluted with 5 mM Tris-HCl buffer (pH 7.4).

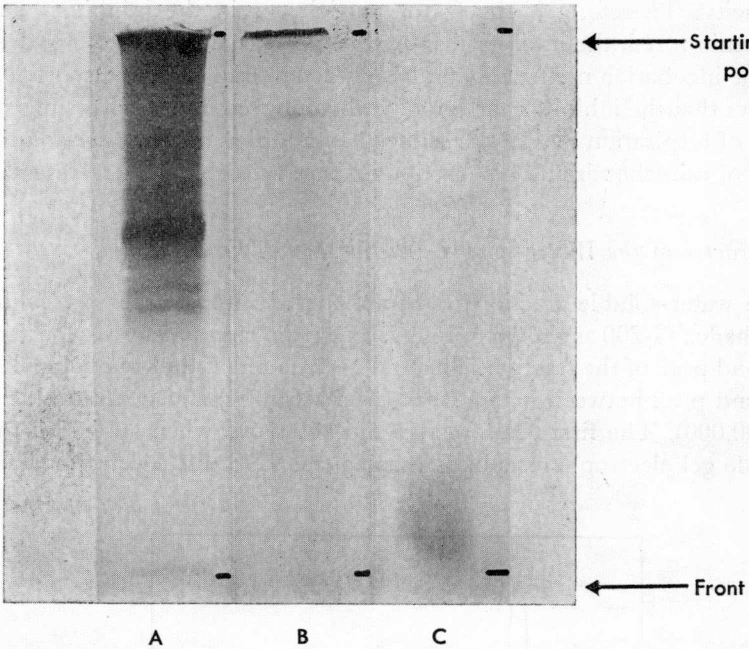

FIG. 8. Separation of the first peak fraction from Sephadex G-200 column chromatography by 0.1% SDS disc electrophoresis. The total (A) and the first fraction (B) in 7% acrylamide gel. The first fraction (C) in 4% acrylamide gel containing 0.1% SDS.

nique, the fraction was separated into more than 6 bands (Fig. 8). The main band moved to the front. The second peak fraction was separated into more than 5 bands by disc electrophoresis, comprising 1 main band and 4 minor components (Fig. 9). The first fraction blocked the inhibitory action of ruthenium red on mitochondrial Ca^{2+} transport and contained sialic acid and hexosamine (56 nmoles and 200 nmoles/mg protein, respectively) and phospholipid. As was expected, the first peak fraction had the capacity to bind ruthenium red (Fig. 10).

As the activity of the first fraction was quite low, we are attempting to obtain a fraction with high activity.

From these data we may summarize as follows: At least part of the specific inhibition by ruthenium red of active Ca^{2+} accumulation by mitochondria results from the inhibition of high-affinity Ca^{2+} binding. The inhibitory ac-

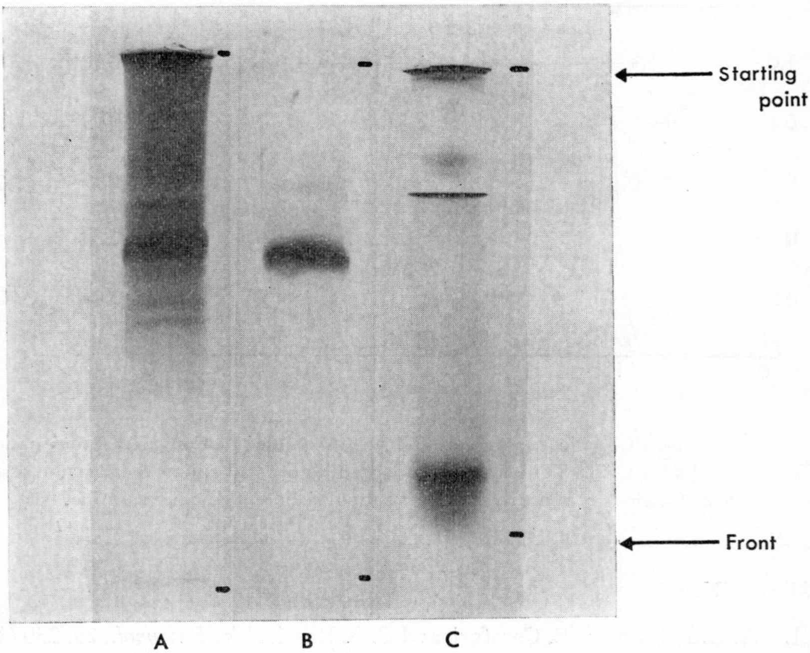

FIG. 9. Separation of the second peak fraction from Sephadex G-200 column chromatography by 0.1% SDS disc electrophoresis. The total (A) and the second fraction (B) in 7% acrylamide gel. The second fraction (C) in 4% acrylamide gel containing 0.1% SDS.

tion of ruthenium red is decreased by the water-soluble fraction extracted from mitochondria by hypotonic shock by the modified method of Lehninger. One of the fractions of the water-soluble factor fractionated on columns of Sephadex G-200 bound ruthenium red and blocked the inhibitory activity of ruthenium red on mitochondrial Ca^{2+} transport. This fraction contained protein, sugars and phospholipid. CPC, a reagent reacting with acid polysaccharides, inhibited not only Ca^{2+} transport but also other energy transfer reactions of mitochondria. However, no conclusions could be drawn concerning the postulated requirement of polysaccharides for Ca^{2+} transport in mitochondria.

In addition, after preparing this article we received information regarding research by Gazzoti *et al.* (*22*) and Gomez-Puyou *et al.* (*23*), which is quite similar to our experiments.

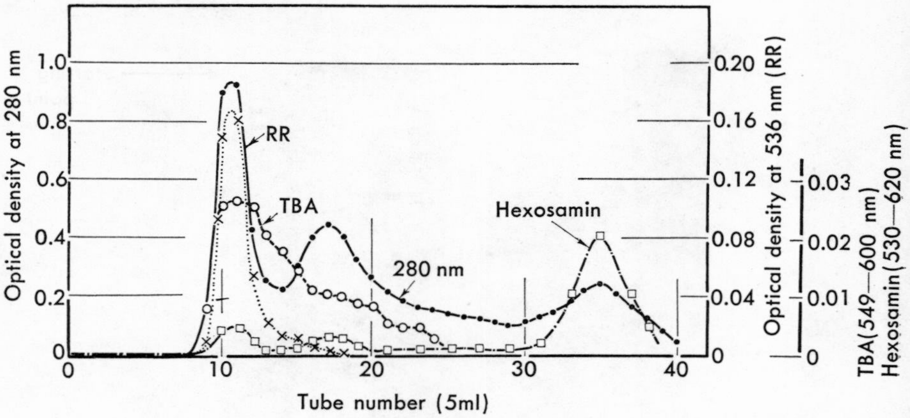

FIG. 10. Chemical nature of the water-soluble: raction after fractionation on Sephadex G-200. The first peak fraction bound ruthenium red. First and second peak fractions contained a small amount of sialic acid and hexosamine.

REFERENCES

1. A. L. Lehninger, E. Carafori, and C. S. Rossi, *Adv. Enzymol.*, **29**, 259 (1967).
2. A. L. Lehninger, *Biochem. J.*, **119**, 129 (1970).
3. B. Reynafarje and A. Lehninger, *J. Biol. Chem.*, **244**, 584 (1969).
4. L. Mela, *Arch. Biochem. Biophys.*, **123**, 286 (1968).
5. B. Chance, A. Azzi, and L. Mela, *in* "Molecular Basis of Membrane function," ed. by G. Tosteson, Prentice Hall, New York, p. 561 (1969).
6. C. L. Moore, *Biochem. Biophys. Res. Commun.*, **42**, 298 (1971).
7. A. L. Lehninger, *Biochem. Biophys. Res. Commun.*, **42**, 312 (1971).
8. K. Utsumi, *Acta Med. Okayama*, **17**, 258 (1963).
9. K. Utsumi, and G. Yamamoto, *Acta Med. Okayama*, **18**, 111 (1964).
10. Z. Dische, *J. Biol. Chem.*, **167**, 189 (1947).
11. I. Werner and L. Odin, *Acta Soc. Med. Upsalienis*, **57**, 230 (1952).
12. L. Warren, *J. Biol. Chem.*, **234**, 1971 (1959).
13. C. J. M. Rondle and W. T. J. Morgan, *Biochem. J.*, **61**, 576 (1955).
14. C. S. Rossi and A. L. Lehninger, *Biochem. Biophys. Res. Commun.*, **11**, 441 (1963).
15. C. S. Rossi and A. L. Lehninger, *J. Biol. Chem.*, **239**, 3971 (1964).
16. C. S. Rossi, J. Bielawaki, and A. L. Lehninger, *J. Biol. Chem.*, **241**, 1919 (1967).
17. C. S. Rossi, A. Azzi, and G. F. Azzone, *Biochem. J.*, **100**, 4C (1967).

18. A. Martinez-Palomo, C. Braislowsky, and Bernhard, *Cancer Res.*, **29**, 925 (1969).
19. I. E. Scott, *Method Biochem. Anal.*, **8**, 145 (1960).
20. F. D. Vasington, P. Gazzoti, R. Tiozzo, and E. Carafori, *Biochim. Biophys. Acta*, **256**, 43 (1972).
21. E. Carafoli, P. Gazzotti, C. S. Rossi, and R. Tiozzo, *in* "Membrane-Bound Enzymes," ed. by G. Porcellati and F. Dijeso, Plenum Press, New York, p. 63 (1971).
22. P. Gazzotti, F. D. Vasington, and E. Carafoli, *Biochem. Biophys. Res. Commun.*, **47**, 808 (1972).
23. A. Gomez-Puyou, M. Gomez-Puyou, G. Becker, and A. L. Lehninger, *Biochem. Biophys. Res. Commun.*, **47**, 814 (1972).

Adenine Nucleotide Control of Heart Mitochondrial Oscillations

Lester Packer and Van D. Gooch

Department of Physiology-Anatomy, University of California, Berkeley, California, U.S.A.

The phenomenon of oscillatory transport of ions in mitochondria is valuable for elucidating aspects of mitochondrial temporal organization, for investigating possible synchronizing mechanisms between mitochondria, and as a sensitive measure of their functional integrity. The operation and participation of the factors which control heart mitochondrial oscillations were studied in this system. The degree of K^+ permeability, as adjusted by the valinomycin concentration, primarily influences the period length. A requirement of ADP, ATP, or 5′AMP is necessary for sustaining the oscillations. Inhibition of the oscillations to one rebound can occur by inhibiting the adenine nucleotide exchange carrier using atractyloside, or by inhibiting the reversible ATPase using oligomycin. Total inhibition occurs if the phosphate carrier is inhibited using the uncoupler 2,4-dinitrophenol (DNP). These experiments indicate that not only the transport of the ions of K^+, H^+, and phosphate but also a functioning reversible ATPase are required and controls heart mitochondrial oscillations.

The ATP synthetase-ATPase system of mitochondria is poised and operates in a manner unlike that in other membrane systems, such as the (Na^+-K^+)-ATPase of the plasma membrane or the (Ca^{2+}-Mg^{2+})-ATPase system of the sarcoplasmic reticulum or the myosin ATPase system. The special features of this system are that it is poised in the direction primarily of ATP synthesis and there is an exchange of adenine nucleotides in a dynamic fashion across the inner mitochondrial membrane, which affords a mechanism whereby energy supply can be regulated in the cell. Thus, the mitochondria distributed throughout the cell provide a means for the energetic communication between cellular compartments and membrane systems on the basis of exchanges of adenine nucleotides and other substances. The special mechanisms for synthesis and exchange of adenine nucleotides are thus of crucial importance in understanding mechanisms of energy transduction and of metabolic control.

The phenomenon of the oscillatory transport of ions in mitochondria is a valuable system for elucidating aspects of mitochondrial temporal organization, for investigating possible synchronizing mechanisms between mitochondria, and as a sensitive measure of their functional integrity in the investigation. The specific system used in this study was first developed for heart mitochondria by Chance and Yoshioka (*1*). It has several significant features; it involves the cyclic transport of H^+ and K^+—both

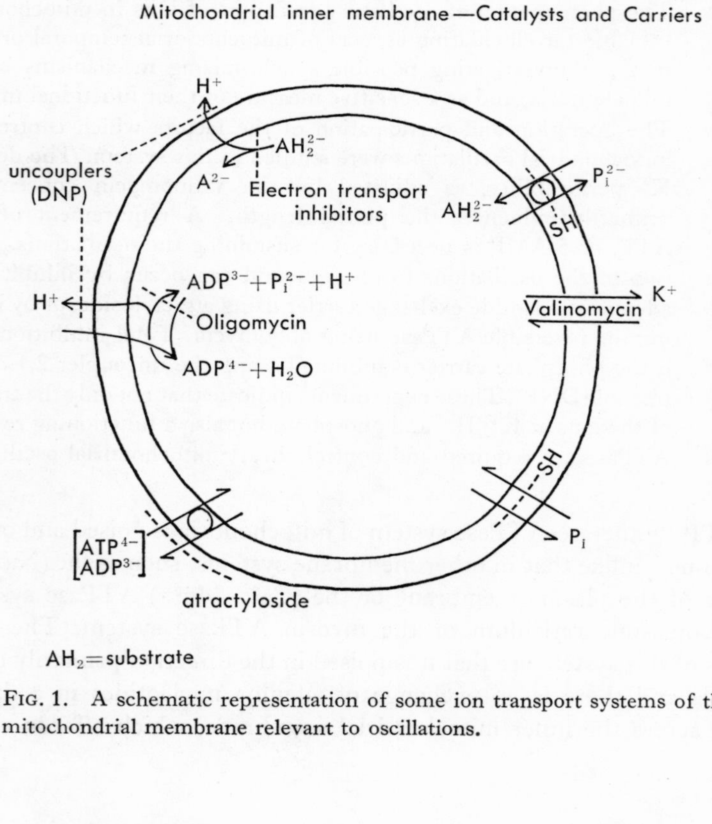

FIG. 1. A schematic representation of some ion transport systems of the inner mitochondrial membrane relevant to oscillations.

parameters can be easily measured and quantified (*cf.* Ref. *2*). Further, this system has very little damping compared to other mitochondrial systems.

Some of the transport systems that we have investigated in mitochondrial inner membranes are depicted schematically in Fig. 1, where the following systems are demonstrated:

Metabolite transport: An energy-dependent mechanism for the transport of substrate anions across the mitochondrial membrane is known (*3*). This carrier-mediated mechanism requires catalytic amounts of phosphate and is inhibited by mercurial reagents which react with SH groups.

The electron transport-dependent production of H^+ outside of the inner membrane: This process is decreased by inhibitors of electron transport or by uncoupling agents which facilitate the equilibration of H^+ across the membranes, an example being 2, 4-dinitrophenol (DNP).

Monovalent cation exchange: Under certain conditions, H^+ gradients can be exchanged for other cations such as K^+. Such a condition occurs in the presence of valinomycin, an antibiotic which induces K^+ transport (*4*).

Phosphorylation mechanisms: The ATP synthetase-ATPase complex is a protein of high molecular weight of around 355,000 (*5*) and is composed of nine subunits. It is involved in the reversible synthesis of ATP. The mechanism of ATP synthesis depends upon the presence of an H^+ gradient, phosphate, and ADP. On the other hand, ATP hydrolysis generates H^+ gradients. The antibiotic oligomycin inhibits ATP synthesis and ATPase activity. The action of oligomycin is thought to involve a phospholipid requirement and/or the association of the enzyme complex with the membrane.

Phosphate transport mechanisms: The transport of inoganic phosphate across the membrane is energy dependent. The phosphate carrier requires sulfhydryl groups and can be inhibited by mercurials.

Adenine nucleotide exchange mechanisms: The adenine nucleotide carrier involves a one-to-one exchange between ATP and ADP. This exchange is specifically blocked by the antibiotic atractyloside which prevents adenine nucleotide binding on the outer surface of the inner mitochondrial membrane.

Evidence in support of operation and participation of these systems in the control of heart mitochondrial oscillations is presented below. Oscillations of heart mitochondria were analyzed by measuring changes in transmittance, H^+, K^+, respiration rate (and morphology, which is not

Homogenize tissue
(0.25 M sucrose, 50 mM EDTA, Nagarse (2 mg/g), pH 7.6)

↓ 750 × g, 10 min

Supernatant | Pellet

↓ 8,000 × g, 10 min

Supernatant | Resuspend pellet (0.25 M sucrose)

CHART 1. Procedures for the isolation of heart mitochondria and assay medium for oscillation. The test medium was maintained at 25°C during experimentation; the concentration of mitochondria was 0.8 mg/ml. Oscillation medium: 0.25 M sucrose, 2.7 mM KCl, 2.0 mM K_2PO_4, 2 mM glutamate, 2 mM malate, pH 6.25 using Tris base.

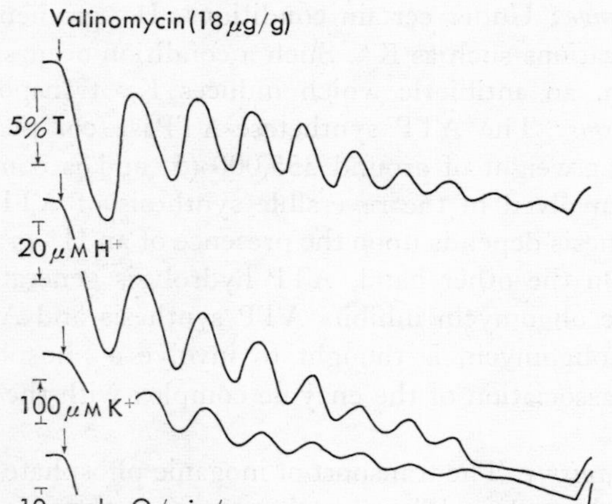

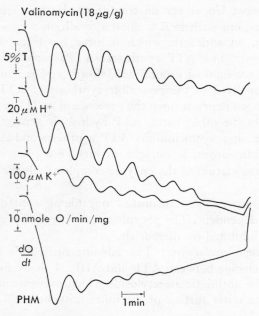

FIG. 2. Oscillation parameters of heart mitochondria. Oscillations of pigeon heart mitochondria (PHM) were measured by changes of transmittance (T), H^+, and K^+ in the medium using glass electrodes and changes of respiration rate (dO/dt) by differentiating a signal of an oxygen electrode. The conditions are as specified in Chart 1 with the addition of 30 μM ADP.

shown here). The transmittance changes are indicative of mitochondrial volume changes; an increase of transmittance is a display of swelling, and a decrease corresponds to contraction. The methods used for isolation of heart mitochondria and the test medium for oscillations are shown in Chart 1, which gives the procedure for PHM. This procedure is substantially the same for beef heart mitochondria (BHM). The results obtained with either PHM or BHM reported in this investigation are virtually identical.

Induction of oscillations and the testing of various oscillation parameters in heart mitochondria are shown in Fig. 2. Oscillations were induced by the addition of valinomycin which initiates oscillatory movement of K^+ and H^+ as well as mitochondrial volume, which is the osmotic response to the

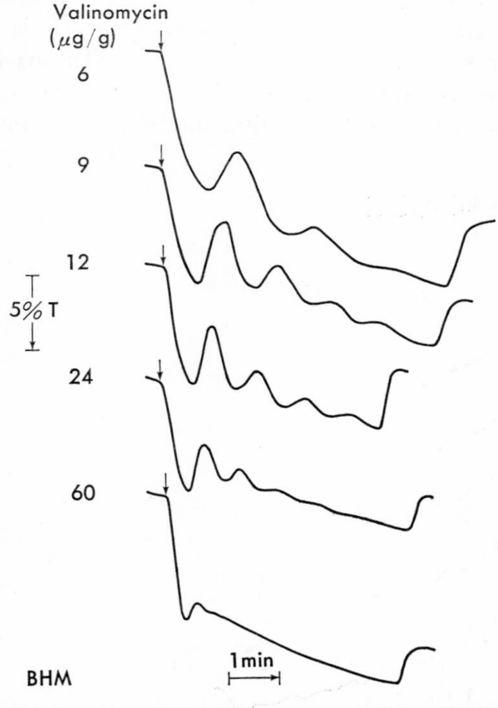

FIG. 3. Effect of various concentrations of valinomycin on the oscillations of BHM. The conditions are as specified in Chart 1 with the addition of 30 μM ADP.

net influx or efflux of substances. Ion transport parameters reflect the coupling of energy of electron transport. This is shown by the oscillation of oxygen uptake, which is particularly evident in differential recordings showing the rate of oxygen consumption. The oscillation is sustained for many minutes, eventually reaching a steady state. When the oxygen supply of the system is exhausted, oxygen uptake ceases, the established gradients of ions collapse, and contraction occurs.

The effect of various concentrations of valinomycin is shown in Fig. 3. It can be seen that the optimal concentration of valinomycin under the basic conditions shown in Chart 1 is about 15 μg/g. These experiments indicate that the oscillation of K^+ is a primary perturbation which initiates the oscillations in this system.

However, it has been found that the concentration of ADP level is critical for preventing the damping and sustaining the heart mitochondrial oscillations. A test for the effects of various ADP concentrations is shown in Fig. 4, where it can be seen that sustaining of the oscillations is best brought about using a concentration over 20 μM. The specificity of adenine nucleotide dependence for sustaining heart mitochondrial oscillations was

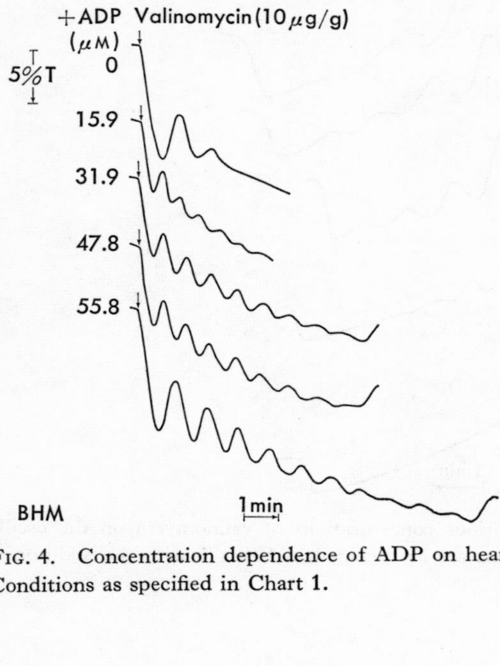

FIG. 4. Concentration dependence of ADP on heart mitochondrial oscillations. Conditions as specified in Chart 1.

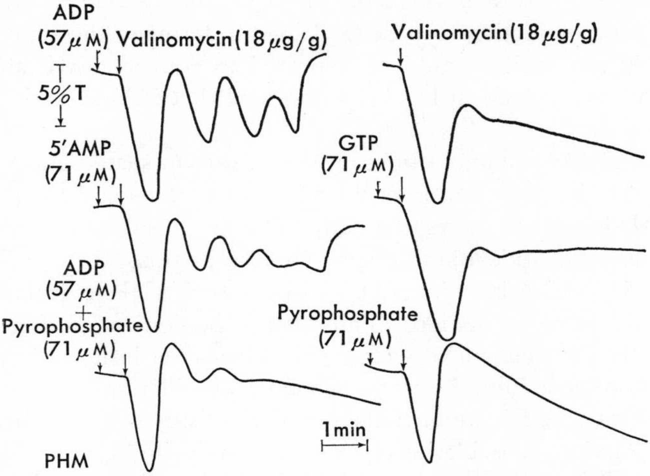

FIG. 5. Adenine nucleotide specificity of oscillations of heart mitochondria.

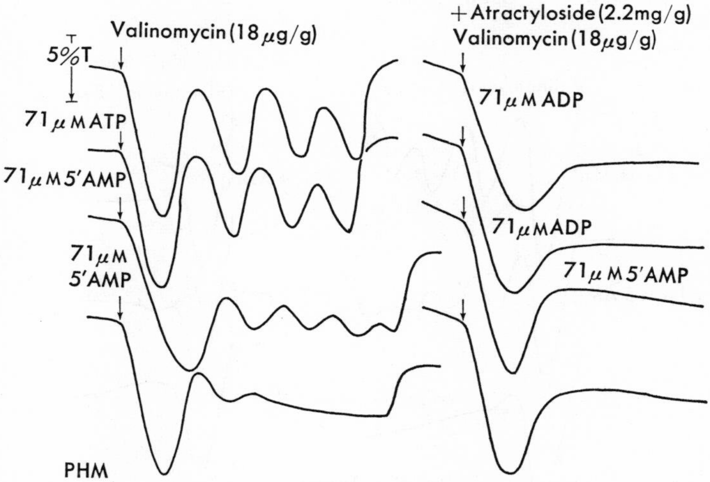

FIG. 6. Influence of atractyloside on the stimulating effects induced by ADP, ATP, and 5'AMP.

further investigated. It is shown in Fig. 5 that among the various nucleotides and pyrophosphate, ADP is the most effective. The action of AMP and ATP in inducing oscillations can be explained in terms of the action of the enzyme myokinase which would be expected to convert AMP and ATP into ADP. For this reason, it has been observed that 5′AMP and ATP act almost equivalently.

The action of adenine nucleotides on the long-term sustaining of oscillations appears to depend upon phosphorylation reactions and the transport of phosphate. Figure 6 shows that atractyloside, a substance which inhibits adenine nucleotide binding, prevents oscillations. Figure 7 shows that oligomycin, which inhibits ATP synthesis and ADP hydrolysis, likewise is inhibitory for oscillations. Thus, it can be concluded that the exchange and phosphorylation of adenine nucleotides is intimately involved in the process of sustaining heart mitochondrial oscillations.

An important role for anion transport mechanisms is also indicated by the inhibitory action on oscillations by mersalyl as shown in Fig. 8. Mersalyl, under the conditions shown, clearly inhibits the oscillations when added before, after, or during induction of the oscillatory state. The action of mersalyl can be interpreted as affecting the phosphate transport system.

The dependence of oscillations upon H^+ gradients is indicated by the

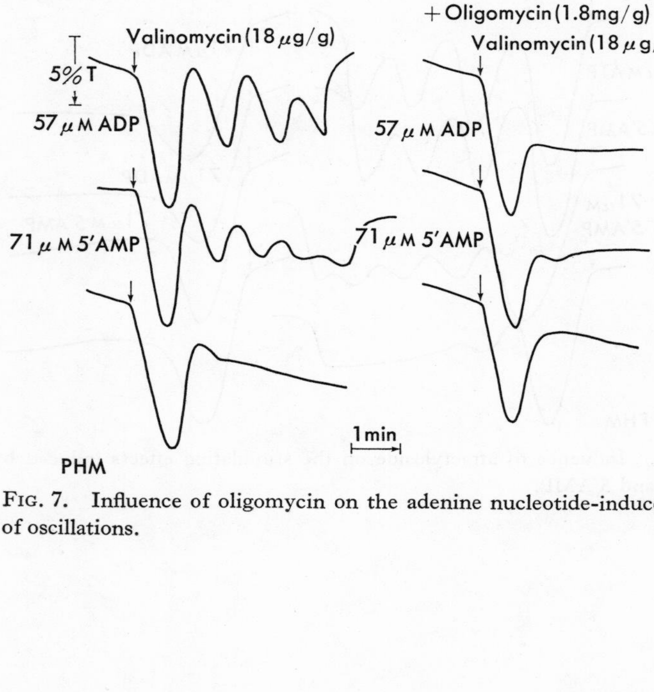

FIG. 7. Influence of oligomycin on the adenine nucleotide-induced stimulation of oscillations.

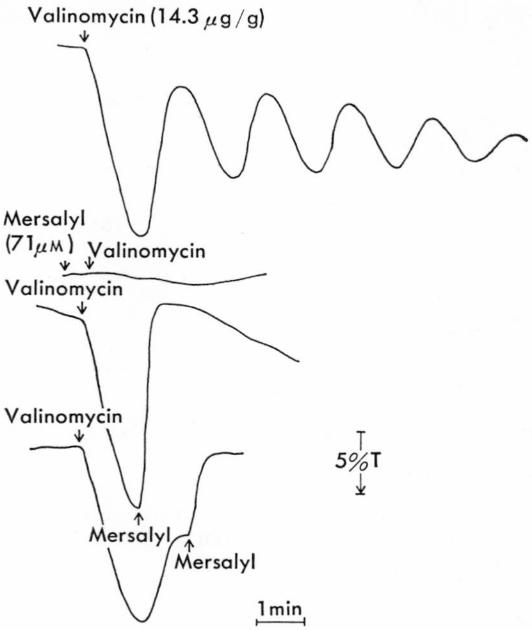

FIG. 8. Inhibition of oscillations by mersalyl. Conditions are as specified in Chart 1 with the addition of 30 μM ADP.

inhibitory effect of uncouplers such as dinitrophenol upon the oscillatory state. 2, 4-Dinitrophenol is known to cause the rapid release of H^+ gradients across mitochondrial membranes (*6*). Figure 9 shows that low concentrations of DNP prevent the oscillatory state.

One of the most important problems revealed by this study is the synchronizing action which ADP and ATP manifest upon the oscillations. The adenine nucleotides thus seem to manifest a synchronizing as well as a controlling factor in the flows of energy in this system. For this reason, it was of interest to test the effect of adding ADP at various times during the oscillatory state. Figure 10 shows the controlling influence of ADP. The addition of this nucleotide at different times during the oscillation always causes a contraction relative to the response without ADP. This response may be commensurate with two possibilities: First, in the presence of ADP, energy from the electron transport chain preferably supports phosphorylation, thus allowing ionic gradients that were being maintained

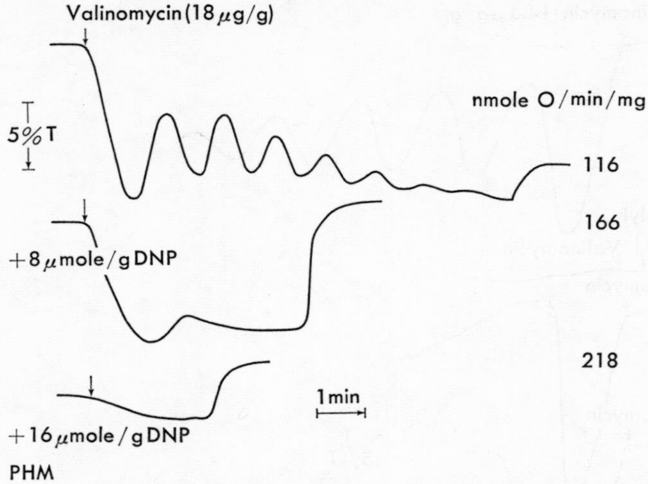

FIG. 9. Influence of the uncoupler DNP on oscillations. Conditions are as specified in Chart 1 with the addition of 30 μM ADP.

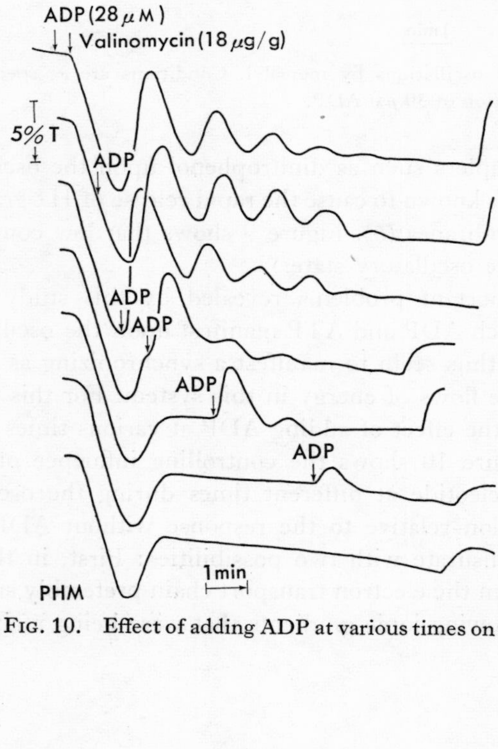

FIG. 10. Effect of adding ADP at various times on the oscillations.

by electron transport to collapse, which would correspond to the observed transmission change shown in Fig. 10. Another possibility involves the known fact that such an ionic gradient in itself is a storage of energy. It has been shown (*7*) that in the presence of ADP such a gradient can collapse, yielding its energy to the formation of ATP. This phenomenon has been reported in a number of different membrane ATPase systems, *viz.* the red blood cell membrane in Glynn's laboratory (*8*), the sarcoplasmic reticulum in Tonomura's (*9*), Deamer and Baskin's (*10*) and Hasselbach's laboratories (*11*). Perhaps both mechanisms play a role. Regardless, it would appear that a major conclusion from this investigation is that since the damping seen in most of the mitochondrial oscillatory systems may be due to the loss of synchrony between individual oscillating mitochondria, it is significant that, in this investigation, there seems to be a strong synchronizing effect manifested by appropriate concentrations of ADP and ATP in the medium. Future studies are being directed toward examining the importance of the interaction and exchange of nucleotides between mitochondria of the population. These studies point to an important regulating and controlling influence—the ability of small molecules such as ATP and ADP to act as a communication system between individual mitochondria, thus synchronizing their behavior. This may have great significance in the cell where concentrations of mitochondria are much higher than those used in *in vitro* experiments.

Acknowledgments

The work reported in this paper was supported by grants from the U.S. Public Health Service (AM-06438-09, S-T 01GM-00829) and the National Science Foundation (GB-20951).

REFERENCES

1. B. Chance and T. Yoshioka, *Arch. Biochem. Biophys.*, **117**, 451 (1966).
2. V. D. Gooch and L. Packer, *Biochim. Biophys. Acta*, **245**, 17 (1971).
3. S. Papa, N. E. Lofrumento, D. Kanduc, G. Paradies, and E. Quagliariello, *Eur. J. Biochem.*, **22**, 134 (1971).
4. E. J. Harris, R. Cockrell, and B. C. Pressman, *Biochem. J.*, **99**, 200 (1966).
5. R. F. Fisher, E. N. Moudrianakis, and D. R. Sanadi, *Fed. Proc.*, **31**, 416 (1972).

6. V. P. Skulachev, A. A. Sharaf, L. S. Yagujzinsky, A. A. Sasaitis, E. A. Liberman, and V. P. Topali, *Curr. Mod. Biol.*, **2**, 98 (1968).
7. R. A. Reid, J. Moyle, and P. Mitchell, *Nature*, **212**, 259 (1966).
8. I. M. Glynn, *Nature*, **216**, 1318 (1967).
9. Y. Tonomura, U. S.-Japan Seminar on Organization of Energy-transducing Membranes, Tokyo, Japan, May (1972).
10. D. W. Deamer and R. J. Baskin, 16th Ann. Meeting of the Biophysical Soc., Toronto, Canada, Feb. (1972).
11. M. Makinose and W. Hasselbach, *FEBS Letters*, **12**, 269 (1971).

Structural and Energy States of Photosynthetic Membranes in Relation to Proton and Cation Gradients

Satoru Murakami

Department of Biology, University of Tokyo, Tokyo, Japan

Under the conditions favorable for electron transport which supports photophosphorylation the chloroplasts display reversible shrinkage, accompanying decrease both in inner space and membrane thickness of the thylakoids. Similar modification of the membrane structure can be induced by proton titration of the chloroplast membrane in the dark. Protonation concept has been proposed to interpret the sequence of observed changes in membrane conformation. Protonation of negatively charged groups of membrane molecules, such as β and γ-carboxyl, secondary phosphate and imidazole leads the membrane to contraction. When 2,4-dinitrophenol (DNP) and valinomycin, which affect, respectively, light-induced H^+ uptake and K^+ translocation, are contained together in the medium, both H^+ and cation gradients are abolished simultaneously, producing swelling and full uncoupling. Differential and synergistic effects of these two chemicals can also be observed on other parameters of structural and energy states of the thylakoids, including electron transport, delayed light emission, 8-anilinonaphthalene-1-sulfonate (ANS) fluorescence and structure as revealed by electron microscopy. The results suggest that H^+ and cation gradients control complementarily the structural and energy states of the thylakoid membranes.

For several years evidence has accumulated which shows that structural and energy states of the inner chloroplast membrane system, the thylakoids, are intimately related to, and are controled by photosynthetic electron transport, phosphorylation, and by the proton and ionic environment of the medium. Isolated chloroplasts manifest H^+ uptake upon illumination (*1*, *2*), accompanied by efflux of K^+ and other cations (*3*, *4*) or influx of Cl^- (*5*) to maintain ionic balance. Under photosynthetic conditions which are favorable for ATP formation and H^+ uptake, the chloroplasts exhibit a reversible shrinkage-swelling cycle (*6–13*). The time courses of light-induced H^+ uptake and of structural changes monitored by 90° light scattering proceed in a similar fashion (*10*). An artificial pH change in the incubation medium also induces structural changes of the chloroplasts in the dark (*10*, *13–15*). If the medium is acidified with a small amount of HCl, the light scattering level of the chloroplasts is increased, and the thylakoids become flattened and tightly packed together. These changes are reversed by alkalinization of the medium. These studies suggest that the H^+ uptake by chloroplasts might be a cause of structural changes in the thylakoids. Accordingly, the present investigation was undertaken to determine the site of H^+ uptake by chloroplasts, and to elucidate the sequence of steps by which a change in the H^+ gradient alters the conformation of the thylakoid membranes (*13*, *16*, *17*).

Protonation and Conformation of Thylakoid Membranes

Illumination of isolated spinach chloroplasts by red actinic light (>650 nm) in the presence of weak acid anions such as acetate causes an increase in 90° light scattering and a decrease in transmission (Fig. 1). Electron microscopic examination revealed that the observed changes in the photometric parameters are always connected with structural changes of the inner membrane system. Upon illumination the inner membrane system becomes stretched, flattened and very tightly packed together, leading to shrinkage of the chloroplasts (Figs. 7 and 8). These changes in membrane structure include 2 phases; gross structural changes caused by a flattening of the thylakoids with a corresponding decrease in the spacing between membranes (configurational change), and reduction in thickness or contraction of the thylakoid membranes (conformational change). Comparison of the time courses of changes in membrane structure and of the photometric parameters disclosed that the change in light scattering is intimately related to the conformational changes in the membranes, and that the spacing or configurational change closely follows the changes in transmission, which are known to correlate very clearly with osmotic volume change. The initial

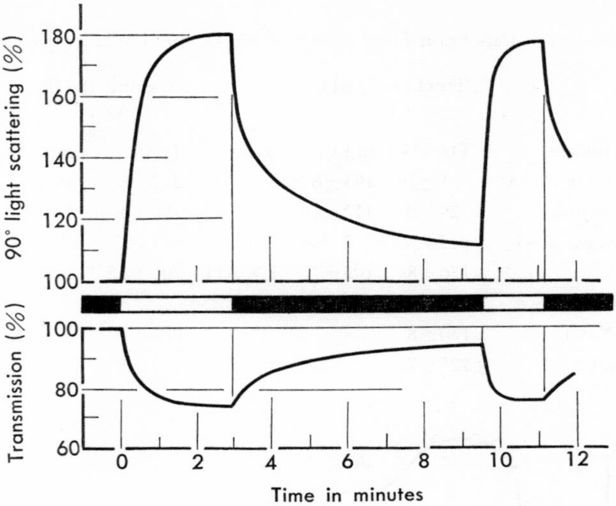

FIG. 1. Light-induced 90° light scattering and transmission changes of isolated spinach chloroplasts accompanying photoshrinkage in a sodium acetate medium. Chloroplasts were isolated in 50 mM Tris-HCl, pH 8.0, containing 175 mM NaCl, and suspended in 150 mM sodium acetate, pH 6.7, plus 15 μM phenazine methosulfate (PMS). The measuring light was filtered at 546 nm with an interference filter (*13*).

osmolarity of the medium in which chloroplasts are incubated is significant in determining the magnitude of the spacing changes. Osmotically contracted chloroplasts exhibit only the spacing change but no change in thickness. Hence it is very likely that the changes in chloroplast volume and configurational changes of the thylakoids are osmotic events which are controlled both by ionic relations inside and outside the thylakoids and by the conformational state of the membrane itself.

Contraction of thylakoid membranes has been studied extensively by various means, including 90° light scattering and transmission measurements, high-resolution electron microscopy and by means of the 8-anilinonaphthalene-1-sulfonate (ANS) fluorescence probe combined with actinic illumination and artificial pH changes of the medium to induce structural changes. An analysis by microdensitometry of electron microscope negatives clearly revealed a 15% reversible decrease in membrane thickness upon illumination in sodium acetate. This decrease was observed under all conditions which induced spacing changes, with the notable

TABLE I
Effect of Incubation Conditions upon Dimensions of the Grana Thylakoid Membrane

Conditions	Thickness (SD) (Å)			Spacing (SD) (Å)		
Light-induced shrinkage	Dark →	light →	dark	Dark →	light →	dark
175 mM NaCl-15 μM PMA	131±10	104±6	130±9	212±8	144±9	214±4
150 mM sodium acetate	129±9	112±10		196±4	144±3	
pH-induced shrinkage (dark)	pH 7.7→	4.7 →	7.3	pH 7.7→	4.7 →	7.3
100 mM NaCl	136±8	105±7	138±11	206±15	143±6	236±24
Osmotic shrinkage						
0.05 M sucrose-NaCl	130±8			252±26		
0.5 M sucrose-NaCl	125±7			172±5		

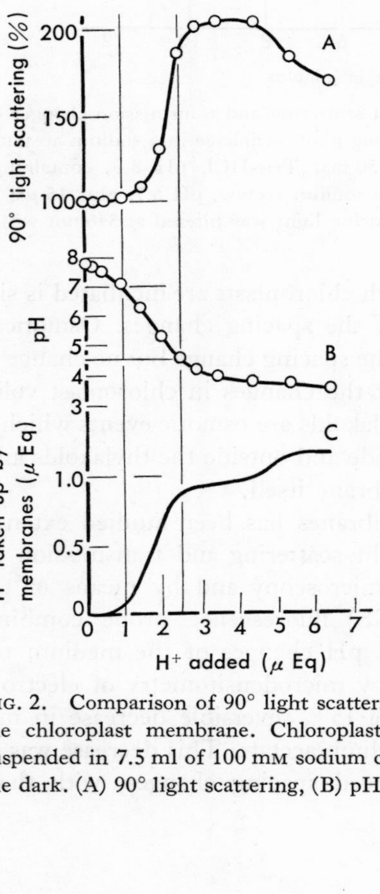

FIG. 2. Comparison of 90° light scattering change with pH and protonation of the chloroplast membrane. Chloroplasts containing 200 μg chlorophyll were suspended in 7.5 ml of 100 mM sodium chloride and titrated with 0.05 N HCl in the dark. (A) 90° light scattering, (B) pH changes, (C) protonation (*13*).

exception of osmotic treatment (Table I). It is of great interest to note that artificial pH changes in the medium caused a 23% decrease in the thickness of thylakoid membranes. When chloroplasts in 100 mM NaCl were titrated in the dark with 0.05 N HCl, light scattering increased on lowering the pH below 6, reached a maximum at pH 4, and then decreased abruptly beyond pH 3.8 (Fig. 2). The increase in light scattering induced by acidification of the suspension was reversed by realkalinization with 0.05 N NaOH. The membrane system has considerable buffering capacity in the pH range between 7 and 4.5. The amount of H^+ taken up was 4.5 μ equivalent per mg chlorophyll. Acidification from pH 7.7 to 4.7 resulted in a 23%

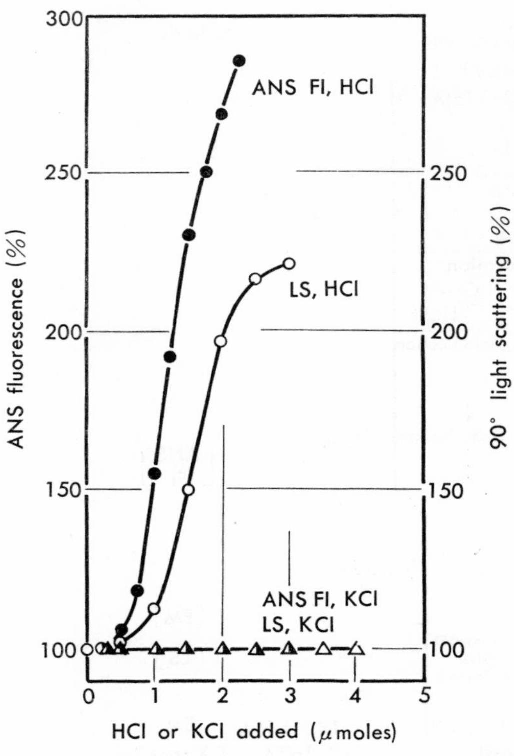

FIG. 3. Effect of H^+ and K^+ on 90° light scattering and ANS fluorescence intensity of chloroplasts. Spinach chloroplasts were isolated in 100 mM sodium chloride, and 90° light scattering (LS) and ANS fluorescence (ANS Fl) were titrated with 0.05 N HCl or 0.05 N KCl (*13*).

decrease in membrane thickness, which was reversed by back titration from pH 4.7 to 7.3. When ANS is added to the chloroplasts, fluorescence emitted by ANS is enhanced by binding to the membrane. Upon either illumination or acidification, a further increase in fluorescence intensity was observed, indicating changes in the number of binding sites for ANS and/or in the microenvironment within the thylakoid membranes in which ANS molecules are embedded (Fig. 3). The kinetics of light-induced ANS fluorescence changes in the presence of weak acid anions (sodium acetate) were similar to those in a medium with strongly dissociated ions (NaCl-phenylmercuric acetate (PMA)). In both systems the speed of the ANS

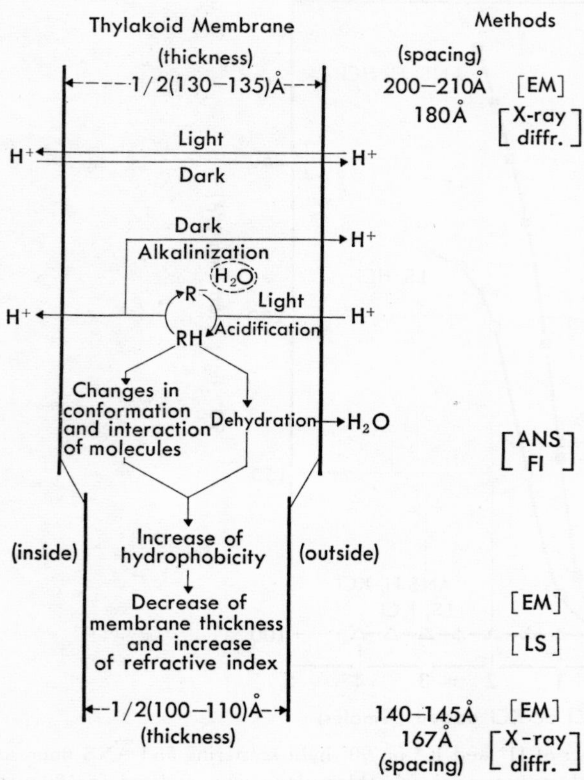

FIG. 4. Proposed mechanism for structural changes in thylakoid membrane caused by protonation (*13*).

fluorescence change is faster than the corresponding light scattering change. The half-time for ANS fluorescence changes taken from first-order plots was 14 sec in NaCl-PMA and 15 sec in sodium acetate, whereas the half-time for light scattering changes was, respectively, 60 and 33 sec. Changes in the thylakoid membranes which can be detected by ANS fluorescence changes must be more primary than those monitored by light scattering changes.

From these results a protonation model can be deduced to interprete the sequence of observed membrane conformational changes (*13*) (Fig. 4). Thus, a decrease in membrane thickness may result from protonation of negatively charged groups which causes changes in the conformation of the molecules, in molecular interactions, and in the hydration of the membrane. The values for the buffer capacity of the thylakoid membranes suggest the participation of β and γ carboxyl and imidazole groups of protein, and secondary phosphate groups of lipid molecules as the most probable protonation sites. This hypothesis provided an experimental basis for a concept of protonation of the membrane which has been also postulated by Dilley and Rothstein (*18*) and by Deamer *et al.* (*10*) as a cause of light-induced changes in chloroplast structure. Furthermore, it was proposed that the temporal sequence of events involved in the hierarchy of structural changes in thylakoid membranes after illumination is probably as follows: (a) protonation, (b) change in the environment within the membrane, (c) change in membrane thickness, (d) change in internal osmolarity of the thylakoids, with accompanying ion movements which result in the collapse and flattening of thylakoids, (e) change in the gross morphology of the inner chloroplast membrane system, and (f) change in the gross morphology of whole chloroplasts.

Uncoupling of Photophosphorylation and Membrane Structure

A thylakoid membrane has net negative charges on its surface and selective permeability to various ions. Hence, when the chloroplasts are suspended in an ionic environment, the electrochemical potential of the thylakoid membranes, which is directly correlated with the energy state and structural states, is influenced both by an electrically neutral concentration gradient of protons and by cation concentration gradients across the membrane. Some of the ionophorous antibiotics such as nigericin make the thylakoid membrane leaky to both K^+ and H^+, and simultaneously collapse the K^+

and H^+ gradients, which results in full uncoupling (*19–22*). In the presence of this uncoupler, light-induced, energy-dependent conformational changes of the thylakoid membranes are also inhibited completely. These facts suggest that the structural state of the thylakoid is supported not only by proton gradient but also by cation gradient.

1. Synergistic effect of valinomycin and DNP

When spinach chloroplasts were suspended in potassium chloride, a high concentration of 2, 4-dinitrophenol (DNP) alone reduced light-induced H^+ uptake to some extent, but did not uncouple photophosphorylation. A proton gradient established upon illumination was dissipated in part by DNP. Valinomycin at low concentration is known to stimulate the dark permeability of the thylakoid to K^+ without influence on the H^+ movement, and to cause no light-induced changes in K^+ concentration. In the presence of valinomycin chloroplasts could still manifest light-induced H^+ uptake. However, when both valinomycin and DNP were present in the medium both the proton and cation gradients were abolished simultaneously, and full uncoupling resulted (Figs. 5 and 16). This synergistic effect, which was first noted by Karlish and Avron (*23*, *24*), is the case in structural response of the thylakoids monitored by changes in light scattering (Fig. 6). Valinomycin or DNP alone was almost ineffective on the

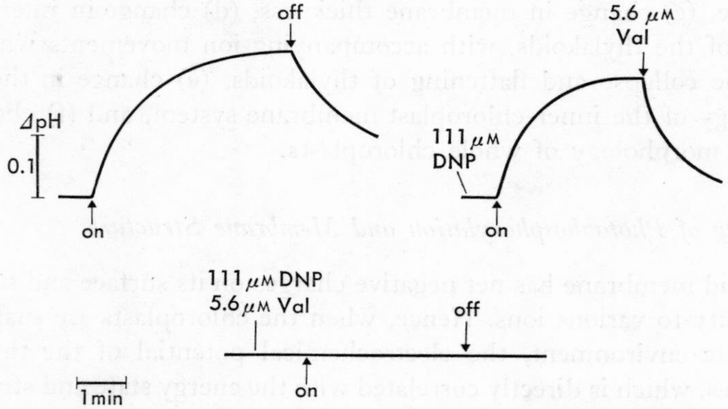

FIG. 5. Effect of valinomycin and DNP on light-induced H^+ uptake by spinach chloroplasts. Chloroplasts were isolated in 50 mM Tris-HCl, pH 7.7, containing 200 mM NaCl, washed in 200 mM KCl and suspended in 50 mM KCl plus 15 μM PMS. Initial pH of the suspension was 6.6.

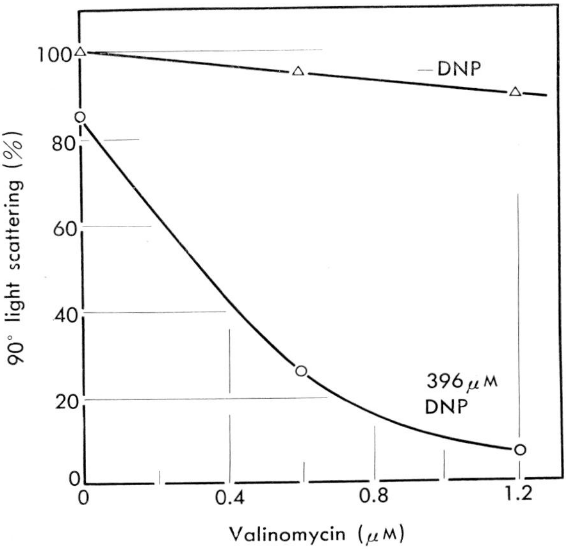

FIG. 6. Synergistic effect of valinomycin and DNP on light-induced 90° light scattering changes of spinach chloroplasts in 150 mM potassium acetate, pH 6.8, containing 15 μM PMS.

light-induced increase in light scattering (*i.e.*, shrinkage) of the chloroplasts in potassium acetate, but when these 2 compounds were present together, no increase in light scattering could be observed upon illumination. Valinomycin added to shrunken chloroplasts in the light did not affect the increased light scattering level, which was abruptly reversed to the initial level by subsequent addition of DNP. The order in which these 2 chemicals were added was not critical, because the light-induced increase in light scattering was also reversed completely when valinomycin was added after DNP.

2. Electron microscopic observations

Isolated spinach chloroplasts observed under the electron microscope are devoid of their outer envelope and stroma components, but have an intact inner membrane system where grana and intergrana thylakoids are relatively loosely stacked because of the rather hypotonic condition of the incubation medium, *i.e.*, 150 mM acetate (pH 6.8) (Fig. 7). After illumination, the thylakoids become very tightly packed together, making it quite difficult

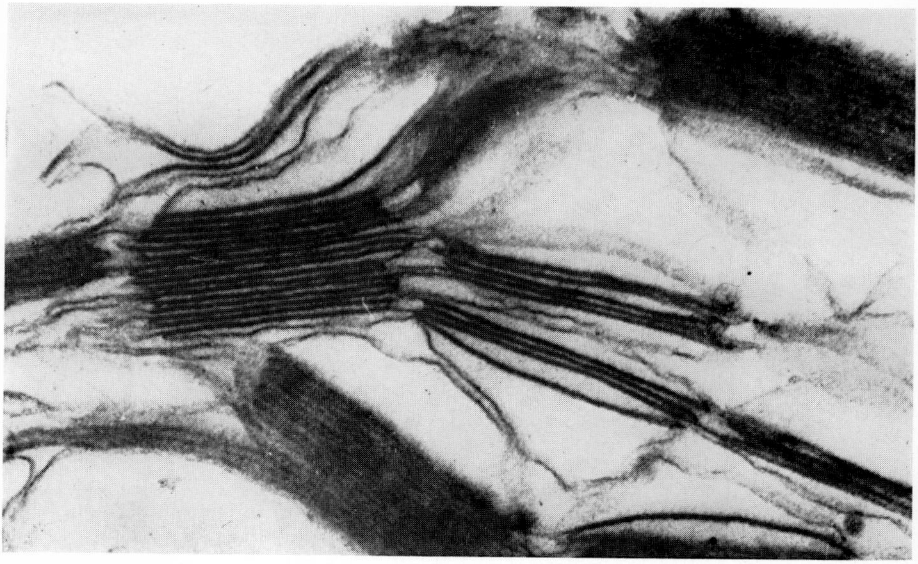

FIG. 7. Chloroplast inner membrane system in 150 mM sodium acetate, pH 6.7, plus 15 μM PMS in the dark before illumination. Both grana and intergrana thylakoids exhibit a slightly swollen configuration. ×56,000.

to discern clear spaces between individual membranes (Fig. 8). It has been found that this structural change is accompanied by a decrease in membrane thickness as well as in the spacing of the thylakoids which amounted to 13 and 27%, respectively (*13*). These changes due to illumination are reversed completely when the chloroplasts are transferred in the dark.

Reversal of light-induced shrinkage of the chloroplast inner membrane system in potassium acetate to the initial swollen state is caused even in the light, if valinomycin and DNP are added together to the medium. Consequently the packing of thylakoid membranes is loosened, and each membrane can be seen clearly at low magnification (Fig. 9). Figure 10 shows details of the membrane system at higher magnification when both valinomycin and DNP were added in the light phase. Many small particles of 90–100 Å diameter spaced at a regular distance from each other can be seen on the outer surface of the thylakoids. These particles are probably those which were observed previously by a negative staining technique

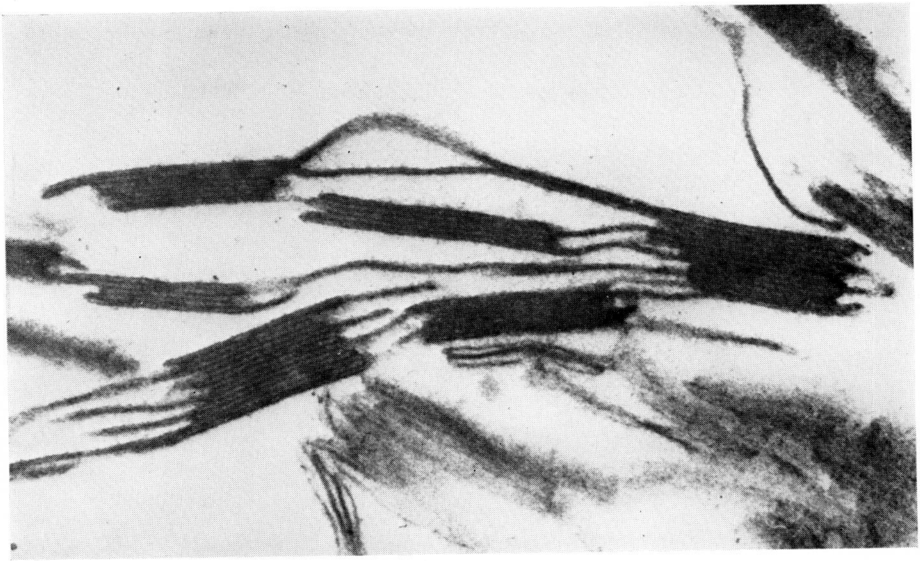

FIG. 8. Chloroplast inner membrane system in 150 mM sodium acetate, pH 6.7, plus 15 μM PMS in the light. Chloroplasts were illuminated by red light for 3 min. Flattening of the thylakoids and contraction of the membranes have occurred. $\times$56,000.

and which were identified as the coupling-factor particle for photophosphorylation (*25*). Separate addition of valinomycin or DNP to shrunken chloroplasts in the light did not change the structure.

In another series of experiments, valinomycin and/or DNP were added first to the medium in the dark and the chloroplasts were then exposed to actinic illumination. Valinomycin or DNP alone caused marked swelling of the chloroplast membrane system in the dark (Fig. 11), but subsequent induction of shrinkage upon illumination was not affected at all. Figure 12 shows chloroplasts illuminated in the presence of valinomycin only. The thylakoids became flattened and tightly packed (Fig. 13), and this was accompanied by transformation of the membrane system from a round shape in the dark to fusiform in the light. The chloroplasts exhibited the same response to DNP treatment as was observed with valinomycin. DNP which was added to the chloroplasts in the dark caused marked swelling of the inner membrane system, and subsequent illumination induced shrinkage. Figure 14 shows a chloroplast membrane system which was first incubated

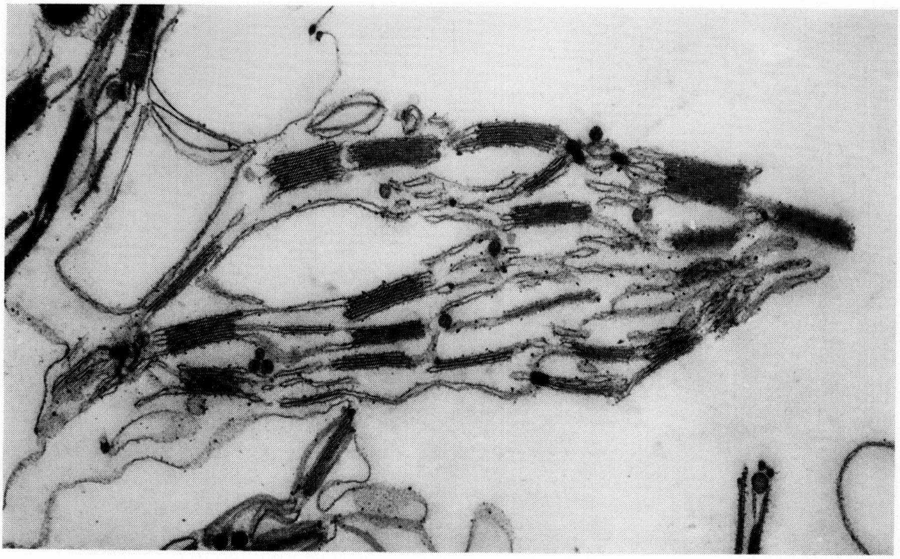

FIG. 9. Effect of valinomycin and DNP on structural state of chloroplasts in the light. Isolated spinach chloroplasts were suspended in 150 mM potassium acetate, pH 6.8, containing 15 μM PMS and were illuminated, and then 1.33 μM valinomycin and 266 μM DNP were added to the suspension in the light. The thylakoids are loosely packed, showing swollen configuration. ×24,000.

with DNP in the dark, illuminated and then treated with valinomycin during illumination. The thylakoids are loosely stacked. When valinomycin and DNP were present together in the medium in the dark before illumination, chloroplast swelling became much more pronounced and no structural changes were induced by illumination.

Thus the synergistic effect of valinomycin and DNP on the structure was also confirmed by electron microscopy.

3. ANS fluorescence

It is well known that ANS emits strong fluorescence when it is bound to biological membranes or when its environment becomes less polar. Fluorescence of ANS bound to thylakoid membranes is further enhanced in light which causes the chloroplasts to contract (*13*). After an initial rapid rise, a slow increase in fluorescence continues for 20–30 sec until a maximum level is reached (Fig. 15). It should be noted that the slow phase of ANS

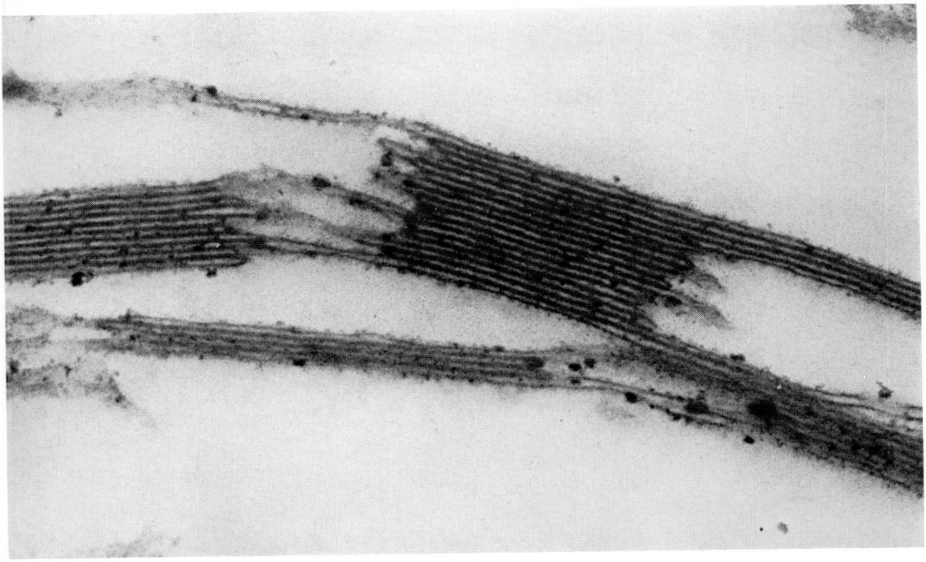

FIG. 10. Effect of valinomycin and DNP on the structure of thylakoids. Conditions as in Fig. 9. ×88,000.

fluorescence change is sensitive to several treatments which severely affect the Hill reaction and photophosphorylation, while the rapid phase of unknown nature is not affected. Light scattering changes and the slow phase in the ANS fluorescence response of chloroplasts to actinic illumination were eliminated completely by 6.5×10^{-7} M 3-(3′, 4′-dichlorophenyl)-1, 1-dimethylurea (DCMU) and 0.5% glutaraldehyde. When nigericin, which uncouples photophosphorylation through dissipating both the proton and cation gradients, was added to the chloroplasts before or during illumination, light-induced changes in ANS fluorescence were also abolished. This led us to carry out experiments which were designed to test whether a synergistic inhibitory effect of valinomycin and DNP on ANS fluorescence could be observed. DNP or valinomycin alone had no effect on ANS fluorescence. Only when both compounds were present together in the medium in the dark or light phase of the experiments, did they act as potent inhibitors of ANS fluorescence, especially of its second slow phase (Fig. 15). Thus it was proved that the structural states of the thylakoid membranes are controlled by both proton and cation gradients not only at the level of gross morphology but also at the molecular level.

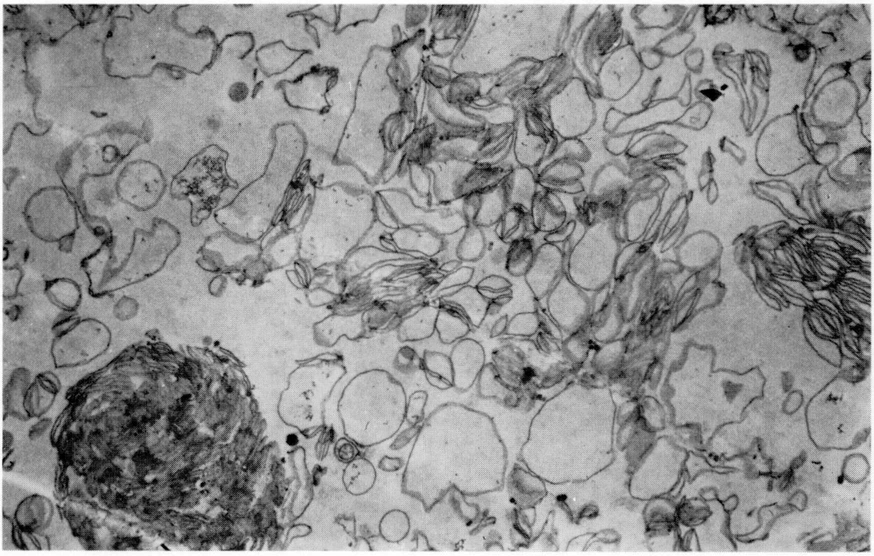

FIG. 11. Effect of valinomycin on chloroplast inner membrane system in the dark. Chloroplasts were suspended in 150 potassium acetate, pH 6.8, containing 15 μM PMS, and 1.33 μM valinomycin was added to the chloroplasts in the dark. Valinomycin caused marked swelling of the membrane system in the dark. ×73,000.

4. *Photosynthetic reactions*

Electron transport mediated by methyl viologen was measured by oxygen consumption, which was determined with an oxygen electrode. The chloroplasts in our experiments have a high coupling capacity of electron transport to photophosphorylation. A photosynthetic control ratio of 3–4 was obtained in the presence of 50 μM ADP. The basic electron flow rate under nonphosphorylating conditions was 32 μmoles O_2 per mg chlorophyll per hr, and was accelerated by addition of 50 μM ADP to 47.5 μmoles O_2 per mg chlorophyll per hr. Valinomycin inhibited basic electron flow by 20–35%, while significant acceleration was observed with DNP. It appears that valinomycin or DNP alone does not affect photophosphorylation since a photosynthetic control ratio of 3.2–3.7 was obtained even after addition of one of these compounds. When valinomycin and DNP were present together in chloroplast incubation in the dark before illumination, or in

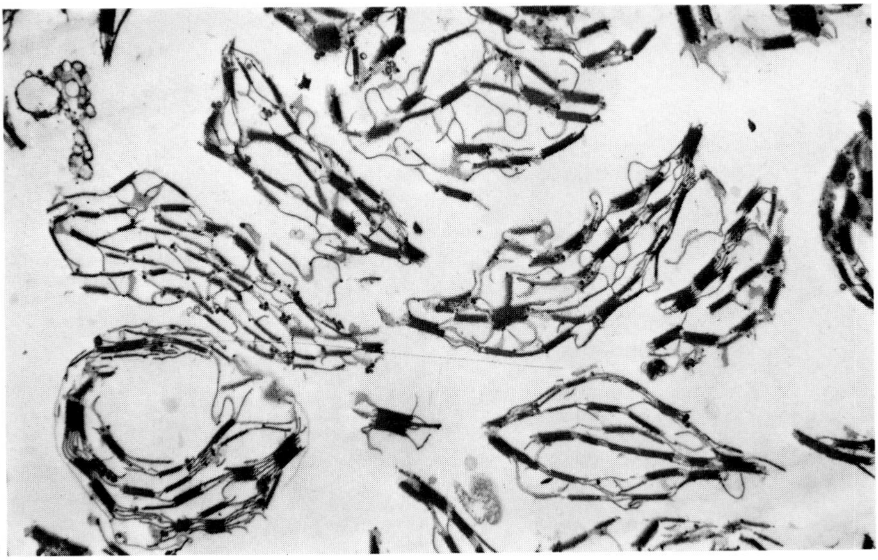

FIG. 12. Light-induced shrinkage of chloroplasts in the presence of valinomycin. Chloroplasts were treated as in Fig. 11, but subsequently illuminated by red light. Transformation of the membrane system from a round shape (swollen configuration) in the dark to fusiform in the light was observed. ×73,000.

the light, electron transport was uncoupled almost completely, giving a maximum flow rate of up to 80 μmoles O_2 per mg chlorophyll per hr (Fig. 16). With 0.3 μM gramicidin, which might abolish both proton and cation gradients, an uncoupled electron flow rate of 95 μmoles O_2 per mg chlorophyll per hr was recorded.

5. *Delayed fluorescence*

A close correlation between the emission of delayed fluorescence from the excited chlorophyll system and the primary photochemical energy conversion process in photosynthesis has been demonstrated by many investigators (*26–31*). It is generally accepted that the intensity of millisecond-delayed light emission depends on the rate of the reverse reaction between the reduced primary electron acceptor (X^-) and the oxidized primary electron donor (Y^+) which are produced by photoreaction system II during excitation. Recent analysis of the kinetics of millisecond-delayed light emission has indicated that it consists of 2 distinct components; an initial

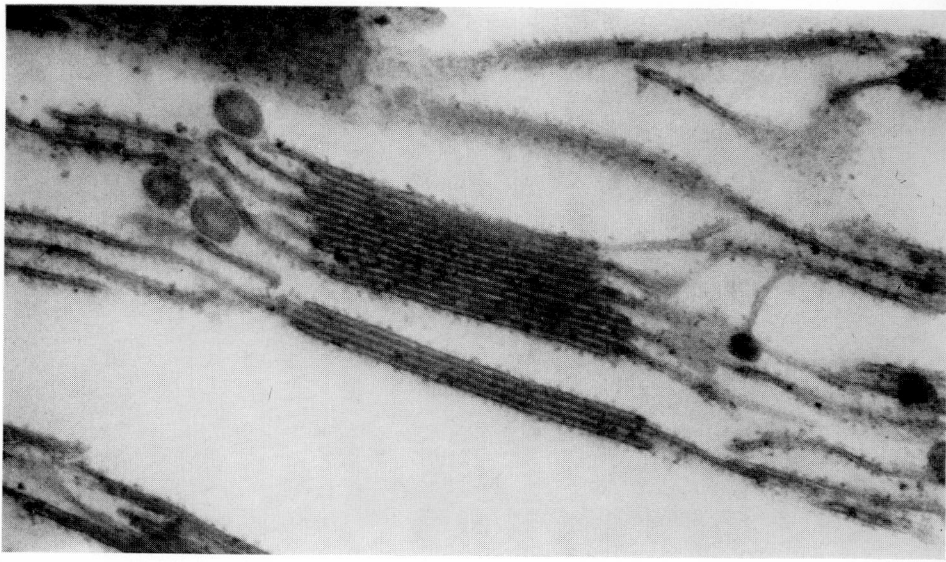

FIG. 13. Ultrastructure of the thylakoids illuminated in the presence of valinomycin. The thylakoids became flattened and tightly packed upon illumination. Conditions as in Fig. 12. ×93,000.

rapid rise and second slow increase (*32*). From studies of the effect of various electron- and energy-transfer inhibitors and uncouplers on delayed light emission it is suggested that the initial rapid rise originates through the reverse process or recombination of oxidized and reduced photoproducts of photoreaction II, while the second slow phase is related to the amount of accumulated high energy state of photophosphorylation (*30*, *32*, *33*). Thus studies of delayed fluorescence seem to provide valuable information concerning the energy state of the thylakoid membranes (*33–36*). Taking account of this, the effect of valinomycin and DNP on the emission of millisecond-delayed fluorescence was examined.

The time course of emission of millisecond-delayed fluorescence of the chloroplasts equivalent to 3.5 μg chlorophyll per ml suspended in 150 mM potassium acetate in the presence of 3.3 μM phenazine methosulfate is shown in Fig. 17. Two components in its response to actinic illumination can be discerned. Intermittent illumination, with light and dark periods of about 2 milliseconds each, induced first a prompt rise in emission with

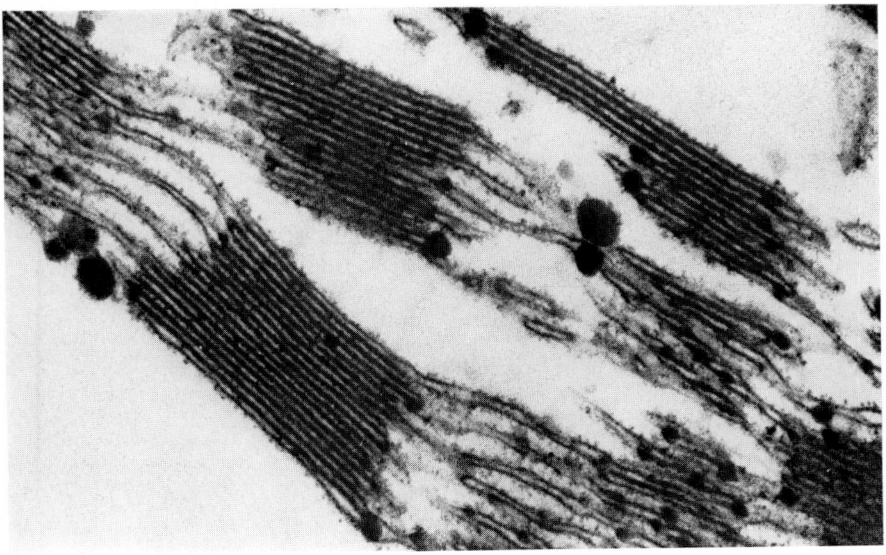

FIG. 14. Effect of valinomycin on contracted thylakoids in the presence of DNP. The chloroplasts were suspended in 150 mM potassium acetate, pH 6.7, containing 15 μM PMS and 266 μM DNP, illuminated, and then treated with 1.33 μM valinomycin during illumination. Tightly packed thylakoids became loosened upon addition of valinomycin. ×88,000.

rapid decay, and then a slow increase which continued for about 30 sec till the maximum level was attained. Under the conditions of our experiments the intensity of the first component was usually higher than that of the second one. It was found that valinomycin and DNP affected the intensity and time course of the emission in different manners (Table II). The intensity of both components was slightly reduced by 13.4 μM DNP. If the concentration of DNP was increased the inhibitory effect also increased. On the other hand, valinomycin severely and selectively inhibited the onset of the rapid component (Fig. 17). Even with 0.33 μM valinomycin the rapid response was reduced to about one-third of that of untreated chloroplasts. The slow component was equally resistant to single treatments with either valinomycin or DNP. Very similar results were obtained when phenazine methosulfate was replaced by different electron acceptors such as methyl viologen and potassium ferricyanide. In Fig. 17 synergistic in-

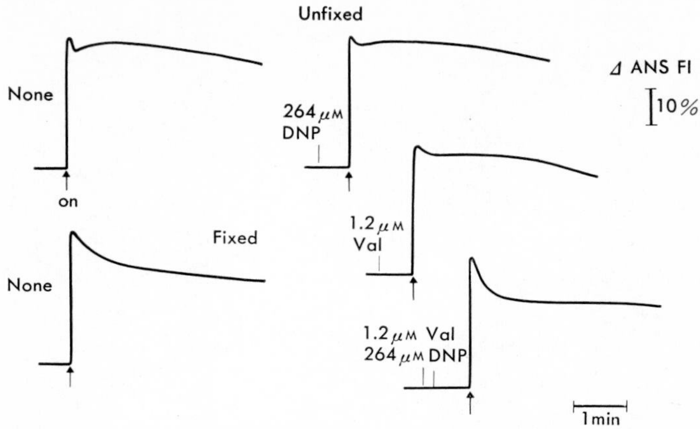

FIG. 15. Light-induced ANS fluorescence changes of chloroplast inner membranes. 100 μg ANS was added to chloroplasts equivalent to 80 μg chlorophyll in 150 mM potassium acetate, plus 15 μM PMS. For the experiments with fixed chloroplasts, chloroplasts were treated with 0.1 M glutaraldehyde in the daık, throughly washed and incubated.

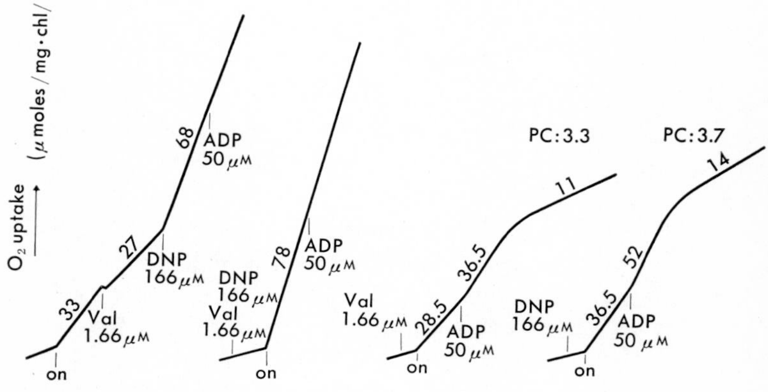

FIG. 16. Effect of valinomycin and DNP on the rate of electron transport from H_2O to methyl viologen. Electron flow rate was monitored by oxygen consumption using methyl viologen as an electron acceptor. The reaction mixture contained 50 mM phosphate, pH 7.8, 10 mM KCl, 1.66 mM $MgCl_2$, 1 mM NaN_3, 50 μM methyl viologen and chloroplasts containing 78 μg chlorophyll in 6 ml.

hibitory effects of valinomycin and DNP on the emission of delayed fluorescence is shown. A combination of 0.66 μM valinomycin and 13.4 μM DNP almost completely eliminated both the rapid and slow components of delayed light emission. As shown in Table II, valinomycin at a concentra-

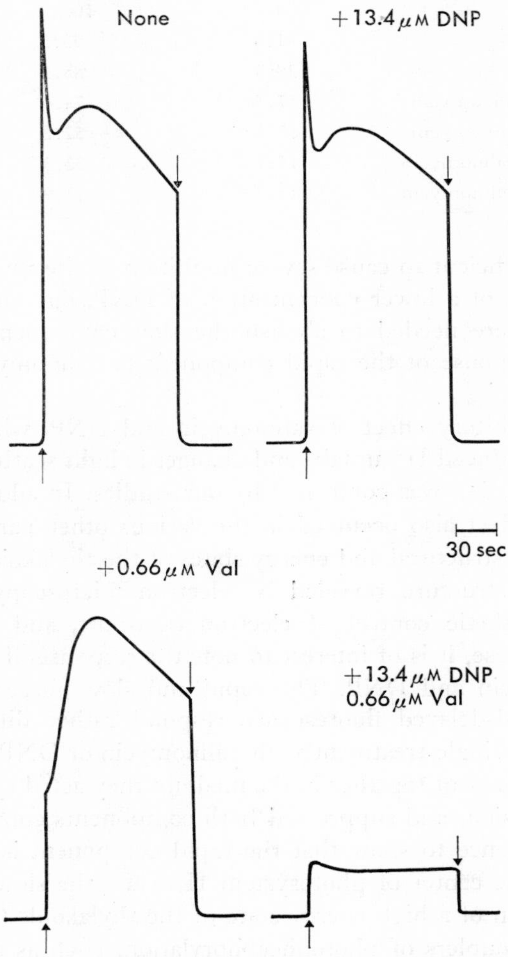

FIG. 17. Effect of valinomycin and DNP on the time course of emission of millisecond-delayed fluorescence of chloroplasts. The chloroplasts equivalent to 10.5 μg chlorophyll were suspended in 150 mM potassium acetate containing 3.3 μM PMS.

TABLE II
Synergistic Effect of Valinomycin and DNP on Delayed Fluorescence

Addition	Relative intensity of delayed fluorescence	
	Rapid component	Slow component
None	100	100
13.4 μM DNP	91.6	93.5
0.66 μM valinomycin	28.3	88.1
6.7 μM DNP + 0.33 μM valinomycin	17.5	74.4
6.7 μM DNP + 0.66 μM valinomycin	17.5	52.2
13.4 μM DNP + 0.33 μM valinomycin	15.4	53.3
13.4 μM DNP + 0.66 μM valinomycin	11.7	11.9

tion of 0.33 μM was sufficient to cause severe inhibition of the rapid component in the presence of a lower concentration of DNP, but much valinomycin and DNP were needed to abolish the slow component. Thus again, the sensitive response of the rapid component to valinomycin was demonstrated.

The synergistic inhibitory effect of valinomycin and DNP, which was first noticed for light-induced H^+ uptake and changes in light scattering by Karlish and Avron (*23*, *24*), was confirmed by our studies. In addition, it was shown that this effect also occurred in the various other parameters employed to follow the structural and energy states of the thylakoid membranes, including the structure revealed by electron microscopy, ANS fluorescence, photosynthetic control of electron transport, and delayed fluorescence. Among these, it is of interest to note the response of the last parameter to valinomycin and DNP. The rapid and slow phases of the emission of millisecond-delayed fluorescence respond rather differently and independently to a single treatment with valinomycin or DNP. When both compounds were present together in the medium they acted synergistically on delayed emission, and suppressed both components completely.

There is ample evidence to show that the rapid component is related to the redox state at the center of photosystem II, while the slow one is more likely an indication of a high energy state of the thylakoids (*34*, *36*). Addition of potent uncouplers of photophosphorylation, such as methylamine and carbonylcyanide-3-chlorophenylhydrazone, selectively and completely eliminate the slow phase of the response (*32*). Nigericin and ammonium chloride, which uncouple photophosphorylation, also abolish the slow component, but not the initial rapid rise (*33*). Electron transport inhibitors such as CMU, DCMU and *o*-phenanthroline were found to suppress completely both the rapid and slow rise in emission. The results presented in this paper enlarged the spectrum of compounds which affect the delayed fluorescence. Valinomycin alone drastically suppresses the initial rapid rise in emission while the slow one is relatively resistant. Since valinomycin reduces electron transport to some extent and increases the permeability of photosynthetic membranes to K^+, resulting in dissipation of the K^+ gradient, the rapid component must be supported to some extent by a cation gradient, or membrane potential, which is established and controlled by the photosynthetic conditions.

On the other hand, the slow component of the delayed fluorescence is reduced only slightly by DNP treatment alone, but is eliminated completely when DNP acts together with valinomycin. Contraction of the thylakoid membranes as revealed by ANS fluorescence and electron microscopy is also suppressed by the combination of valinomycin and DNP. From the studies in the literature it is most likely that the slowly increasing component of the delayed light emission is an indication of a high energy state of the membranes under photosynthetic conditions. Hence it is evident that the energized state and contracted conformation of the thylakoid membranes are related to the light-induced H^+ gradient which is supported by operation of the light-dependent cation gradient. This leads us to the conclusion that proton and cation gradients control complementarily the structural and energy states of the thylakoid membranes. The differential effect of valinomycin and DNP on cation and proton gradients, respectively, suggests that these 2 gradients operate at different sites in the electron transfer pathway.

REFERENCES

1. A. T. Jagendorf and G. Hind, *in* " Photosynthetic Mechanisms of Green Plants," ed. by B. Kock and A. T. Jagendorf, Natl. Acad. Sci., Natl. Res. Council, Publ. 1145, Washington, D.C., pp. 599–610 (1963).
2. J. Neumann and A. T. Jagendorf, *Arch. Biochem. Biophys.*, **107**, 109–119 (1964).
3. R. A. Dilley, *Biochem. Biophys. Res. Commun.*, **17**, 716–722 (1964).
4. R. A. Dilley and L. P. Vernon, *Arch. Biochem. Biophys.*, **111**, 365–375 (1965).

5. D. W. Deamer and L. Packer, *Biochim. Biophys. Acta*, **172**, 539–545 (1969).
6. L. Packer, *Biochem. Biophys. Res. Commun.*, **9**, 355–360 (1962).
7. L. Packer, *Biochim. Biophys. Acta*, **75**, 12–22 (1963).
8. M. Itoh, S. Izawa, and K. Shibata, *Biochim. Biophys. Acta*, **66**, 319–327 (1963).
9. R. A. Dilley and L. P. Vernon, *Biochemistry*, **3**, 817–824 (1964).
10. D. W. Deamer, A. R. Crofts, and L. Packer, *Biochim. Biophys. Acta*, **131**, 81–96 (1967).
11. A. R. Crofts, D. W. Deamer, and L. Packer, *Biochim. Biophys. Acta*, **131**, 97–118 (1967).
12. R. A. Dilley, R. B. Park, and D. Branton, *Photochem. Photobiol.*, **6**, 407–412 (1967).
13. S. Murakami and L. Packer, *J. Cell Biol.*, **47**, 332–351 (1970).
14. Y. Mukohata, M. Mitsudo, and T. Isemura, *Ann. Rep. Biol. Works, Fac. Sci., Osaka Univ.*, **14**, 107–119 (1966).
15. Y. Mukohata, *Ann. Rep. Biol. Works, Fac. Sci., Osaka Univ.*, **14**, 121–134 (1966).
16. S. Murakami and L. Packer, *Biochim. Biophys. Acta*, **180**, 420–423 (1969).
17. S. Murakami and L. Packer, *Plant Physiol.*, **45**, 289–299 (1970).
18. R. A. Dilley and A. Rothstein, *Biochim. Biophys. Acta*, **135**, 427–443 (1967).
19. N. Shavit and A. San Pietro, *Biochem. Biophys. Res. Commun.*, **28**, 277–283 (1967).
20. N. Shavit, R. A. Dilley, and A. San Pietro, *Biochemistry*, **7**, 2356–2366 (1968).
21. L. Packer, *Biochem. Biophys. Res. Commun.*, **28**, 1022–1027 (1967).
22. P. Mitchell, Chemiosmotic Coupling and Energy Transduction, Glynn Res. Ltd., Bodmin, pp. 111 (1968).
23. S. J. D. Karlish and M. Avron, *FEBS Letters*, **1**, 21–24 (1968).
24. S. J. D. Karlish and M. Avron, *Eur. J. Biochem.*, **9**, 291–298 (1969).
25. S. Murakami, *in* "Comparative Biochemistry and Biophysics of Photosynthesis," ed. by K. Shibata *et al.*, University of Tokyo Press, Tokyo, pp. 82–88 (1968).
26. W. E. Arthur and B. L. Strehler, *Arch. Biochem. Biophys.*, **70**, 507–526 (1957).
27. J. C. Goedheer, *Biochim. Biophys. Acta*, **66**, 61–71 (1963).
28. W. Arnold, J. R. Azzi, *Proc. Natl. Acad. Sci. U.S.*, **61**, 29–35 (1968).
29. G. P. B. Kraan, J. Amesz, B. R. Velthuys, and R. G. Steemers, *Biochim. Biophys. Acta*, **223**, 129–145 (1969).
30. R. K. Clayton, *Biophys. J.*, **9**, 60–76 (1969).
31. S. Itoh, S. Katoh, and A. Takamiya, *Biochim. Biophys. Acta*, **245**, 121–128 (1971).

32. S. Itoh, N. Murata, and A. Takamiya, *Biochim. Biophys. Acta*, **245**, 109–120 (1971).
33. C. A. Wraight and A. R. Crofts, *Eur. J. Biochem.*, **19**, 386–397 (1971).
34. A. R. Crofts, C. A. Wraight, and D. E. Fleischmann, *FEBS Letters*, **15**, 89–100 (1971).
35. J. Barber and W. J. Varley, *Nature*, **234**, 188–189 (1971).
36. J. Barber, *FEBS Letters*, **20**, 251–254 (1972).

Composition and Possible Energy Transformation of Cytochromes in the Respiratory System in Relation to the Study on *b*-Type Cytochromes in Mitochondria

Bunji Hagihara,[*1] Nobuhiro Sato,[*1] Kenji Takahashi,[*2] and Saburo Muraoka[*2]

*Department of Biochemistry, School of Medicine, Osaka University, Osaka, Japan,[*1] and Faculty of Pharmaceutical Sciences, Tokushima University, Tokushima, Japan[*2]*

This communication describes the composition of the mitochondrial respiratory system under various physiological and pathological conditions as well as a brief history of studies on *b*-type cytochromes, and also discusses the heterogeneity, distribution, and complex behavior of these cytochromes in mitochondria. The change of redox states of *b*-type cytochromes was studied in relation to the change of the energy states of mitochondrial membrane, using difference spectrum of visible region taken at room and low (77°K) temperatures with fairly rapid response. The redox change of cytochrome b_T (b_{565}) induced by ATP and by respiration was accompanied by the redox change of not only cytochrome c_1 but also cytochromes a, a_3, c, and b_K (b_{561}). From these studies a tentative scheme for the mechanism of coupled respiration and energy-linked reversal of electron transfer is proposed.

Based on the work of Keilin (*1*), Keilin and Hartree (*2*), and Yakushiji and Okunuki (*3*), it was long believed that the respiratory chain of mammalian mitochondria contained 5 cytochromes, namely cytochromes a, a_3, b, c, and c_1. However, several workers have noted that there were differences in the extent of reduction of cytochrome b with different reducing agents such as succinate, NAD-linked substrates, NADH and dithionite (*4–7*). Recently, the existence of at least 2 distinct spec es of mitochondrial *b*-type cytochromes in a wide variety of sources has been established (*8–23*). It is the purpose of this communication to present a brief history of studies on mitochondrial *b*-type cytochromes, and also to discuss the heterogeneity, distribution and behavior of these cytochromes.

History

In 1958, Chance (*24*) described the possible existence of 3 *b*-type cytochromes in the mitochondrial fragments from beef heart; the first component was characterized as a succinate-reducible pigment with an α maximum at 562 nm, the second as succinate-reducible in the presence of antimycin A, having an α maximum at 566 nm, and the third as reducible only by dithionite and having a broad α maximum from 556 to 566 nm. Slater and Colpa-Boonstra (*25*) supposed the third pigment to be contaminant myoglobin, and concluded that their own data and those of Chance provided evidence for the presence of 2 *b*-type cytochromes. The succinate-reducible one was referred to as classical cytochrome *b*, and the other, reducible with succinate only in the presence of antimycin, was referred to as cytochrome *b′*. Pumphrey also described the possible presence of the former 2 moieties (*26*). From studies on the reactivity of cytochrome *b* with substrates in the presence and absence of various inhibitors and chelating agents, Wainio and co-workers (*27–29*) suggested that cytochrome *b* is present in 3 different compartments, or as 3 moieties in heart mitochondria.

In 1966 Chance and Schoener (*30*) reported the presence of a new cytochrome *b* moiety showing an α peak at 555 nm at liquid nitrogen temperature (around 560 nm at room temperature) in pigeon heart mitochondria. This absorption band appeared when ATP was added to the mitochondria (in the sulfide-inhibited state), and disappeared on the addition of uncoupling agents, such as dicumarol or dinitrophenol. From the above behavior, this band was proposed to be due to the existence of a new type of cytochrome *b*, or a modified form of the classical one, which was tentatively designated as cytochrome b_{555}.

In 1968, we reported the existence of 2 hemeproteins additional to the established cytochromes in various mammalian mitochondria so far tested (*31*, *36*). They are also contained in mitochondria from a wide variety of cancer cells (*31*, *32*). The above pigments were identified from the difference spectra (of mitochondrial and cell suspensions) obtained in the presence and absence of dithionite after anaerobiosis was attained by succinate respiration and endogenous respiration, respectively. At room temperature, these spectra exhibited 2 peaks at 559 and 565 nm and at the temperature of liquid nitrogen they peaked at 556 and 563 nm in the α region. These 2 peaks were supposed to be derived from 2 pigments. Although these difference spectra appeared to be those of typical cytochromes, the pigments were tentatively named hemeproteins 559 and 565 (HP_{559} and HP_{565}) (*31*), since their electron donors and acceptors were not known at that time.

In 1970, Slater *et al.* (*15–17*) observed an absorption peak at 565 nm in beef heart mitochondria when ATP was added to a mitochondrial suspension containing succinate in anaerobic conditions or in the presence of antimycin A. They considered that this peak was produced by a spectral shift of the 562 nm peak of cytochrome *b* or of the 558 nm peak of cytochrome *b*′.

In the same year, Wilson and Dutton (*8*) presented evidence for the existence of 2 distinct species of *b*-type cytochromes in rat liver mitochondria, based on measurement of the midpoint oxidation-reduction potentials (E_m) of the cytochromes of mitochondria ***in situ***. One has variable E_m values depending on the energy states (the (ATP)/(ADP)(P_i) ratio) in mitochondria, and has a double α peaks at 565 (α_1) and 558 nm (α_2) in the reduced minus oxidized difference spectrum (*12*). This component was

TABLE I
Mitochondrial *b*-Type Cytochromes in the Literature Other than Classical Cytochrome *b*[a]

Authors	Designation and α peak (nm)		Conditions of reduction	Ref. No.
Chance		566	Succinate+AM[b]	(*24*)
		556–566	Dithionite	
Slater and Colpa-Boonstra	*b*′	565	Succinate+AM Dithionite	(*25*)
Chance and Schoener	b_{555}	560 (555 at 77°K)	Succinate+ATP	(*30*)
Wainio *et al.*	*b*′		Succinate+AM	(*27–29*)
	b″		NADH	
Sato and Hagihara	HP_{565}	565 (563 at 77°K)	Dithionite	(*31, 32*)
	HP_{559}	559 (556 at 77°K)	Dithionite	
Slater *et al.*	b_i	558 (565 by AM)		(*15–17*)
Wilson *et al.*	b_T	565 and 558 (562.5 and 555 at 77°K)	Succinate+ATP	(*8, 12*)

[a] The classical cytochrome *b* (=b_K, α peak 562 nm) is reduced easily by succinate. [b] Succinate+AM means the component is reduced with succinate only in the presence of antimycin A.

supposed to be the cytochrome directly responsible for energy transduction and was named cytochrome b_T (*8*). The presence of such a cytochrome was confirmed by Chance's group in a wide variety of mitochondria, including those of pigeon heart (*9–12*), beef heart (*11*), ascites cells (*14*), and yeast (*13*). The other component of *b*-type cytochromes is the classical cytochrome *b* having a fixed E_m value, which was named cytochrome b_K in honor of D. Keilin. Evidence has also been obtained by Chance and Erecińska (*18*, *21*) for the existence of 2 kinetically distinct species of *b*-type cytochromes in pigeon heart mitochondria. Wikström also showed independently the presence of energy-dependent and -independent cytochromes *b* in rat liver mitochondria (*22*, *23*).

Table I shows a summary of the work on the *b*-type cytochromes contained in mitochondria. The cytochrome with α absorption at 566 nm reported by Chance (*24*) and cytochrome b' of Slater and Colpa-Boonstra (*25*) might be identical with cytochrome b_T of Wilson and Dutton (*8*). The 555 nm absorption of cytochrome b_{555} at 77°K observed by Chance and Schoener (*30*) is also identical with the α_2 peak of cytochrome b_T.

Relation among Cytochrome b_T, HP_{565}, HP_{559}, and b_{559}

Considering the experimental conditions of analysis, it is clear that our HP_{565} (*31*, *32*) is identical with the low-energy form of cytochrome b_T. Our HP_{559} is entirely or partly, depending on the type of mitochondria, due to the α_2 absorption maximum of cytochrome b_T. As shown in Fig. 1 and Table II, the ratio of the absorption at 559 nm to that at 565 nm is variable with different sources of mitochondria. It is less than 1 in mitochondria of brain and kidney, and more than 1 in those of liver, lung and various tumors. In the former cases, the slightly lower ratio of extinction at 559 nm to that at 565 nm can be explained by the contribution of cytochromes b_T, c_1, and b_K. The extinction at 559 nm is the result of the sum of roughly 2/3 of the extinction of cytochrome b_T at 565 nm, 1/4 of that of cytochrome c_1 at 554 nm and 4/5 of that of cytochrome b_K at 562 nm. The presence of cytochrome c_1 has a negligible effect on the absorbance at 565 nm, and b_K has a similar effect on the absorbance at both 559 and 565 nm. In addition to the above 2 *b*-type cytochromes, cytochrome b_5 is contained in liver mitochondria (*35*). However, this cytochrome was found only in negligible amounts in the other mitochondria tested (*40*). Taking the effect of the above cytochromes in consideration, the large absorbance in liver and tumor mitochondria at 559 nm as compared to that at 565 nm, suggests the presence of a pigment having an absorption maximum around 559 nm (556 nm at 77°K) in addition to cytochrome b_T. This pigment, which appears to be one of the b-type cytochromes from its absorption spectrum, is tentatively

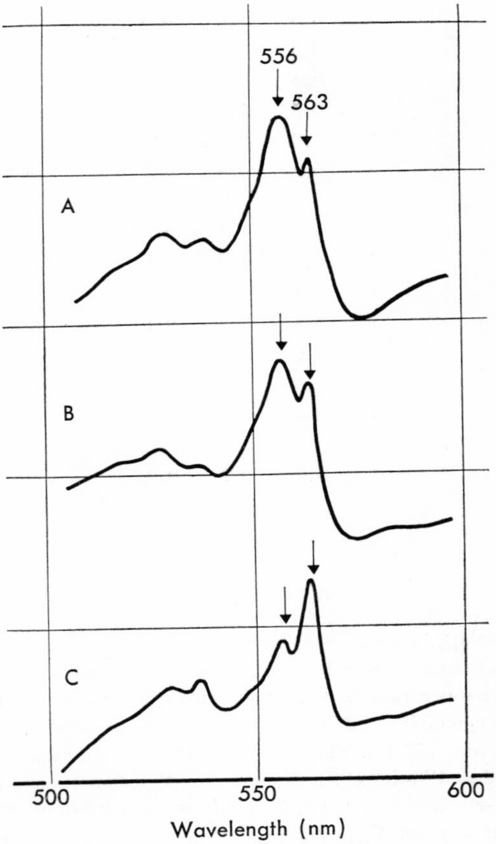

FIG. 1. Low-temperature spectra representing the difference between the presence and absence of dithionite in anaerobic mitochondria supplemented with 2.5 μM rotenone, 5 mM succinate and 0.1 mM NADH. Curve A, normal rat liver mitochondria (1.8 mg protein/ml); Curve B, Morris 7316 B hepatoma mitochondria (1.6 mg protein/ml); Curve C, normal dog kidney mitochondria (0.8 mg protein/ml). Arrow at 563 nm, α_1 band of cytochrome b_T; arrow at 556 nm, α_2 band of cytochrome b_T plus α band of cytochrome b_{559}.

TABLE II
Relative Concentration of Cytochromes in Mammalian Mitochondria (*cf.* Ref. *36*)

Tissue	Concentration of cytochrome *c* (nmoles/mg protein)	Relative concentration[a] of cytochromes					
		c	c_1	b_K [b]	$HP_{565}(b_T)$[b]	b_{559}	$a+a_3$
Kidney cortex							
Beef	0.44	1	0.3	1.0	0.3	—	1.2
Dog	0.61	1	0.3	0.8	0.4	—	0.9
Pig	0.46	1	0.3	0.9	0.3	—	1.2
Rabbit	0.51	1	0.3	1.0	0.2	—	1.2
Mouse	0.53	1	0.3	0.8	0.2	—	1.1
Rat	0.65	1	0.3	0.6	0.2	—	0.8
Liver							
Beef	0.20	1	0.6	1.1	0.1	0.2	1.2
Dog	0.21	1	0.5	1.0	0.1	0.2	1.1
Rabbit	0.19	1	0.4	1.0	0.1	0.2	1.0
Mouse	0.19	1	0.5	1.0	0.1	0.3	1.2
Rat	0.18	1	0.6	1.0	0.1	0.3	1.2
Brain							
Rat	0.19	1	0.4	0.7	0.3	—	0.8
Skeletal muscle							
Rat	0.47	1	0.3	0.7	0.3	—	0.9
Tumor							
Ehrlich	0.41	1	0.3	0.3	0.1	0.1	0.3
Sarcoma 180	0.33	1	0.4	0.4	0.1	0.1	0.4
AH130	0.20	1	0.5	0.7	0.2	0.2	0.8

The contents of cytochromes $(a+a_3)$, b_K, and $(c+c_1)$ were determined according to Chance (*44*) using the difference spectra between anaerobic conditions in the presence of succinate and aerobic conditions, at room temperature. Separate values for cytochromes *c* and c_1 were calculated using the ratio of *c* to c_1 which was determined from the same difference spectra taken at liquid nitrogen temperature (77°K). The content of HP_{565} was determined from $E_{565}-E_{575}$ in the "dithionite-reduced" minus "succinate-reduced" difference spectra (*31*). Under these conditions, the low-energy form of cytochrome b_T is supposed to be measured. The content of b_{559} was determined from $(E_{559}-2/3\ E_{565})-E_{575}$ in the same spectra. The 2/3 value of E_{565} is nearly equal to the absorbance of b_T at 559 nm. For calculations, a difference extinction coefficient $(\varepsilon_{\lambda max}-\varepsilon_{575})$ of 18 $mM^{-1}\,cm^{-1}$ was tentatively used for all *b*-type cytochromes. [a] Since the content of cytochrome *c* in tissues (or cells) is least variable, the relative concentrations were expressed using *c* (or $c+c_1$ when separate values were not determined) as a standard in this and other tables. [b] The values for b_K are somewhat overestimated and those for b_T are underestimated since a part of b_T is reduced by succinate.

referred to here as cytochrome b_{559}. The name HP_{559} will be no longer used since this seems not to be a single component but to be composed of cytochrome b_T and b_{559}.

Cytochrome Composition in the Respiratory System

Several workers have vaguely believed in the presence of a stoichiometric relationship of the cytochrome components in the respiratory systems, and supposed the existence of a minimum unit or set of the respiratory assembly composed of a small number of electron transport and phosphorylating components. Green's group (*33, 34*), for instance, reported that beef heart mitochondria contain cytochrome $a(+a_3)$, b, c_1, and flavoproteins in the ratio of 6: 3: 1: 2. As the minimum respiratory assembly, the name "elementary particle" was proposed by Green's group (*33, 34*) and "oxisome" by Chance's group (*35*).

We have noticed that even mitochondria in the same kind of tissue or cells contain cytochromes in extremely variable ratios according to the physiological conditions. In the case of 2 kinds of yeast, for instance, the relative concentration of *a*-, *b*-, and *c*-type cytochromes in the respiratory system changed markedly and continuously during the different stages of cultivation (Table III) (*37*). It is very difficult to postulate a unit set of the respiratory assembly which could account for the above change. Changes were also observed in the relative concentration of cytochromes in the respiratory chain of mammalian liver following change in the hormonal conditions (Table IV) (*38, 39*). The relative concentration of respiratory cytochromes in liver also changed during the regeneration of liver, during

TABLE III
Relative Contents of Respiratory Cytochromes in Yeast Cells at Various Stages of Cultivation (*37*)

Kind of yeast	*Saccharomyces cerevisiae*			*Candida roubsta*		
Cytochrome	$c+c_1$ [a]	*b*-type	$a+a_3$	$c+c_1$ [a]	*b*-type	$a+a_3$
Lag phase	1	0.13	0.13	1	0.22	0.09
Log phase	1	0.42	0.30	1	0.44	0.19
Stationary phase	1	0.50	0.38	1	0.61	0.28

[a] Absolute concentration of cytochrome $(c+c_1)$ changed from roughly 0.6 to 1.0 nmoles/mg nitrogen of cells in both strains.

treatment with carcinogenic amino-azo dyes and even in the presence of tumors in the other part of the body (Table IV) (*40*).

Generally speaking, the composition of the respiratory cytochromes including the *b*-type cytochromes, as well as their absolute concentration, in given tissues or cells are not constant but are readily changeable depending on the physiological conditions. From the above considerations, it may

TABLE IV
Relative Concentrations of Cytochromes in Mitochondria of Rat Livers under Various Conditions (*38–40*)

Treatment or conditions	Relative concentration of cytochromes			
	$c+c_1$	b_K	b_{others} [a]	$a+a_3$
Normal rat				
Donryu	1	0.43	0.90	0.87
Buffalo	1	0.42	0.74	0.96
Wistar	1	0.38	0.79	0.82
Hypophysectomized rat				
after 3 days	1	0.47		0.77
after 9 days	1	0.43		0.75
after 48 days	1	0.37		0.80
Thyroidectomized (7 days)	1	0.50		0.70
Regenerating liver (26 hr)	1	0.41	0.49	0.91
Embryonal liver (−2 days)	1	0.56	0.91	0.65
3′-DAB feeding (22nd week)				
Noncancerous part	1	0.32	0.51	0.40
Hepatoma part	1	0.33	0.47	0.39
Liver of tumor-bearing rat				
7794 A	1	0.46		0.58
7316 B	1	0.37		0.54
7793	1	0.46		0.67
Hepatoma				
7794 A	1	0.44		0.55
7316 B	1	0.37		0.47
7793	1	0.46		0.66

Wistar rats were used in the analysis of the cytochromes in hypophysectomized-, thyroidectomized-, regenerating, and embryonic livers, while buffalo rats were used in the study of the Morris hepatoma 7794A, 7316B, and 7793. [a] Assay methods, see Table II. All values are a mean of 5 to 10 rats.

be concluded that the respiratory components are distributed randomly in the membrane instead of composing unit sets.

Change between High- and Low-Energy States

Wilson *et al.* and many other workers observed, on the addition of ATP, transition of cytochrome b_T from the low- to the high-potential form (E_m, at pH 7.2≒−40 and 240 mV, respectively) and of cytochrome a_3 from the high- to the low-potential form (about 390 and 190 mV, respectively) in rat liver (*8*) and pigeon heart (*9*) in the anaerobic state. The same transitions were also produced by the energy generated by electron transport through the respiratory chain from substrates to oxygen (*18*, *19*) or to other acceptors (*41*, *42*).

In most of the above studies, the double-beam technique was used for the detection of the transitions. In the case of cytochrome b_T, the extinction at 565 nm (measurement wavelength, λ_2) minus that at 575 nm (reference wavelength, λ_1) was usually recorded. In order to detect the cytochrome component whose oxidation-reduction state is dependent on the energy state, the experiment was repeated using a different λ_2 and constant λ_1, and the difference spectrum between the 2 different energy states was drawn consecutively point by point using the various ($E_{\lambda_2}-E_{\lambda_1}$) traces. Since this type of spectra does not show the detailed shape over a wide range of wavelength, we directly measured the difference spectrum of mitochondria in different energy states, using a split-beam spectrophotometer (*43*, *44*) with a fairly rapid response (7.5 sec per 100 nm within 1 nm delay) equipped with a 60-cycle vibration mirror and a double monochromator (*45*).

When ATP was added to mitochondria made anaerobic by addition of an equimolar quantity of succinate and fumarate, a prominant peak appeared at 565 nm, indicating the reduction of cytochrome b_T, which gradually disappeared with time (Fig. 2). This could be well explained by supposing that the addition of ATP transformed cytochrome b_T from the low- to the high-potential form, which has a greater affinity for electrons (see Fig. 4). The gradual disappearance of the resulting peak is due to the limited supply of energy, which is reduced with time. In the same spectrum, a comparatively big trough was observed around 605 nm. This is presumed to be due to the oxidation of cytochromes *a* and a_3 which was supposed to be the result of electron transfer from cytochrome a toward cytochrome b_T.

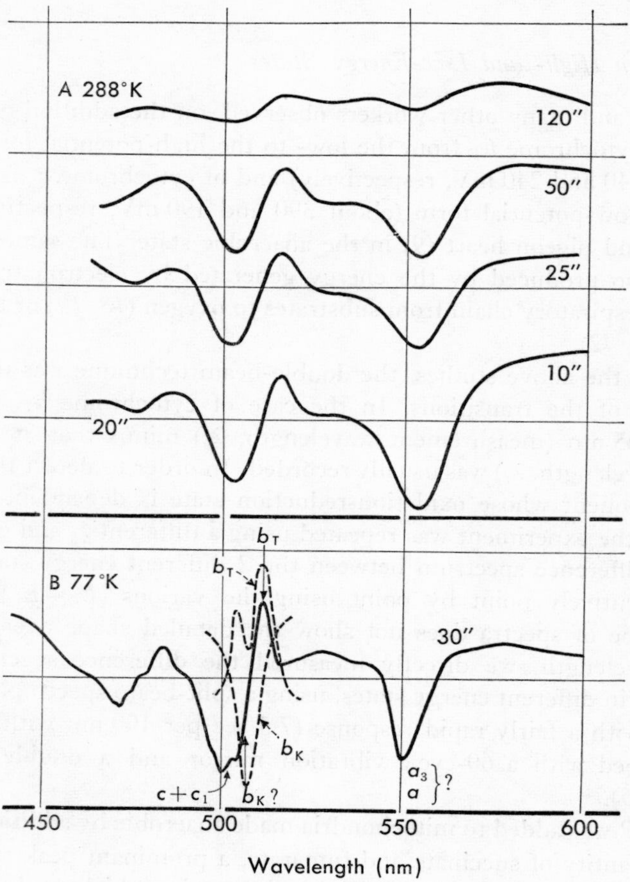

FIG. 2. Effect of ATP on the oxidation-reduction level of cytochromes of rat liver mitochondria in anaerobic conditions. A: The mitochondrial suspensions (containing 10 mM succinate and 10 mM fumarate) in both the sample and reference cuvettes were kept at 15°C for a few minutes after anaerobiosis and then 0.2 mM ATP was added only to the sample cuvette. The difference spectra were taken 10, 25, 50, and 120 sec after the addition of ATP. B: Reaction was carried out as above, but the difference spectrum was taken at 77°K after the mitochondrial suspensions were frozen for about 30 sec after the ATP addition.

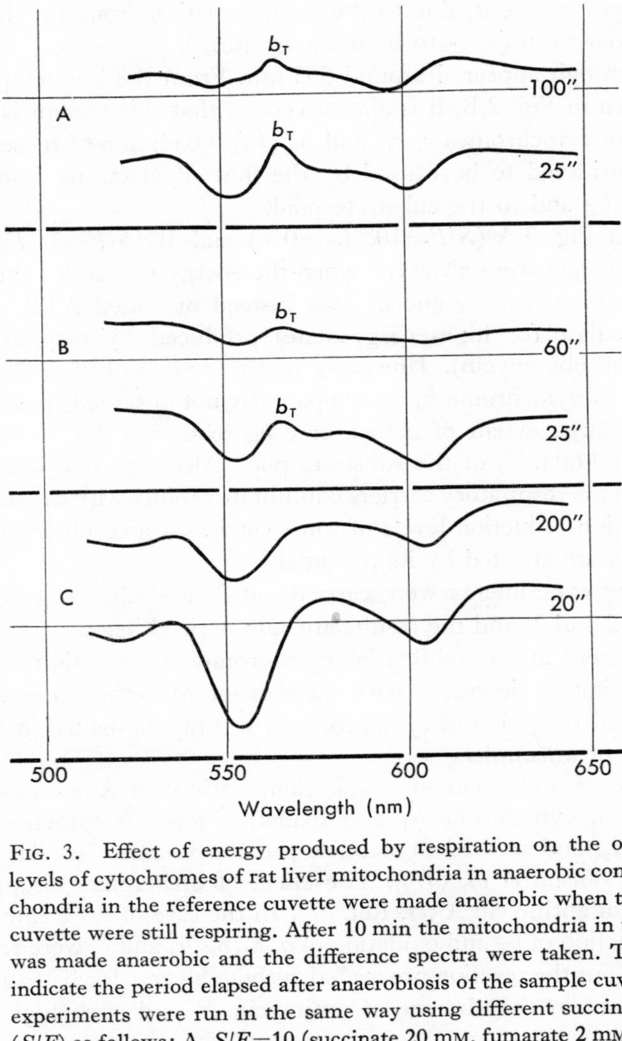

FIG. 3. Effect of energy produced by respiration on the oxidation-reduction levels of cytochromes of rat liver mitochondria in anaerobic conditions. The mitochondria in the reference cuvette were made anaerobic when those in the sample cuvette were still respiring. After 10 min the mitochondria in the sample cuvette was made anaerobic and the difference spectra were taken. The specified times indicate the period elapsed after anaerobiosis of the sample cuvette. Three sets of experiments were run in the same way using different succinate/fumarate ratios (S/F) as follows: A, $S/F=10$ (succinate 20 mM, fumarate 2 mM); B, $S/F=1$ (succinate 10 mM, fumarate 10 mM); C, $S/F=0.1$ (succinate 2 mM, fumarate 20 mM). In all cases, sufficient amounts of oligomycin were included.

This had been transformed by ATP from the low- to the high-potential form (low- to high-energy form), through cytochrome c_1 and c and the oxidation of cytochrome a_3 due to the transformation from the high- to the low-potential form (low- to high-energy form).

One more trough appeared around 555 nm. From the low-temperature spectrum shown in Fig. 2,B, it is almost certain that this trough is due to the oxidation of cytochromes c, c_1, and b_K (*19*). Oxidation of these 3 cytochromes is considered to be caused by the flow of electrons from them to cytochrome b_T and to the substrate pool.

As shown in Fig. 3,A ($S/F=10$, $E_h=0$ V) and B ($S/F=1$, $E_h=+30$ mV), similar changes were observed when the energy source for the transformation of cytochrome b_T and a_3 was, instead of added ATP, a high-energy intermediate (or high-energy state) produced by respiration (in the presence of oligomycin). However, in the case of Fig. 3,C ($S/F=0.1$), reduction of cytochrome b_T was apparently not observed, presumably because of the large extent of cytochrome b_K oxidation due to the high potential level (+60 mV) of the substrate pool. Although it is difficult to consider that all the respiratory carriers equilibrate rapidly with the substrate pool, the oxidation-reduction levels of some carriers, especially those close to the substrate, are affected by its potential.

Various other experiments were carried out in a similar way to those shown in Figs. 2 and 3, and the results are summarized below.

1) Both the extent and durability of cytochrome b_T reduction by ATP addition or respiration decreased with the damage of mitochondria.

2) ATP-induced reduction of cytochrome b_T was rapidly halted (oxidized) by the addition of uncouplers.

3) In the case of induction of respiration, antimycin A increased the reduction level of cytochrome b_T and oxidation level of cytochromes c, c_1, b_K, and $a+a_3$, and seemed to stabilize the above states (similar to the observations of Wilson *et al.* (*41*)). The effect of uncouplers disappeared in the presence of antimycin A (*cf.* Ref. *41*). In the case of ATP induction, the rates of reduction of b_T and oxidation of a, a_3, b_K, c, and c_1 were reduced by antimycin A, but the reaction proceeded, although very slowly, to a much higher extent than in the absence of antimycin A, and such high levels were stable.

4) In the presence of antimycin A at a concentration which largely inhibits respiration, cytochrome b_T was in a more reduced form under aerobic conditions (State 4) than in anaerobic conditions (State 5) in the absence

of antimycin A, while other cytochromes including b_K were in a more oxidized form under the former conditions.

In order to explain the above observations, a tentative scheme for the mechanism of coupled respiration and energy-linked reversal of electron transfer is proposed in Fig. 4, together with a modification of Wilson and Dutton's scheme for the respiratory chain (*8*). The above mechanism is based on the scheme of Wilson *et al.* (*8*, *10*, *46*) and that of Chance (*18*) in several respects.

In the above mechanism, coupled respiration proceeds according to the following reactions:

$$b_T^{3+} + b_K^{2+} \longrightarrow b_T^{2+} + b_K^{3+} \quad (1)$$

$$b_T^{2+} + c_1^{3+} \longrightarrow b_{T}{\sim}^{3+} + c_1^{2+} \quad (2)$$

$$b_{T}{\sim}^{3+}\,X + I \longrightarrow b_T^{3+} + X{\sim}I \quad (3)$$

$$X{\sim}I + ADP + P_i \longrightarrow ATP \quad (4)$$

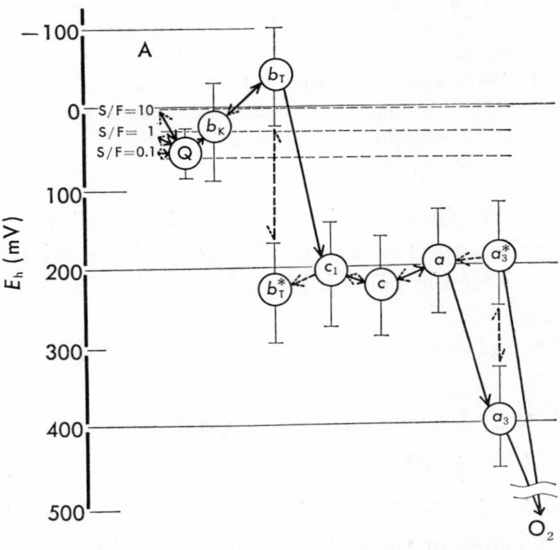

FIG. 4. Respiratory chain and mechanism of coupling. A: Modification of Wilson and Dutton's scheme (*8*, *10*) for the respiratory chain showing the oxidation-reduction potential of the components. Each circle indicates the position of E_m and the upper and lower lines of the respective components indicate their potential when the ratios of reduced to oxidized form are 1: 10 and 10: 1, respectively. The 3 horizontal dotted lines at the upper part of the figure indicate the potentials of solutions containing succinate and fumarate in ratios of 10: 1 ,1: 1, 1: 10.

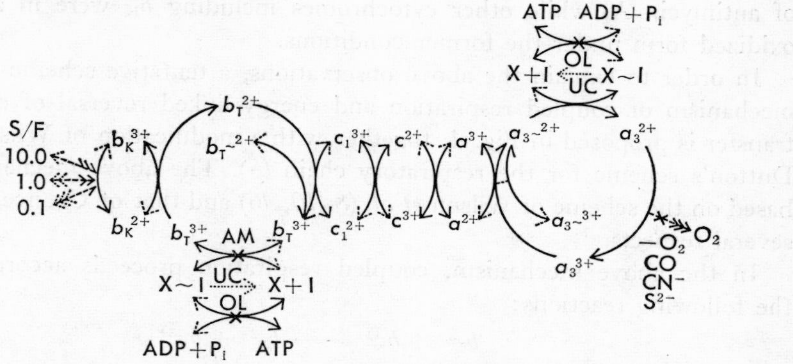

B: Proposed scheme for the mechanism of coupled respiration and reversed electron transfer by energy supplied by added ATP or respiration. *S/F*, succinate-fumarate ratio of substrate pool; b_K^{3+}, c_1^{3+}, *etc.*, oxidized form of cytochrome b_K, c_1, *etc.*; b_K^{2+}, c_1^{2+} *etc.*, reduced form; b_T^{3+} or b_T^{2+}, low-energy form (low-potential form) of b_T; $b_T{\sim}^{3+}$ or $b_T{\sim}^{2+}$, high-energy form (high-potential form); a_3^{3+} or a_3^{2+}, low-energy form (high-potential form) of a_3; $a_3{\sim}^{3+}$, $a_3{\sim}^{2+}$, high-energy form (low-potential form); X~I, high-energy intermediate (or high-energy state); AM, antimycin A; OL, oligomycin; UC, uncouplers; *S/F*, succinate fumarate ratio. → directions of reaction in the case of respiration; ⇢directions of reaction in the case of energy-linked reversal; × inhibition; ⇢ induction.

In Reaction (2) above, cytochrome b_T is transformed from a low- to a high-energy form utilizing the large free energy change of the reaction between cytochrome b_T and c_1. Reaction (3) is inhibited by antimycin A and Reaction (4) by oligomycin.

The reduction of cytochrome b_T by the addition of ATP is explained by the reactions below:

$$X{\sim}I + b_T^{3+} \longrightarrow b_T{\sim}^{3+} + X + I$$
$$b_T{\sim}^{3+} + c_1^{2+} \longrightarrow b_T{\sim}^{2+} + c_1^{3+}$$

The oxidation and reduction levels of other cytochromes are modified by the above reactions. Change of the redox levels of cytochrome a_3 may be explained in similar ways according to the scheme.

REFERENCES

1. D. Keilin, *Proc. Roy. Soc.*, **B98**, 312 (1925).
2. D. Keilin and E. F. Hartree, *Proc. Roy. Soc.*, **B127**, 167 (1939).

3. E. Yakushiji and K. Okunuki, *Proc. Imp. Acad. Tokyo*, **16**, 299 (1940).
4. E. G. Ball, *Z. Biochim.*, **295**, 262 (1938).
5. E. C. Slater, *J. Biochem.*, **46**, 484 (1950).
6. B. Chance, *Nature*, **169**, 215 (1952).
7. E. C. Slater, *Adv. Enzymol.*, **20**, 147 (1958).
8. D. F. Wilson and P. L. Dutton, *Biochem. Biophys. Res. Commun.*, **39**, 59 (1970).
9. B. Chance, D. F. Wilson, P. L. Dutton, and M. Erecińska, *Proc. Natl. Acad. Sci. U.S.*, **66**, 1175 (1970).
10. P. L. Dutton, J. G. Lindsay, and D. F. Wilson, *Int. Symp. Mitochondrial Membranes*, Academic Press, in press.
11. P. L. Dutton, D. F. Wilson, and C. P. Lee, *Biochemistry*, **9**, 5077 (1970).
12. N. Sato, D. F. Wilson, and B. Chance, *Biochim. Biophys. Acta*, **253**, 88 (1971).
13. N. Sato, T. Ohnishi, and B. Chance, *Biochim. Biophys. Acta*, **275**, 288 (1972).
14. A. Cittadini, T. Galleotti, B. Chance, and T. Terranova, *FEBS Letters*, **15**, 133 (1971).
15. E. C. Slater, C. P. Lee, J. A. Berden, and H. J. Wegdam, *Nature*, **226**, 1248 (1970).
16. E. C. Slater, C. P. Lee, J. A. Berden, and H. J. Wegdam, *Biochim. Biophys. Acta*, **223**, 354 (1970).
17. H. J. Wegdam, J. S. Berden, and E. C. Slater, *Biochim. Biophys. Acta*, **223**, 365 (1970).
18. B. Chance, *FEBS Letters*, **23**, 3 (1972).
19. M. Erecińska, B. Chance, D. F. Wilson, and P. L. Dutton, *Proc. Natl. Acad. Sci. U. S.*, **69**, 50 (1972).
20. A. Boveris, M. Erecińska, and M. Wagner, *Biochim. Biophys. Acta*, **256**, 223 (1972).
21. M. Erecińska and B. Chance, *in* " Energy Transduction in Respiration and Photosynthesis," ed. by E. Quagliariello *et al.*, Adriatica Editrice, Bari, p. 729 (1971).
22. M. K. F. Wikstrom, *Biochim. Biophys. Acta*, **253**, 332 (1971).
23. M. K. F. Wikstrom, *in* " Energy Transduction in Respiration and Photosynthesis," ed. by E. Quagliariello *et al.*, Adriatica Editrice, Bari, in press.
24. B. Chance, *J. Biol. Chem.*, **233**, 1223 (1958).
25. E. C. Slater and J. P. Colpa-Boonstra, *in* " Haematin Enzymes," ed. by J. E. Falk *et al.*, Pergamon, London, p. 575 (1961).
26. A. M. Pumphrey, *J. Biol. Chem.*, **237**, 2384 (1962).
27. J. D. Shore and W. W. Wainio, *J. Biol. Chem.*, **240**, 3165 (1965).
28. J. Kirschbaum and W. W. Wainio, *Biochim. Biophys. Acta*, **113**, 27(1966).

29. W. W. Wainio, J. Kirschbaum, and J. D. Shore, *in* "Structure and Function of Cytochromes," ed. by K. Okunuki *et al.*, University of Tokyo Press, Tokyo, p. 713, (1968).
30. B. Chance and B. Schoener, *J. Biol. Chem.*, **241**, 4567 (1966).
31. N. Sato, and B. Hagihara, *J. Biochem.*, **64**, 723 (1968).
32. N. Sato and B. Hagihara, *Cancer Res.*, **30**, 2061 (1970).
33. P. V. Blair, T. Oda, D. E. Green, and H. Fernandez-Moran, *Biochemistry*, **2**, 756 (1963).
34. H. Tisdale, D. C. Wharton, and D. E. Green, *Arch. Biochem. Biophys.*, **102**, 114 (1963).
35. B. Chance, D. F. Parsons, and G. R. Williams, *Science*, **143**, 136 (1964).
36. N. Sato and B. Hagihara, *Seikagaku* (*J. Jap. Biochem. Soc.*), **41**, 297 (1969) (in Japanese).
37. R. Oshino and B. Hagihara, *J. Ferment. Technol.*, in press.
38. T. Matsubara, T. Hasegawa, Y. Tochino, A. Tanaka, Y. Uemura, N. Sato, K. Hirai, and B. Hagihara, *J. Biochem.*, **72**, 1379 (1972).
39. N. Sato, T. Matsubara, and B. Hagihara, unpublished.
40. B. Hagihara, unpublished.
41. D. F. Wilson, M. Koppelman, M. Erecińska, and P. L. Dutton, *Biochem. Biophys. Res. Commun.*, **44**, 759 (1971).
42. J. S. Rieske, *Arch. Biochem. Biophys.*, **145**, 179 (1971).
43. C. C. Yang, and V. Legallais, *Rev. Sci. Instr.*, **25**, 801 (1954).
44. B. Chance, *in* "Methods in Enzymology," ed. by S. P. Colowick and N. O. Kaplan, Academic Press, New York, Vol. IV, p. 273 (1957).
45. B. Hagihara, *Protein Nucleic Acid Enzyme*, **13**, 30 (1968) (in Japanese).
46. P. L. Dutton, M. Erecińska, N. Sato, Y. Mukai, M. Pring, and D. F. Wilson, *Biochim. Biophys. Acta*, **267**, 15 (1972).

Redox Changes of Longer-Wavelength Cytochrome *b* (b_{566}) in Rat Liver Mitochondria

Saburo Muraoka and Masafumi Okada

Faculty of Pharmaceutical Sciences, Tokushima University, Tokushima, Japan

The redox changes of longer-wavelength cytochrome *b* (b_{566}) were divided to 2 types: Energy-dependent reduction and oxidant-induced reduction. Oxidant-induced reduction of b_{566} was observed in rat liver mitochondria in State 5 on addition of ferricyanide. Addition of menadione caused oxidation of b_{566} and reduction of cytochrome $c_1(+c)$. In the presence of menadione, ferricyanide induced reduction of b_{566} and oxidation of cytochrome c_1+c, and exhaustion of the added ferricyanide resulted in oxidation of b_{566} and reduction of cytochrome c_1+c. Antimycin A increased the extent of the ferricyanide-induced redox change of b_{566} from 20 to 68% of the total absorbance at 566 nm. The energy-dependent type of reduction was abolished by uncouplers. However, uncouplers hardly affected the oxidant-induced reduction in the presence of antimycin A.

The functional role of longer-wavelength cytochrome *b* is under extensive study in several laboratories. This cytochrome was named cytochrome b_T by Chance's group and cytochrome b_i by Slater and his co-workers. In this paper cytochrome b_{566} or b' is used for the longer-wavelength cytochrome *b*, and cytochrome b_{560} or simply cytochrome *b* for the classical cytochrome *b*. The redox changes of cytochrome b_{566} seem to be divided to 2 types, which may be described as energy-dependent reduction and oxidant-dependent reduction, though it remains uncertain whether they are based on different mechanisms or essentially on the same mechanism.

Abbreviations used: EDTA, ethylenediaminetetraacetic acid; PCP, pentachlorophenol.

The energy-dependent reduction of cytochrome b_{566} was discussed by Chance *et al.* (*1*, *2*). They reported that ATP, added to pigeon heart mitochondria and submitochondrial particles, caused a shift of the absorption peak of cytochrome *b* to 555 nm (at 77°K). Slater and his co-workers (*3–5*), on the other hand, reported that ATP added to phosphorylating submitochondrial particles in the presence of substrate caused a shift of the *b* band to 565 nm (at 20°C), *i.e.*, to the red. The discrepancy has recently been solved by Sato *et al.* (*6*, *7*), since they found that the 555 nm band previously observed is a low-wavelength shoulder of a *b* species absorbancy at high wavelength.

The oxidant-dependent reduction of cytochrome *b* components was first described by Chance (*8*). He found that oxygen induced the reduction of cytochrome *b* in antimycin A-treated yeast cells. In the presence of antimycin A a similar phenomenon was observed by Pumphrey (*9*) in electron transport particle by Rieske (*10*) in the *b*-c_1 complex, and by Wilson *et al.* (*11*) in succinate-cytochrome *c* reductase and in a cholate-treated preparation of chicken heart mitochondria. In the light of recent findings on *b*-type cytochromes made in several laboratories (*3–5*, *12*, *13*), Wilson *et al.* (*11*) reported that the cytochrome *b* reduced under these experimental conditions was longer-wavelength cytochrome *b* (b_{566}) and that the reduction of b_{566} required activation of electron transport through cytochrome c_1. More recently, Erecińska *et al.* (*14*) measured the rate of reduction of b_{566} and that of oxidation of cytochrome c_1 by oxygen in anaerobic, uncoupled pigeon heart mitochondria. They postulated that the aerobic reduction of cytochrome b_{566} is directly related to energy conservation at Site II.

In studies of the effects of various oxidants on the redox state of cytochrome components in rat liver mitochondria oxidizing succinate, we obtained spectral evidence indicating that ferricyanide caused reduction of cytochrome b_{566} and oxidation of cytochrome c_1+c and that menadione caused oxidation of cytochrome b_{566} and reduction of cytochrome c_1 $(+c)$ under conditions where the effects of energy coupling at Sites I and III were eliminated by rotenone and KCN, respectively.

Spectrophotometric Analysis

Rat liver mitochondria were isolated by the method of Hogeboom (*15*), as described by Myers and Slater (*16*). Protein was determined by the biuret method, as described by Cleland and Slater (*17*). All reactions were carried out in one of the following media: Medium A; 25 mM Tris-HCl buffer,

50 mM sucrose, 5 mM $MgCl_2$, 2 mM EDTA, and 15 mM KCl; medium B; 30 mM Tris-HCl buffer, 70 mM sucrose, 200 mM mannitol, and 20 mM KCl. The other components used are indicated in the legends to the relevant figures. The final volume of the mixture was 3 ml and the pH was 7.4. Measurements of the absorbance changes of cytochrome components were made with a Hitachi 2-wavelength spectrophotometer, Model 356, using the following wavelength pairs: Cytochrome *b* (classical cytochrome *b*, b_{560}), 560 nm minus 575 nm; cytochrome b_{566}, 566 nm minus 575 nm; cytochrome c_1+c, 550 nm minus 540 or 575 nm. Difference spectra were obtained by one of the following 2 procedures. Procedure A was the 2-wavelength/double-beam method described previously (*18*). In procedure B difference spectra were obtained by scanning the wavelengths, taking 575 nm as a reference wavelength.

Effect of External Oxidants

A highly reduced state of the respiratory-chain components in mitochondria was induced by addition of succinate with rotenone via State 2→3→4→5 (*25*). On addition of ferricyanide to State 5 mitochondria, cytochrome c_1+c was nearly 100% oxidized in the presence of KCN, whereas a considerable part of the cytochrome *b* components remained in the reduced state (37% at 560 nm and 62% at 566 nm) as shown in Fig. 1. Figure 2 shows that the difference spectrum between mitochondria treated with ferricyanide and mitochondria in State 2 has a peak at 565 nm (Curve B) and the absorption due to b_{566} vanishes on adding uncouplers, such as flufenamic acid (*19*) (Curve C).

Subsequently, the effect of menadione on the redox state of cytochrome components was tested using various wavelength pairs, since menadione is known as a mediator of electron flow from various dehydrogenases by interaction with respiratory-chain components on the substrate side of the antimycin site (*20*, *21*). In the presence of rotenone, antimycin A and KCN, menadione added to anaerobic mitochondria oxidizing succinate induced 3 phases in the redox state of b_{566} after the rapid oxidation process, as shown in B, C, and D in Fig. 3A. In phases B and C, cytochrome b_{566} was oxidized, whereas cytochrome c_1+c, measured at 550–575 nm, was highly reduced. In phase D, reduction was observed of all components, such as b_{566}, b_{560} (not shown), cytochrome c_1+c and pigment 558 (*1*, *2*, *6*, *7*). The results obtained by the 2-wavelength method were confirmed by the

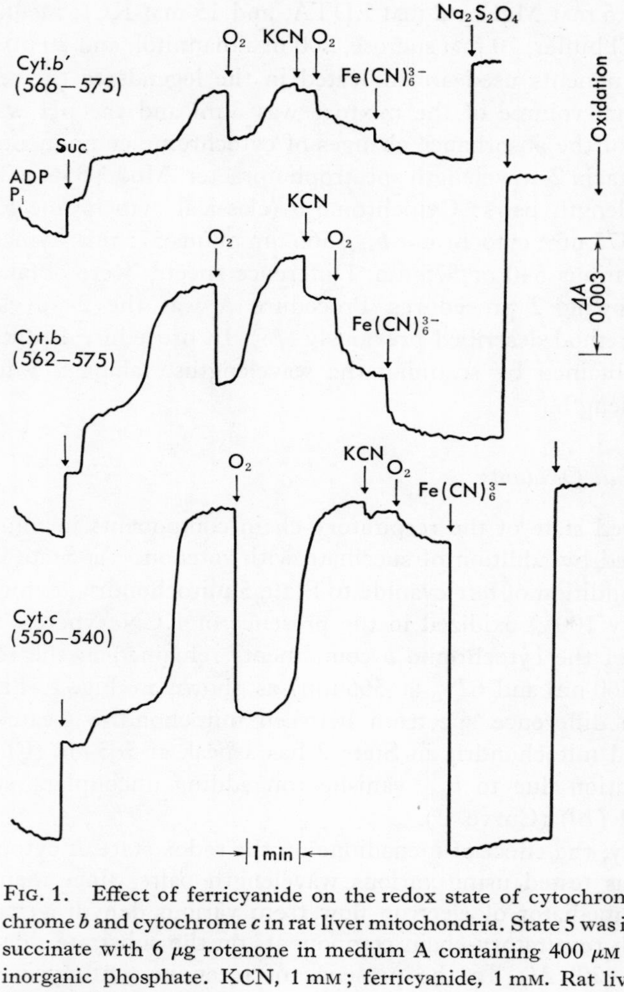

FIG. 1. Effect of ferricyanide on the redox state of cytochrome b' (b_{566}), cytochrome b and cytochrome c in rat liver mitochondria. State 5 was induced by 10 mM succinate with 6 μg rotenone in medium A containing 400 μM ADP and 10 mM inorganic phosphate. KCN, 1 mM; ferricyanide, 1 mM. Rat liver mitochondria, 1.6 mg/ml.

difference spectrum shown in Fig. 3B. The difference spectrum before and after addition of menadione in phase C indicated the oxidation of b_{566} and reduction of cytochrome c_1+c, as shown in Curve C-A, though little change was observable in cytochrome b_{560}. Curve D-C indicates the reduction of b_{566}, b_{560} and pigment 558 during the transition from phase C to phase D.

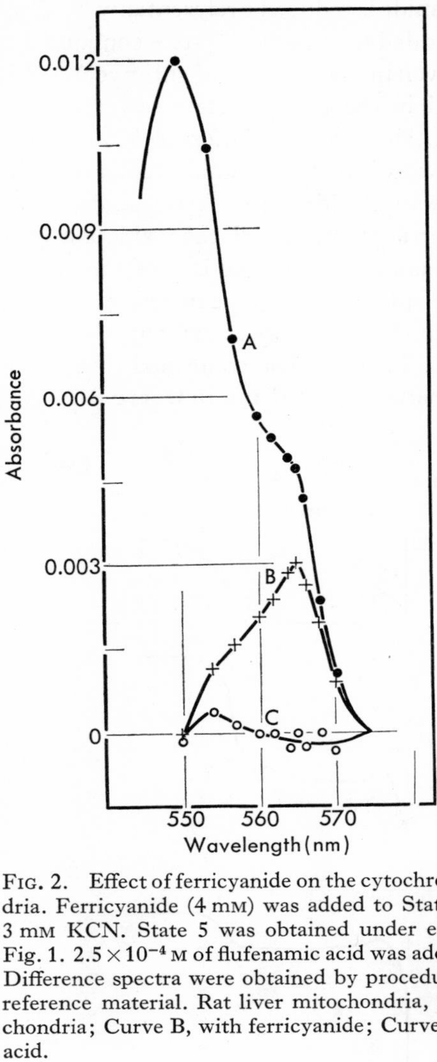

FIG. 2. Effect of ferricyanide on the cytochrome *b* spectrum in rat liver mitochondria. Ferricyanide (4 mM) was added to State 5 mitochondria in the presence of 3 mM KCN. State 5 was obtained under essentially the same conditions as in Fig. 1. 2.5×10^{-4} M of flufenamic acid was added after treatment with ferricyanide. Difference spectra were obtained by procedure A taking State 2 mitochondria as reference material. Rat liver mitochondria, 2.46 mg/ml. Curve A, State 5 mitochondria; Curve B, with ferricyanide; Curve C, with ferricyanide and flufenamic acid.

The oxidation of b_{566} may be caused by the transfer of reducing equivalents from b_{566} to menadione and the simultaneous reduction of cytochrome c_1+c is caused by the acceptance of reducing power from succinate and also from b_{566} *via* menadione through an antimycin-insensitive path.

It is well known that ferricyanide is reduced preferentially by cytochrome *c* (*22*). When ferricyanide was added to a reaction system containing menadione, an electron flow occurred even in the presence of antimycin A. Figure 4A shows that ferricyanide added in the reduced state scarcely affected b_{566} in the absence of antimycin A. However, oxidation of b_{566} occurred when ferricyanide (yellow) was converted to ferrocyanide (colorless). The difference spectrum immediately after addition of ferricyanide and after its conversion to ferrocyanide (ferri=0) indicated the oxidation of b_{566} and reduction of cytochrome c_1+c, as shown in Curve B-A of Fig. 4B. Classical cytochrome *b* did not show an anomalous change in redox state under the same conditions, but changed in the same way as cytochrome c_1+c, though to a smaller extent (not shown). This was also confirmed from the fact that the point of zero absorbance change is at 561 nm in Curve B-A. Antimycin

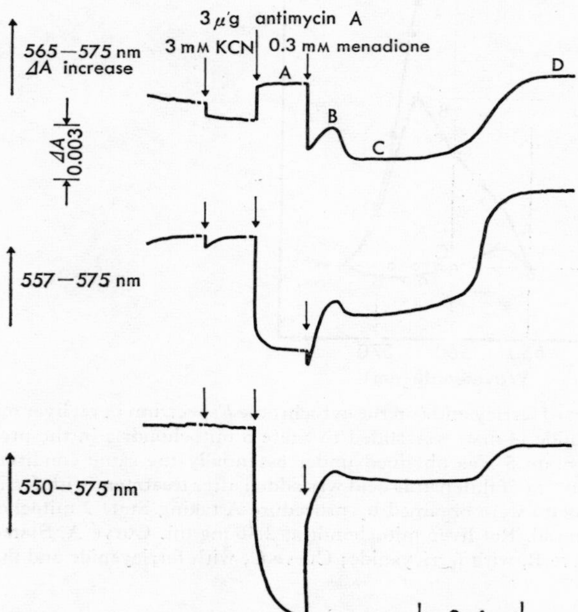

FIG. 3A. Effect of menadione on the redox state of cytochrome components. Rat liver mitochondria (7.0 mg/ml) were suspended in medium B and State 5 was induced by 10 mM succinate with 4 μg rotenone. Menadione was added to State 5 mitochondria after addition of KCN and antimycin A.

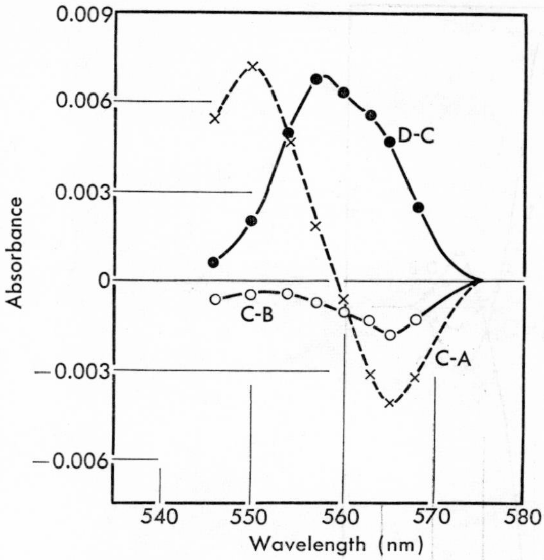

FIG. 3B. Absorption spectra of cytochrome components. The wavelength scanning was performed point by point with separate incubation for each wavelength couple, taking 575 nm as reference wavelength. Curve C-B, b_{566} was oxidized after addition of menadione in phase B; Curve C-A, b_{566} was oxidized in phase C($+\Delta A_{550}/-\Delta A_{565}=1.7$; point of zero absorbance change, 559 nm); Curve D-C, cytochrome components were reduced in phase D.

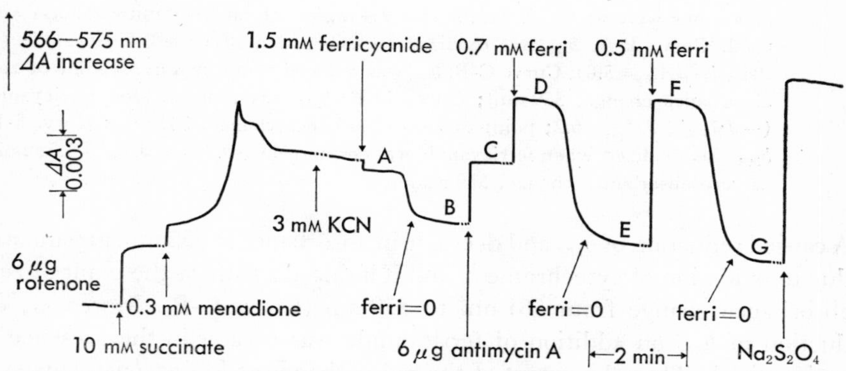

FIG. 4A. Effect of ferricyanide on the redox state of cytochrome b_{566} (measured at 566 nm minus 575 nm) in the presence of menadione. Rat liver mitochondria (6.8 mg/ml) were suspended in medium B.

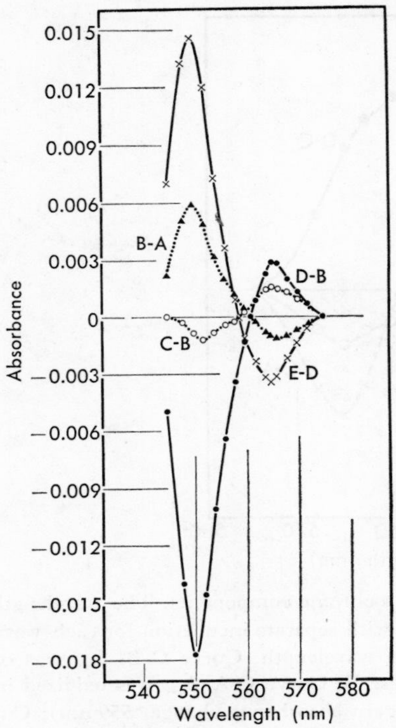

FIG. 4B. Absorption spectra of cytochrome components. Difference spectra were obtained by procedure B, with 575 nm as the reference wavelength. The conditions were as for A except that 3.4 mg/ml of rat liver mitochondria were used. Curve B-A, b_{566} was oxidized on exhaustion of added ferricyanide ($+\Delta A_{550}/-\Delta A_{565}=5.0$); Curve C-B, b_{566} was reduced by antimycin A (point of zero absorbance change, 559 nm); Curve D-B, b_{566} was reduced after ferricyanide ($-\Delta A_{550}/+\Delta A_{565}=6.2$; point of zero absorbance change, 561 nm); Curve E-D, b_{566} was oxidized when ferricyanide was consumed ($+\Delta A_{550}/-\Delta A_{565}=4.3$; point of zero absorbance change, 559 nm).

A caused reduction of b_{566} and decrease in absorbance at 552 nm, presumably due to oxidation of cytochrome c_1 and it induced a shift of the point of zero absorbance change from 561 nm to 559 nm (Curve C-B). Moreover, reduction of b_{566} on addition of ferricyanide was greater in the presence of antimycin A. Thus the extent of the redox change of b_{566} as a percentage of the total difference between the absorbance on reduction with $Na_2S_2O_4$ and on oxidation with rotenone in the absence of substrate, was 20% in the absence of antimycin A and 68% in its presence. The redox change induced by ferricyanide was reproducible as indicated in D, E, F, and G in Fig. 4A and the reduction and oxidation of b_{566} were usually accompanied by oxidation and reduction, respectively, of cytochrome c_1+c (Curve D-B and Curve E-D in Fig. 4B). The redox change of b_{566} is probably not controlled by the ferricyanide/ferrocyanide ratio, since the amount of ferricyanide added did not influence the level of reduced b_{566} but only the duration of its reduced state, and since addition of ferrocyanide to b_{566} in the reduced state did not cause any change in the degree of reduction. Thus the transition of b_{566} from the reduced to the oxidized state is caused by a stoppage of electron flow resulting from exhaustion of the electron acceptor. In other words, the reduction or oxidation of b_{566} depends strictly on whether there is an electron flow through cytochrome c_1+c to ferricyanide.

Oxidant-induced Type and Energy-dependent Type

There was a distinct difference between the absorbance change of b_{566} in the presence of PCP (+PCP, solid line in Fig. 5) and in its absence (—PCP, dotted line in Fig. 5) when antimycin A was absent. In the presence of the uncoupler, the absorbance at 566 nm became much lower than in the system without uncoupler. This may be due to inhibition by the uncoupler of the energy-dependent reduction of b_{566}, as reported by several workers (*23*, *24*). Ferricyanide induced a decrease in absorbance at 566 nm and this was followed by an increase in absorbance on exhaustion of ferricyanide (A, B in Fig. 5). This change was different in direction from that in the uncoupler-free system, as shown by the dotted curve in Fig. 5, and was similar to those of b_{560} and cytochrome c_1+c. Considering the contribution of the classical cytochrome *b* to the absorbance at 566 nm (*7*, *14*), it seems that this change is mainly due to the absorbance change of b_{560} at 566 nm, although spectral evidence for this is not yet available. As shown in D, E, F, and G in Fig. 4A, however, PCP did not influence the ferricyanide-induced reduction of b_{566} in the presence of antimycin A. It was also noted that the oxidant-dependent reduction of b_{566} was hardly affected by previous addition of ATP or oligomycin to reaction systems containing menadione and antimycin A.

The oxidant-induced type was characterized by a larger redox change (not shown) at 566 nm compared with energy-dependent reduction. Moreover, the former was usually accompanied by a redox change in the opposite direction of a counter-component such as cytochrome c_1+c, but little is known about the behavior of the counter-component(s) *e.g.*, in ATP-dependent reduction of b_{566} (*24*). Coexistence of menadione and antimycin A was essential for demonstrating typical oxidant-induced reduction in the present experiments. Using these agents, the redox change of b_{566} and

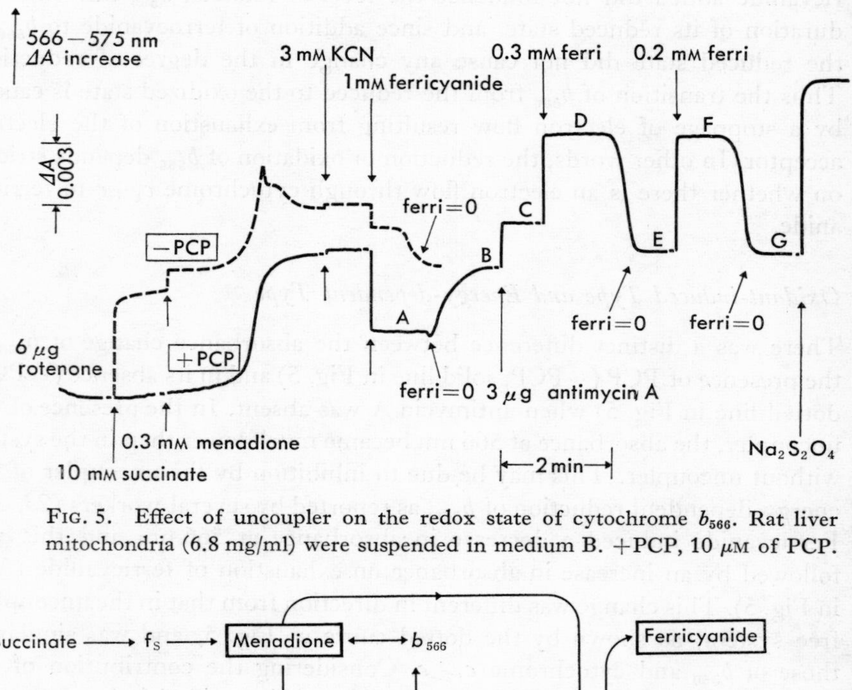

FIG. 5. Effect of uncoupler on the redox state of cytochrome b_{566}. Rat liver mitochondria (6.8 mg/ml) were suspended in medium B. +PCP, 10 μM of PCP.

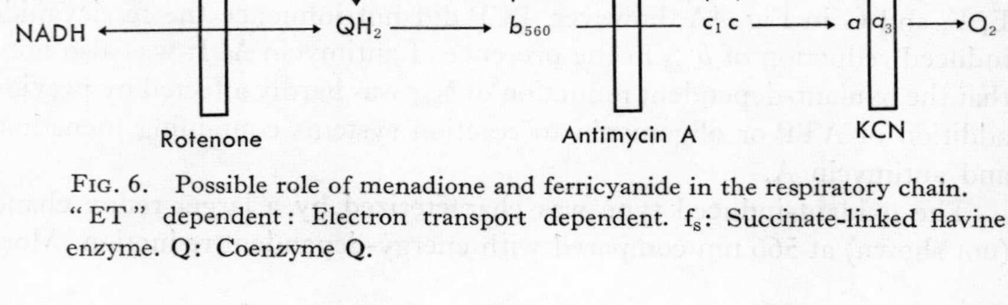

FIG. 6. Possible role of menadione and ferricyanide in the respiratory chain. " ET " dependent: Electron transport dependent. f_S: Succinate-linked flavine enzyme. Q: Coenzyme Q.

cytochrome c_1+c became reproducible on addition of a limited amount of ferricyanide, since menadione induced an electron flow from substrate to ferricyanide in the presence of antimycin A, as shown in Fig. 6. The energy-dependent change in b_{566} may be differentiated from the oxidant-dependent one by the use of uncoupler in some cases, as in Fig. 5, since uncoupler halted the energy-dependent type but not the ferricyanide-induced reduction. However, it was rather difficult to distinguish both types in systems without menadione and antimycin A, as seen in Curves B and C in Fig. 2. In these spectra, the identification of b_{566} by ferricyanide may probably be explained as the oxidant-induced type. The disappearance of a peak of b_{566} caused by uncoupler may be due to a release of the energy coupling at Site II resulting in slight reduction of cytochrome c_1 and oxidation of b_{566}, though the possibility of an uncoupler effect on energy-dependent reduction could not be ruled out.

REFERENCES

1. B. Chance and B. Schoener, *J. Biol. Chem.*, **241**, 4567–4573 (1966).
2. B. Chance, C. P. Lee, and B. Schoener, *J. Biol. Chem.*, **241**, 4574–4576 (1966).
3. E. C. Slater, C. P. Lee, J. A. Berden, and H. J. Wegdam, *Nature*, **226**, 1248–1249 (1970).
4. E. C. Slater and J. A. Berden, *in* " Wenner-Gren Symp. on Structure and Function of Oxidation-Reduction Enzymes," ed. by Å. Åkeson and A. Ehrenberg, Pergamon Press, Oxford, pp. 291–301 (1972).
5. E. C. Slater, C. P. Lee, J. A. Berden, and H. J. Wegdam, *Biochim. Biophys. Acta*, **223**, 354–364 (1970).
6. N. Sato, D. F. Wilson, and B. Chance, *FEBS Letters*, **15**, 204–213 (1971).
7. N. Sato, D. F. Wilson, and B. Chance, *Biochim. Biophys. Acta*, **253**, 88–97 (1971).
8. B. Chance, *in* " Abstr. 2nd Int. Congr. Biochem.," Paris, p. 32 (1952).
9. A. M. Pumphrey, *J. Biol. Chem.*, **237**, 2384–2390 (1962).
10. J. S. Rieske, *Arch. Biochem. Biophys.*, **145**, 179–187 (1971).
11. D. F. Wilson, M. Koppelman, M. Erecińska, and P. L. Dutton, *Biochem. Biophys. Res. Commun.*, **44**, 759–766 (1971).
12. P. L. Dutton, D. F. Wilson, and C. P. Lee, *Biochemistry*, **26**, 5077–5082 (1970).
13. D. F. Wilson and P. L. Dutton, *Biochem. Biophys. Res. Commun.*, **39**, 59–64 (1970).

14. M. Erecińska, B. Chance, D. F. Wilson, and P. L. Dutton, *Proc. Natl. Acad. Sci. U.S.*, **69**, 50–54 (1972).
15. G. H. Hogeboom, *in* " Methods in Enzymology," ed. by S. P. Colowick and N. O. Kaplan, Academic Press, New York, Vol. 1, p. 16 (1955).
16. D. K. Myers and E. C. Slater, *Biochem. J.*, **67**, 558–572 (1957).
17. K. W. Cleland and E. C. Slater, *Biochem. J.*, **53**, 547–556 (1953).
18. S. Muraoka, K. Takahashi, and M. Okada, *Biochim. Biophys. Acta*, **267**, 291–299 (1972).
19. H. Terada and S. Muraoka, *Mol. Pharmacol.*, **8**, 95–103 (1972).
20. J. P. Colpa-Boonstra, and E. C. Slater, *Biochim. Biophys. Acta*, **27**, 122–133 (1958).
21. E. J. De Haan and R. Charles, *Biochim. Biophys. Acta*, **180**, 417–419 (1969).
22. R. W. Estabrook, *J. Biol. Chem.*, **236**, 3051–3057 (1961).
23. H. J. Wegdam, J. A. Berden, and E. C. Slater, *Biochim. Biophys. Acta*, **223**, 365–373 (1970).
24. M. K. F. Wikström, *Biochim. Biophys. Acta*, **253**, 332–345 (1971).
25. B. Chance and G. R. Williams, *Adv. Enzymol.*, **17**, 65–134 (1956).

MOLECULAR BASIS OF ANION TRANSPORT

Chloride and Hydroxyl Ion Conductance of Sheep Red Cell Membranes

D. C. Tosteson, R. B. Gunn, and J. O. Wieth*

*Department of Physiology and Pharmacology, Duke University Medical Center, Durham, North Carolina, U.S.A., and Department of Biophysics, University of Copenhagen, Copenhagen, Denmark**

In this paper we will present evidence in support of the conclusion that membrane conductance of Cl^- (G_{Cl}) in sheep (and presumably human) red cell membranes is about 10^{-6} ohm^{-1} cm^{-2} at 37°C. This value is about 10^2 times greater than membrane conductance of K^+ and Na^+ (G_K and G_{Na}) computed from normal passive fluxes, but 10^{-6} times that computed from the Cl^- equilibrium exchange flux. Thus, only about 10^{-6} of normal Cl^- transport occurs by ionic diffusion, the remainder presumably involving some sort of facilitated exchange. Membrane conductance of OH^- (G_{OH}) estimated by a similar method is about 10^{-7} ohm^{-1} cm^{-2}.

The high Cl^- permeability of the plasma membranes of red blood cells has long been recognized to play an important role in the capacity of red blood cells to transport CO_2 between tissues and lung. However, the mechanism of Cl^- transport has remained obscure. For example, in 1959 we reported that the Cl^- equilibrium exchange flux in human and ox erythrocytes suspended in normal saline at 20°C is about 10^{-8} moles cm^{-2} sec^{-1} (*1*). If all of this flux were by ionic diffusion, we computed that the Cl^- conductance would be about 0.1 ohm^{-1}cm^{-2}. However, at that time, we pointed out that it was impossible to decide from these measurements whether Cl^- transport occurred by ionic diffusion or by some kind of facilitated exchange requiring interaction of Cl^- with components of the membrane. During the past several years, work in many laboratories (*2–13*) has suggested that a substantial fraction of the Cl^- equilibrium exchange flux occurs by facilitated exchange. For example, the Cl^- equilibrium exchange flux has a large temperature coefficient (*3*), is independent of Cl^- concentration above 60 mM (*13*), is maximal at pH 8.0 (*4*), and is poisoned by inhibitors such as trinitrocresolate (*5*, *13*). The goal of the investigation described in this paper is to estimate the Cl^- conductance of the red cell membrane, *i.e.*, that moiety of Cl^- transport which occurs by ionic diffusion rather than by some facilitated process.

The most direct way to make such an estimate would be to measure net Cl^- transport at different known values of the electrical potential difference across the membrane (V_m). Unfortunately, measurements of membrane potential by microelectrode techniques are not possible because of the small size of mammalian red cells. Measurements of membrane potential and resistance have been made in the large red cells of amphiuma by Lassen and Hoffman (*14*) and extended by Lassen and Vestergard-Bogind. Because of the technical problem of measuring V_m, we have been forced to resort to an indirect method to estimate the Cl^- conductance in sheep red cell membranes. We have made use of the fact that the antibiotic valinomycin produces highly selective K^+ permeability in red cell membranes (*15*). As pointed out by Hunter (*9*), this offers the possibility to change V_m by changing the concentration ratio of K^+ in the presence of valinomycin.

The argument is shown in more detail in Table I. Equation 1 states that the sum of the membrane currents of K^+ and Cl^- (I_K and I_{Cl}) is zero, and that these are the only significant ionic currents. Equations 2 and 3 state that I_K and I_{Cl} are equal to the product of a driving force ($E_K - V_m$) and ($E_{Cl} - V_m$) (where E_K and E_{Cl} are the equilibrium potentials of K^+ and Cl^-, and V_m is the membrane potential) and a chord conductance (G_K and

TABLE I
Estimation of Cl^- and K^+ Conductances

Eq. No.	Equation
1	$I_K + I_{Cl} = 0$
2	$I_K = G_K(E_K - V_m) = Z_K F M_K$
3	$I_{Cl} = G_{Cl}(E_{Cl} - V_m) = Z_{Cl} F M_{Cl}$
4	$\dfrac{{}^iM_K}{{}^oM_K} = \dfrac{(K)_o}{(K)_i} \exp \dfrac{-Z_K F V_m}{RT}$

Known: E_K, E_{Cl}, M_K, M_{Cl}, ${}^iM_K$, ${}^oM_K$, $(K)_o$, $(K)_i$. Unknown: V_m, G_K, G_{Cl}.

G_{Cl}). Equation 4 states that the ratio of K^+ influx to K^+ outflux is equal to the ratio of the electrochemical activities of K^+ in external and internal solutions. If this system of equations describes red cells, the electrical potential difference (V_m) and the ionic conductances (G_K and G_{Cl}) can be computed from measurements of the external and internal concentrations and the net fluxes of K^+ and Cl^-, as well as K^+ influx and outflux even in the presence of a high exchange flux of Cl^-. This paper describes experiments which show that Eqs. 1, 2 and 3 are adequate descriptions of sheep red cells exposed to high concentrations of valinomycin. We then use the assumptions contained in equation 4 to calculate V_m. Knowing V_m, G_K and G_{Cl} are easily obtained from Eqs. 2 and 3.

Figure 1 shows the results of a typical experiment designed to test the relevance of Eq. 1. High K^+ (HK) sheep red cells were washed in 0.17 M

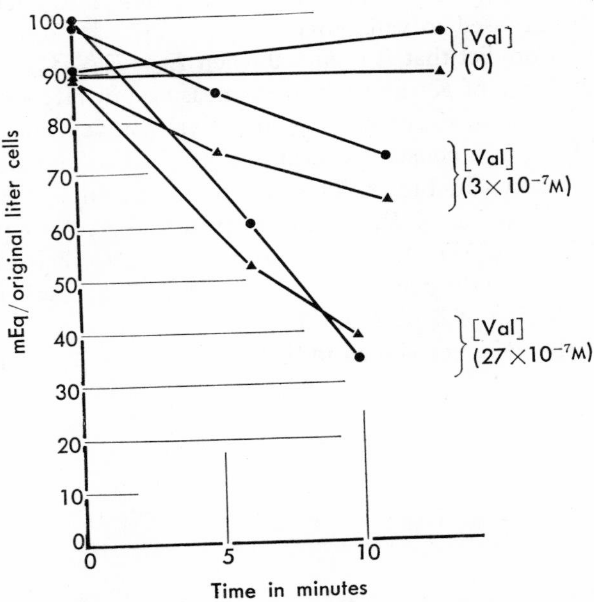

FIG. 1. K^+ and Cl^- contents of HK sheep red cells. At zero time, ethanol sufficient to make a final concentration of 0.01 v/v was added to the suspension which contained red cells in a concentration of 0.01 v/v. The ethanol contained sufficient valinomycin (Val) to make final concentrations of 3×10^{-7}, or 27×10^{-7} M. Temperature: 37°C. ▲ K^+ content; ● Cl^- content.

NaCl and suspended in an unbuffered medium containing known concentrations of KCl and NaCl to make a volume fraction of cells of 0.01. In all cases, the CO_2 concentration in the medium was reduced by constant bubbling with base-washed N_2, and the temperature was 37°C. At time zero, valinomycin dissolved in ethanol was added. The final concentration of ethanol in each flask was 0.01 (v/v). Cell contents of K^+ and Cl^- plotted on the ordinate were determined on cells washed once in ice-cold 0.34 M sucrose. This washing procedure did not lead to loss of intracellular K^+ or Cl^-. Note in Fig. 1 that loss of K^+ and Cl^- occurred at about the same rate even though both processes were more rapid at higher concentrations of valinomycin. This point is made more quantitatively in Table II. Measured values of I_{Cl} and I_K as well as their ratio are shown for several concentrations of valinomycin in a typical experiment. The ratio is not significantly different from 1 at all valinomycin concentrations tested. Furthermore, net Na^+ fluxes were always zero. Therefore, Eq. 1 describes adequately HK sheep red cells exposed to valinomycin.

Equations 2 and 3 predict that $I_K = I_{Cl} = 0$ when $E_K = E_{Cl} = V_m$. To test this prediction, net fluxes of K^+ and Cl^- were measured in HK sheep red cells suspended at 37°C in media containing 10^{-6} M valinomycin and various concentrations of K^+ but a constant concentration of Cl^-. The results of the experiment are shown in Fig. 2. The membrane currents of K^+ and Cl^- are plotted as a function of E_K calculated from the Nernst equation, $E_K = (RT/F)(\ln [(K^+)_o/(K^+)_i])$. Note that both I_K and I_{Cl} become zero when E_K equals the similarly computed value of E_{Cl}. This result is consistent with the prediction of Eqs. 2 and 3.

On the basis of the evidence shown in Table II, and Figs. 1 and 2, we

TABLE II
HK Sheep Red Cells

Valinomycin (M × 10^7)	M_K (10^{-3} moles l^{-1} min^{-1})	M_{Cl}	M_K/M_{Cl}
0	0.0	0.0	—
3	−3.0	−2.7	1.1
9	−4.2	−3.3	1.3
27	−6.0	−6.3	0.9
81	−6.8	−8.4	0.8

Temperature: 37°C.

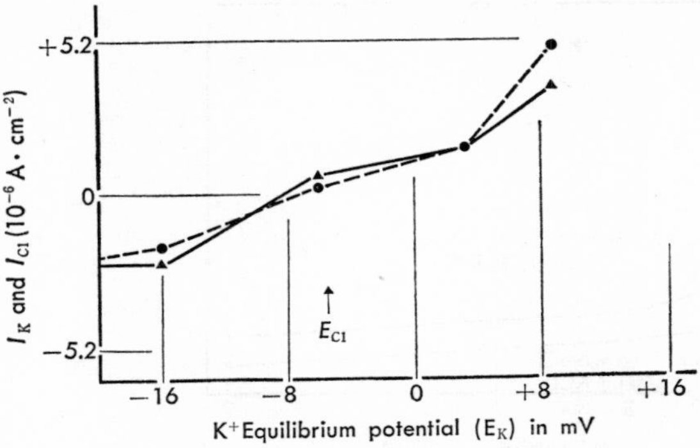

FIG. 2. I_K and I_{Cl} of HK sheep red cells. The value of E_{Cl} was constant in all experiments and is indicated by the arrow. E_K and E_{Cl} were calculated from measured concentrations of the ions and the Nernst equation. Valinomycin: 10^{-6} M. Temperature: 37°C.

TABLE III
HK Sheep Red Cells

Valinomycin (M × 10^7)	${}^iM_K$ (10^{-3} moles l^{-1} min^{-1})	${}^oM_K$	${}^oM_K/{}^iM_K$	$(K)_i/(K)_o$	V_m (10^{-3}V)
3	0.28	3.3	12	13.5	−3.6
9	0.68	4.9	7.2	13.4	−16
27	1.5	7.4	5.1	13.5	−26
81	2.0	8.9	4.4	13.2	−27

Temperature: 37°C.

decided to apply this argument to the estimation of V_m, G_K and G_{Cl} in HK sheep red cells exposed to valinomycin. The results of a typical experiment are shown in Table III. In addition to K^+ and Cl^- concentrations and net fluxes, K^+ influx was measured with $^{42}K^+$, and K^+ outflux from the difference between influx and net flux. From these data and the assumptions expressed in equation 4, V_m was calculated. Note that the ratio of K^+ outflux (${}^oM_K$) to influx (${}^iM_K$) decreased progressively as valinomycin concentration increased. Since the ratio of internal to external K^+ concentrations $[(K^+)_i/(K^+)_o]$ was the same in all experiments, Eq. 4 requires that the reduction in the flux ratio be due to a change in V_m.

Figure 3 shows calculated values for V_m as a function of valinomycin

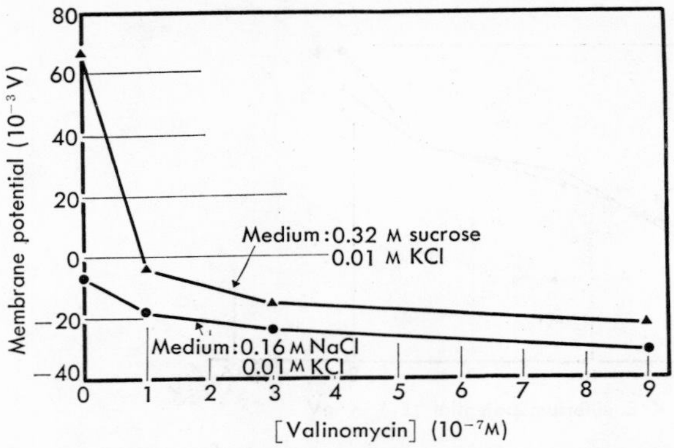

FIG. 3. Calculated values for membrane potential of HK sheep red cells. Results are shown for cells suspended in 2 different media. In both cases, the concentration of cells was 0.01 v/v. Temperature: 37°C.

concentration for 2 different media. In one case, the medium contained 0.16 M NaCl, 0.01 M KCl, in the other case 0.32 M sucrose, 0.01 M KCl. In the former, E_{Cl} was −7 mV and in the latter +70 mV, while E_K was −70 mV in both cases. In both cases, $V_m = E_{Cl}$ in the absence of valinomycin and moves toward E_K as the concentration of the compound increases because of the resultant specific elevation of K^+ permeability. Note, however, that this effect saturates so that V_m reaches values between −30 and −40 mV at the highest valinomycin concentrations despite the fact that E_K is about −70 mV.

Figure 4 shows computed values for G_K and G_{Cl} in a medium containing 0.16 M NaCl and 0.01 M KCl plotted on the ordinate as a function of valinomycin concentration. G_K increases progressively, but G_{Cl} remains relatively constant as valinomycin concentration increases. The extrapolated value for G_{Cl} at zero valinomycin is about 2×10^{-6} ohm^{-1} cm^{-2}. Comparable results for experiments carried out in 0.32 M sucrose, 0.01 M KCl are shown in Fig. 5. In this case, the extrapolated value of G_{Cl} is about 1×10^{-6} ohm^{-1} cm^{-2}, consistent with the reduced Cl^- concentration. These values may be compared with the maximum value of 5×10^{-4} ohm^{-1} cm^{-2} for G_{Cl} estimated by Lassen and Hoffman from direct measurements of the electrical resistance of the amphiuma red cell membrane at 18°C (*14*). Thus, in both cases, G_{Cl} estimated by this method in HK sheep red cells at 37°C is about 10^{-5} times that calculated from the Cl^- equilibrium exchange flux in human

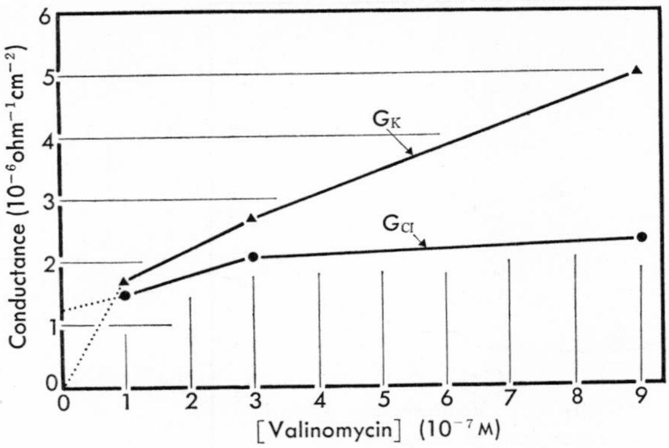

FIG. 4. Calculated values for G_K and Cl^- (G_{Cl}) of HK sheep red cells. These results are from the same experiments as the lower curve in Fig. 3. Medium: 0.32 M sucrose, 0.01 M KCl. Temperature: 37°C.

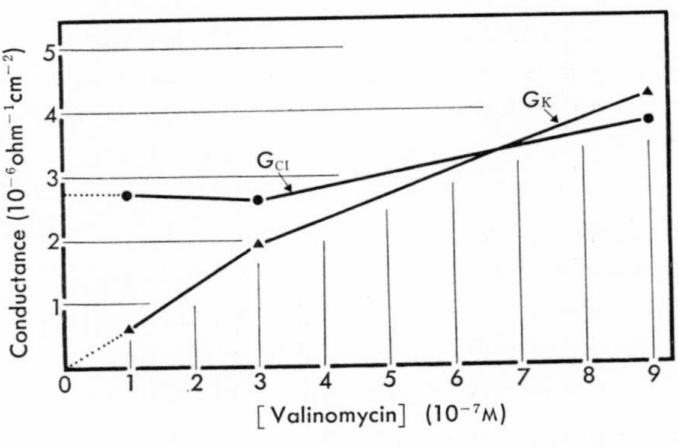

FIG. 5. Same as Fig. 4. For the experiments shown in the upper curve on Fig. 3. Medium: 0.16 M NaCl, 0.01 M KCl. Temperature: 37°C.

and ox red cells at 20°C. From the known temperature dependence of Cl^- equilibrium exchange flux, this means that only about 10^{-6} of the total Cl^- flux occurs by a process of ionic diffusion. On the other hand, G_{Cl} is about 10^2 times greater than G_K estimated from the passive K^+ flux in the absence of valinomycin. Thus, the electrical properties of the normal red cell membrane appear to be dominated by Cl^- but with a permselectivity of 10^2 rather than 10^8.

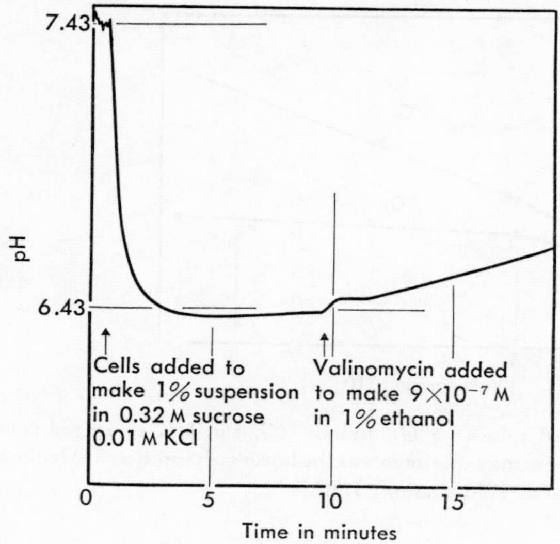

FIG. 6. External pH of HK sheep red cells. At zero time, HK sheep red cells which had been washed in 0.17 M NaCl were added to unbuffered medium in which the CO_2 concentration was reduced by bubbling with washed N_2. Temperature: 37°C.

This method can also be used to define both facilitated exchange and ionic diffusion processes for OH^- transport. Figure 6 shows the measured pH of a CO_2-free, unbuffered medium containing 0.32 M sucrose, 0.01 M KCl when normal HK sheep red cells are added. The sudden change in E_{Cl} and V_m from -7 to $+70$ mV results in a rapid Cl^--OH^- exchange which acidifies the medium until E_{OH} once more equals E_{Cl}. The half-time for this process in our experiments was 20 sec and undoubtedly was determined largely by the kinetics of mixing. Addition of sufficient valinomycin to return V_m to -70 mV fails to produce a comparably rapid return of E_{OH} to its original value but rather leads to a much slower linear rate of change of pH in the alkaline direction. It seems reasonable to conclude that the former process occurs by facilitated exchange and the latter by ionic diffusion. If this interpretation is correct, G_{OH} can be calculated from I_{OH} and V_m as shown in Table IV. I_{OH} was measured by removing the cells and back-titrating the medium to the pH which obtained at the moment of addition of valinomycin. The resultant values of G_{OH} are about one-tenth

TABLE IV
HK Sheep Red Cells: Valinomycin OH^- (H^+) Conductance

Valinomycin ($M \times 10^7$)	I_{OH} ($A\ cm^{-2} \times 10^8$)	$E_{OH} - V_m$ (10^{-3} V)	G_{OH} ($ohm^{-1}\ cm^{-2} \times 10^6$)
1.0	0.64	72	0.09
3.0	1.0	88	0.11
9.0	1.3	105	0.12
27.0	1.8	103	0.18

Medium: 0.32 M sucrose, 0.01 M KCl. Temperature: 37°C.

those computed for G_{Cl}. However, since the concentration of OH^- is only about 10^{-6} that of Cl^- in the medium, it would appear that, if these estimates of G_{OH} and G_{Cl} are correct, either the partition coefficient (membrane/medium) or the mobility or both of OH^- is much greater than for Cl^-. It is possible that our estimate of G_{OH} is too high because of residual CO_2 in the system permitting Cl^--HCO_3^- exchange which would also lead to alkalinization of the medium.

REFERENCES

1. D. C. Tosteson, *Acta Physiol. Scand.*, **46**, 19–41 (1959).
2. J. B. Chappell and A. R. Crofts, *in* "Biochim. Biophys. Library," ed. by J. M. Tager *et al.*, Vol. 7, p. 293 (1966).
3. M. Dalmark and J. O. Wieth, *Biochim. Biophys. Acta*, **219**, 525 (1970).
4. R. B. Gunn, *in* "Oxygen Affinity of Hemoglobin and Red Cell Acid Base Status," ed. by M. Rørth and P. Astrup, Munksgaard, Copenhagen, p. 823 (1972).
5. R. B. Gunn and D. C. Tosteson, *J. Gen. Physiol.*, **57**, 593–609 (1971).
6. P. J. F. Henderson, J. D. McGivan, and J. B. Chappell, *Biochem. J.*, **111**, 521–534 (1969).
7. F. R. Hunter, *Biochim. Biophys. Acta*, **135**, 784–787 (1967).
8. F. R. Hunter, *J. Gen. Physiol.*, **51**, 579–587 (1968).
9. M. J. Hunter, *J. Physiol.*, **218**, 49 (1971).
10. E. J. Harris, and B. C. Pressman, *Nature*, **216**, 918–920 (1967).
11. A. Scarpa, A. Cecchetto, and G. F. Azzone, *Nature*, **219**, 529–531 (1968).
12. A. Scarpa, A. Cecchetto, and G. F. Azzone, *Biochim. Biophys. Acta*, **219**, 179–188 (1970).
13. J. O. Wieth, *in* "Oxygen Affinity of Hemoglobin and Red Cell Acid Base Status," ed. by M. Rørth and P. Astrup, Munksgaard, Copenhagen, p. 265 (1972).

14. U. V. Lassen and P. G. Hoffman, *in* " Oxygen Affinity of Hemoglobin and Red Cell Acid Base Status," ed. by M. Rørth and P. Astrup, Munksgaard, Copenhagen, p. 291 (1972).
15. D. C. Tosteson, P. Cook, T. Andreoli, and M. Tieffenberg, *J. Gen. Physiol.*, **50**, 2513–2525 (1967).

Phosphate Transport and Identification of a Binding Protein of Phosphate in Mitochondria*

Osamu Hatase and Takuzo Oda

Department of Biochemistry, Cancer Institute, Okayama University Medical School, Okayama, Japan

The mechanism of phosphate translocation in mitochondria was investigated by a mitochondria-submitochondrial particle system and a specific inhibitor, NEM, and a possible phosphate-binding protein was identified. P_i was transported energetically, and the translocation system was specifically inhibited by NEM. As the sites of oxidative phosphorylation, headpiece-stalk secter, project into reaction medium in sonicated submitochondrial particles, it is not necessary for P_i to be translocated from the outside of mitochondria into the matrix space for oxidative phosphorylation. NEM completely inhibited energy-dependent P_i translocation and P_i-requiring processes in mitochondria, but it showed no inhibitory effects in ETP_H and P_i-independent functions in mitochondria. The inhibitory effects of NEM in mitochondria were overcome by high levels of P_i. This protection of P_i-requiring functions by high levels of P_i against NEM effects strongly suggests the possibility of identifying certain component(s) that may perform an important role in P_i transport. The data to be shown by P_i-^{14}C-NEM system and polyacrylamide gel electrophoresis identified a protein fraction to be labeled specifically, that is considered a possible P_i-binding protein.

The mechanism of ion translocation in biomembrane systems has been investigated in bacteria, red blood cells, cultured cells, mitochondria,

* Full papers have been and will be submitted to certain journals.

Abbreviations used: ETP_H, phosphorylating electron transfer particles prepared from (cont'd)

artificial membranes, *etc.*, and some models have been proposed. The models basically depend on the kind of ions transported and the model structures of the biomembranes. Among these, the macromolecular models are more relevant to the mechanisms of ion transport, because we can search for possible binding proteins for ions specifically transported by the carrier system.

Some binding proteins from bacteria for sugars, sulfate, and other ions have been reported but in mitochondria the calcium binding protein is the only example. In mitochondria, Na^+ and K^+ would be transported by ionophores, endogenous and exogenous (valinomycin and gramicidin), while anions, such as tricarboxylic acids, adenine nucleotides, and P_i would be translocated by exchange diffusion and macromolecular carrier systems. The proton translocation system is a specific transport system in mitochondrial membranes that would be very important for energization of the inner membrane. There are suggestive communications that indicate the possible presence of a carrier system for P_i, but there is not yet enough evidence.

In this article we shall consider the translocation of P_i in mitochondria and identification of a binding protein for phosphate by using a specific inhibitor, NEM. Inorganic phosphate as a substrate for oxidative phosphorylation is transported energetically, and the transport system is specifically inhibited by NEM. The inhibition pattern is a competitive one, and the effects of NEM on mitochondrial functions that require P_i are overcome by high levels of P_i. By using an NEM-phosphate system, we have identified a protein that may play an important role as a binding protein in the P_i translocation system from the inner mitochondrial membrane.

Mitochondria show configurational change in respose to change in energy level, osmotic pressure, ionic conditions, and so on. Green and his colleagues proposed a configuration, energized-twisted as the high-energy precursor state for ATP synthesis (*1*) (Fig. 1). Though the twisted configuration was proved not to be an artifact, but a natural configuration that is observed in State 4 respiration, the interpretation of the twisted configuration has been modified by ourselves (*2–4*). It is not a cause and a high-energy precursor state, but represents an effect that is induced by energy-dependent P_i uptake, changes in ionic conditions in matrix space and

HBHM; FMA, fluoromercuriacetate; HBHM, heavy beef heart mitochondria; NEM, N-ethylmaleimide; PCMB, *p*-chloromercuribenzoate; SDS, sodium dodecyl sulfate; TMPD, tetramethyl-*p*-phenylenediamine.

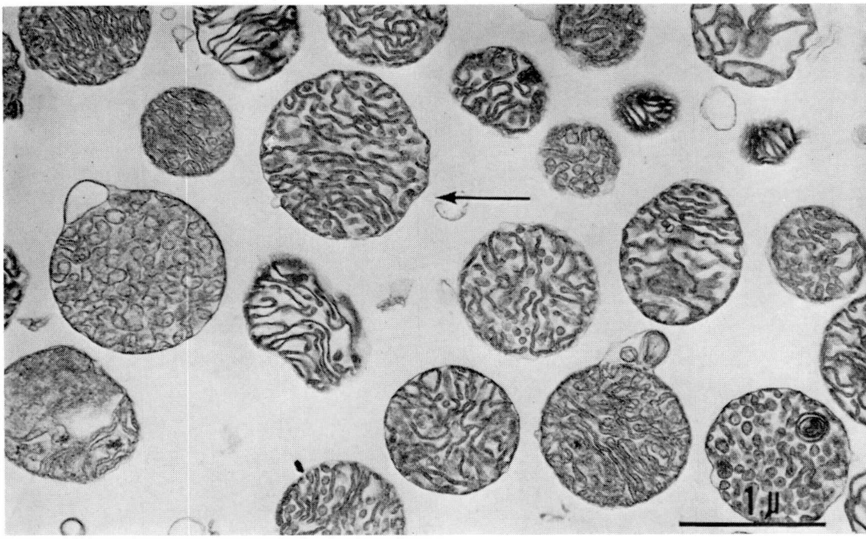

FIG. 1. Energized-twisted configuration of beef heart mitochondria. This configuration is induced by addition of P_i to respiring mitochondria (*4*).

conformation of matrix proteins, and configurational and conformational changes in the inner mitochondrial membrane. The mitochondrial configuration and conformation may be regulated, energetically or nonenergetically, by the ionic conditions in the matrix space and location of charged groups in the inner mitochondrial membrane. As a feedback phenomenon, it is quite possible that the mitochondrial functions may be controlled by ultrastructural changes. It is very important to study the mechanism of ion transport in mitochondria for the investigation of oxidative phosphorylation as well as for the study of the fundamental structure and function of biomembranes.

In ion translocation in mitochondria, there is no special mechanism compared with other biomembranes, except for the presumed electron-proton translocation mechanism (*5*). Transport of adenine nucleotides in mitochondria would be performed by a postulated carrier system (*6*), and Na^+ and K^+ by ionophores, endogenous (*7*) and exogenous (valinomycin and gramicidin). Calcium would be transported by a macromolecular carrier system (*8*). However, translocation of P_i may be different, and there are possibly 2 mechanisms; one is energy-dependent and rapid, and the other is energy-independent and slow. The former may be performed by a carrier system, and the latter by exchange diffusion in salts media.

In the present communication we shall present evidence that: 1) NEM inhibits all P_i-dependent processes that require energy supply. 2) In comparable concentrations NEM has little or no effect on processes which are not P_i-dependent. 3) NEM has no inhibitory effect on P_i-requiring processes in a submitochondrial particles. 4) The inhibitory effects of NEM are released by high concentrations of P_i. 5) Using high levels of P_i, it is possible in theory to identify the site or sites at which NEM reacted to cause specific inhibition of P_i translocation. On the basis of these results we have invoked a transmembrane phosphate carrier system (transphosphorylase) in the inner membrane, the activity of which is modulated by sulfhydryl groups.

First, the effect of NEM on P_i-requiring processes in mitochondria and ETP_H will be discussed. NEM completely inhibits ADP-induced State 3 respiration with 20 nmoles of NEM per mg protein of mitochondria, but has slight or no effect on electron transfer at the same concentration (Fig. 2 and Table I). Next, the specificity of NEM effects on P_i-requiring functions of mitochondria is indicated in Table II–IV and Fig. 2. Tables II to IV show that the effects of NEM are very marked at the above concentration (20 nmoles of NEM per mg protein) on P/O ratio, ATP-$^{32}P_i$ exchange, K^+ translocation with P_i as anion, and energized calcium translocation in presence of P_i (Tables II and III). However, ATPase and K^+ translocation

TABLE I
Inhibition by NEM of Electron Transfer, in Mitochondria and ETP_H

Particle	Substrate	Rate of respiration (nmoles $\frac{1}{2}O_2$/min/mg protein)		Inhibition (%)
		−NEM	+NEM	
Mitochondria	Pyruvate+malate	29.0	20.0	38
Mitochondria	Succinate	110.0	83.0	25
Mitochondria	Ascorbate+TMPD	100.0	106.0	0
ETP_H	NADH	156.0	69.0	49
ETP_H	Succinate	94.0	79.0	16
ETP_H	Ascorbate+TMPD	92.0	95.0	0

The measurement of respiratory rates was carried out at 30°C. Samples were preincubated with 20 nmoles of NEM per mg protein at 0°C for 20 min. The concentrations of the substrates used were: Pyruvate, succinate, and ascorbate (5 mM); NADH (2.5 mM); malate (0.5 mM); and TMPD (0.1 mM).

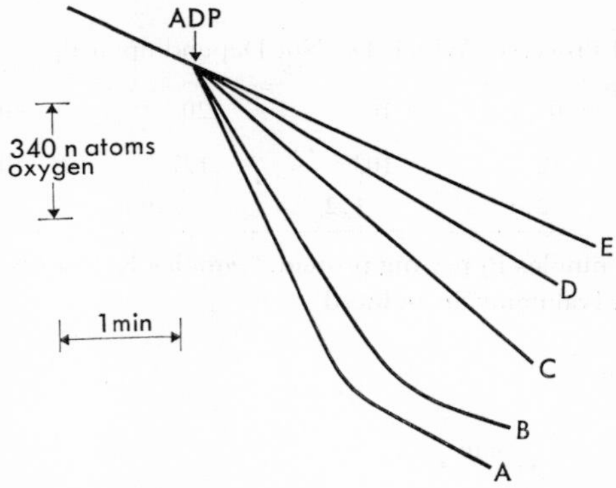

FIG. 2. Effect of NEM on State 3 respiration of mitochondria. The assay was performed in a sucrose medium containing P_i (10 mM), ADP (0.8 or 1.6 μmoles), NEM (0–24 nmoles per mg protein), and mitochondria (4 mg). A–E represent concentrations of NEM. A, 0; B, 8; C, 16; D, 32; E, 48 nmoles of NEM.

TABLE II

Inhibition of P_i-requiring Mitochondrial Functions by NEM

NEM[a]	0	10	20	40
P/O ratio	1.7	1.0	0.15	0
ATP-$^{32}P_i$ exchange[b]	61.7	34.6	6.2	1.2
K^+ translocation (P_i)[c]	2.0	0.10	0.10	0.10

[a] nmoles NEM per mg protein. [b] nmoles P_i per min per mg protein. [c] μmoles K^+ translocated per mg protein with P_i as anion (valinomycin-induced).

TABLE III

Effect of NEM on Energized Translocation of Ca^{2+} plus P_i

System	Rate of translocation of Ca^{2+} (nmoles Ca^{2+}/min/mg protein)		Inhibition (%)
	−NEM	+NEM	
Complete	226	62	73
P_i omitted	33	38	0

The experiments were carried out at 30°C for 60 sec. The mitochondrial suspensions were exposed to NEM (20 nmoles per mg protein) for 10–20 min at 0°C before initiation of the experiment. The system (ascorbate+TMPD) was used as substrate at the same concentrations specified in the legend of Table I.

TABLE IV
Effect of NEM on Mitochondrial Processes Which Do Not Depend upon P_i

NEM[a]	0	10	20	40
ATPase activity[b]	82	103	127	107
K^+ translocation (AC)[c]	2.3	1.2	0.90	0.05

[a] nmoles NEM per mg protein. [b] nmoles P_i per mg protein. [c] μmoles K^+ translocated per mg protein with acetate as anion (valinomycin-induced).

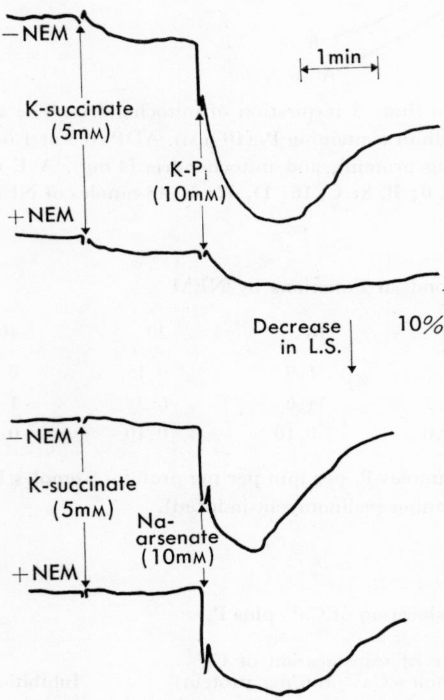

FIG. 3. Effect of NEM on 90° light scattering. The concentration of NEM was 10 nmoles per mg protein.

with acetate as anion, are not inhibited (Table IV). The change in 90° light scattering induced by P_i is inhibited by NEM, but the change induced by arsenate is less sensitive to NEM (Fig. 3). Energy-independent swelling in the presence of acetate is not inhibited by NEM (not shown). On the contrary PCMB inhibits both of the changes in light scattering induced by P_i and acetate, and FMA induces swelling (decrease in light scattering) by itself. These suggest a certain specificity of NEM compared with other sulfhydryl reagents on P_i-requiring processes (*9–16*). It may depend on differences in the chemical properties of NEM. NEM is different from other sulfhydryl reagents; NEM is noncharged, making covalent bonds with sulfhydryl groups that are stable against dithiothreitol treatment. These differences may be the cause of the specificity of NEM.

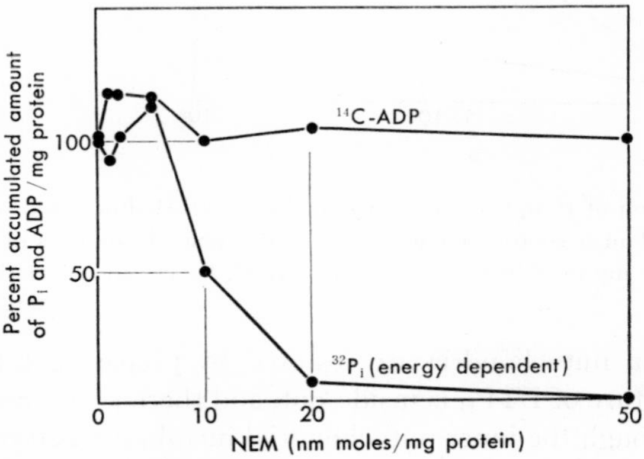

FIG. 4. Effect of NEM on energized uptake of P_i and ADP. The 100% values for P_i uptake and ADP binding (values in the absence of added NEM) were 33.4 or 9.2 nmoles per mg protein, respectively.

Energy-dependent P_i accumulation in mitochondria is completely suppressed by NEM, but ADP uptake is not inhibited at all (Fig. 4). The inhibition pattern of NEM on energy-dependent P_i uptake is competitive (Fig. 5), and high concentrations of P_i overcome the effect of NEM on oxidative phosphorylation and energy-dependent P_i uptake. However, sulfate does not have such an effect (Table V). This strongly suggests a specific correlation between the effect of NEM and energy-dependent P_i translocation.

Table VI is extremely important in demonstrating the presence of a P_i carrier system and the specificity of NEM on it. Other sulfhydryl reagents, such as organic mercurials, inhibit oxidative phosphorylation both in mitochondria and submitochondrial particles, but NEM shows no inhibition on coupling functions in ETP_H. Moreover, the coupling processes

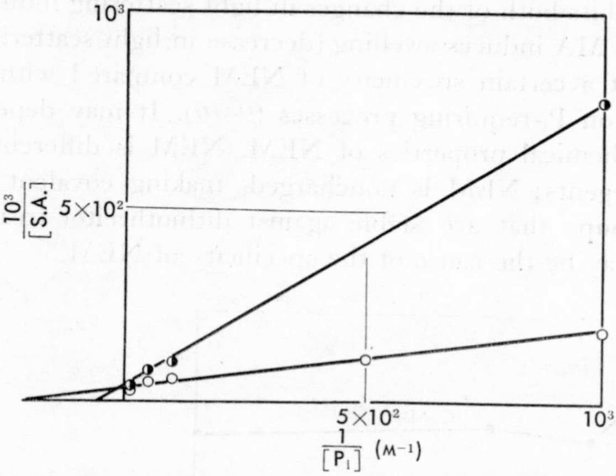

FIG. 5. Inhibition of P_i uptake in mitochondria by NEM. Energized uptake of P_i was measured in a sucrose medium containing graded amounts of P_i with (◐, 5 nmoles per mg protein) or without (○) NEM. S. A.: Specific activity.

inhibited by NEM in mitochondria are restored by preparing ETP_H by sonication. The structure of ETP_H is inside-out, and there is no need for P_i to be transported through the inner mitochondrial membrane before reaching the active site of ATPase, the head-piece of the tripartite repeating unit of mitochondria. These results lead us to the next step of the investigation. To identify the site or sites at which NEM reacted to cause specific inhibition of P_i translocation, we used a high level of phosphate (or sulfate as a control) to protect the specific sulfhydryl groups that would be constructed in a certain component of the inner mitochondrial membrane.

In the first incubation, mitochondria were incubated in the presence of nonradioactive NEM (100 nmoles per mg protein) and either potassium phosphate (150 mM) or potassium sulfate (150 mM). After washing out nonreacted NEM and salts, a second treatment with ^{14}C-NEM (50 nmoles per mg protein) was performed in absence of salts. In the second incubation, the specific sulfhydryl groups that were protected by high levels of P_i and a certain amount of nonspecific sulfhydryl groups would be labeled by radioactive NEM. The labeled mitochondria were sonically irradiated for preparation of ETP_H. Table VII shows the specific protection of sulfhydryl groups by a high level of P_i; the difference between P_i system and no-addition system represents the amount of sulfhydryl groups specifically protected

TABLE V
Inhibition of Oxidative Phosphorylation and Energized Phosphate Uptake in Mitochondria by NEM and Release of the Effect by High Levels of P_i

Additions	P/O ratio		Energized P_i uptake[b]	
	−NEM	+NEM[a]	−NEM	+NEM[a]
Potassium phosphate (10 mM)	1.9	0.24	13	2
Potassium phosphate (100 mM)	2.0	1.6	18	11
Potassium sulfate (100 mM)	0.91	0.15	16.5	0

[a] 20 nmoles NEM/mg protein were added to the incubation medium. [b] Expressed as nmoles P_i accumulated/mg protein. This figure was calculated as the difference between the total system and a control in which succinate was omitted.

TABLE VI
Inhibition of Oxidative Phosphorylation and ATP-P_i Exchange in Mitochondria and Release of the Effect by Sonication of NEM-Exposed Mitochondria

Preparation	P/O ratio with		ATP-$^{32}P_i$ exchange[a]
	Succinate	Ascorbate-TMPD	
HBHM	1.8	0.90	61.7
NEM-HBHM	0.19	0.09	4.1
ETP_H from HBHM	1.4	0.35	62.0
ETP_H from NEM-HBHM	1.3	0.30	59.1
ETP_H+NEM[b]	1.4	0.37	60.5

[a] nmoles P_i exchanged per min per mg protein. [b] ETP_H from HBHM was assayed in presence of 20 nmoles NEM per mg protein.

TABLE VII
Labeling of Mitochondrial Inner Membrane by ^{14}C-NEM[a]

Addition	nmoles ^{14}C-NEM bound/mg ETP_H protein
None	4.0
Potassium phosphate (150 mM)	11.6
Potassium sulfate (150 mM)	6.4

[a] Mitochondria were labeled with ^{14}C-NEM as described in the text.

by P_i (7.6 nmoles per mg ETP_H protein). Sulfate shows slight but non-specific protection as an anion at high concentrations. The labeled ETP_H in the P_i-protecting system were fractionated through a column of Sephadex G-200 with a solution (0.1 M in NH_4HCO_3 and 1% in sodium dodecyl

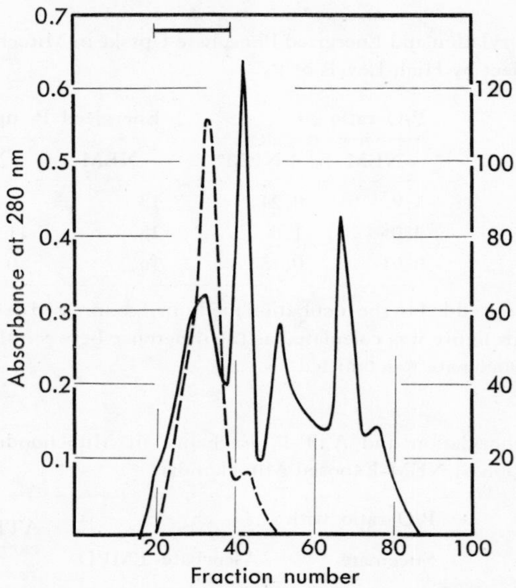

FIG. 6. Gel filtration of ^{14}C-NEM-labeled ETP_H after treatment with SDS to dissolve the protein. The fractions in the radioactive peak were pooled. - - - radioactivity; ——absorbance at 280 nm.

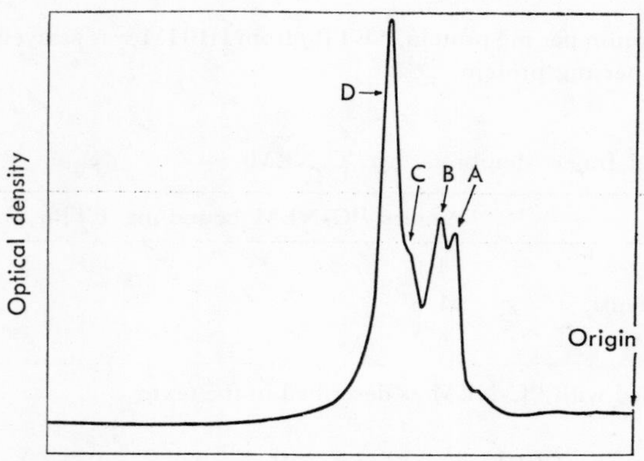

FIG. 7. Polyacrylamide gel electrophoresis of labeled proteins from ETP_H. The radioactive peak obtained in gel filtration (Fig. 6) was examined on 5% gels in SDS. This figure is a typical densitometric trace of a gel; 4 protein bands (A–D) were present.

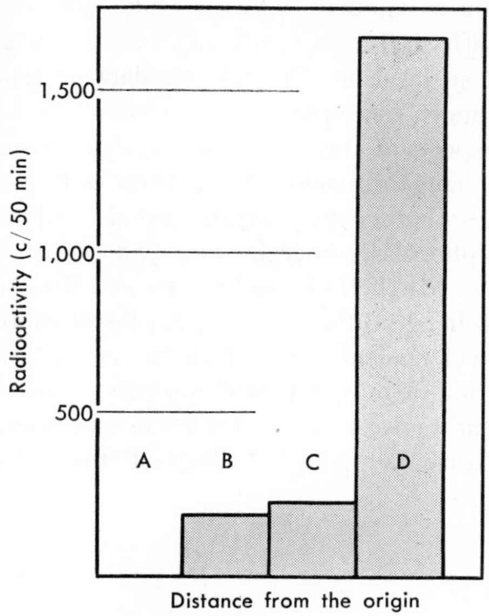

FIG. 8. Radioactivity of labeled proteins from ETP_H. The protein bands A–D were cut out of the polyacrylamide gels (Fig. 7) and examined for radioactivity. Controls were run using slices of unstained gel. This histogram shows a typical experiment in which 50 μg of protein were applied in 20 μl.

sulfate) after SDS treatment in the presence of β-mercaptoethanol and NH_4HCO_3. The proteins in the most highly labeled fraction in Fig. 6 (broken line) were examined by gel electrophoresis in SDS-phosphate buffer (5% acrylamide gels). The proteins were fractionated into 4 bands in the gel, and the highest one, D in Fig. 7, was labeled most strongly (Fig. 8). The molecular weight of this protein was found to be 23,500.

This protein was the specific target of NEM, and it plays a crucial role as a binding protein in energy-dependent P_i translocation.*

The experimental data now available on the selective suppression of P_i-dependent mitochondrial functions by NEM provide powerful support for the notion of a system in the inner mitochondrial membrane which facilitates the transmembrane transport of inorganic phosphate. Such a

* The existence of a phosphate-binding protein has been identified by affinity chromatography and will be published in a different paper.

system may well have a protein component with molecular weight 23,500, and this protein and some sulfhydryl groups may play an important role in the process of phosphate translocation. Though the data presented are very powerful evidence to support a complex carrier system, 2 other problems remain; the energetic aspects of the P_i translocation mechanism and nonenergized P_i transport, exchange diffusion. Thus, there is no way with the available evidence to decide whether the energized uptake of P_i involves energization of the transphosphorylase, energization of a redox loop, or the transprotonase proposed by Mitchell (*5*), and Green and Brucker (*17*), respectively. Nonenergized exchange diffusion of P_i in phosphate media is not inhibited by any sulfhydryl reagents (*18*). This is another problem, but the most important function of mitochondria, oxidative phosphorylation, is performed in coupled and energized conditions in which energized uptake of P_i is the main mechanism supplying sufficient P_i for P_i-requiring processes.

REFERENCES

1. D. E. Green, J. Asai, R. A. Harris, and J. T. Penniston, *Arch. Biochem. Biophys.*, **125**, 684–705 (1968).
2. T. Wakabayashi, O. Hatase, D. W. Allmann, J. M. Smoly, and D. E. Green, *J. Bioenergetics*, **1**, 527–549 (1970).
3. T. Wakabayashi, J. M. Smoly, O. Hatase, and D. E. Green, *J. Bioenergetics*, **2**, 167–182 (1971).
4. O. Hatase, T. Wakabayashi, and D. E. Green, *J. Bioenergetics*, **2**, 183–195 (1971).
5. P. Mitchell, *Nature*, **191**, 144–148 (1961).
6. M. Klingenberg and E. Pfaff, *in* " Regulation of Metabolic Processes in Mitochondria," ed. by J. M. Tager, S. Papa, E. Quagliariello, and E. C. Slater, Elsevier, Amsterdam, pp. 180–201 (1966).
7. G. A. Blondin, A. F. Decastro, and A. E. Senior, *Biochem. Biophys. Res. Commun.*, **43**, 28–35 (1971).
8. A. L. Lehninger, *Biochem. J.*, **119**, 129–138 (1970).
9. R. G. Hansford and J. B. Chappell, *Biochem. Biophys. Res. Commun.*, **30**, 643–648 (1968).
10. A. Fonyo, *Biochem. Biophys. Res. Commun.*, **32**, 624–628 (1968).
11. D. D. Tyler, *Biochem. J.*, **107**, 121–123 (1968).
12. S. Papa, N. W. Lofrumento, M. Loglisci, and E. Quagliariello, *Biochim. Biophys. Acta*, **189**, 311–314 (1969).

13. A. J. Meyer and J. M. Tager, *Biochim. Biophys. Acta*, **189**, 136–139 (1969).
14. A. Tulp and K. Van Dam, *Biochim. Biophys. Acta*, **189**, 337–341 (1969).
15. N. Haugaard, N. M. Lee, R. Kostrzewa, R. S. Harris, and E. S. Haugaard, *Biochim. Biophys. Acta*, **172**, 198–204 (1969).
16. M. J. Lee, R. A. Harris, and D. E. Green, *J. Bioenergetics*, **2**, 13–31 (1971).
17. D. E. Green and R. F. Brucker, *BioScience*, **22**, 13–19 (1972).
18. G. A. Blondin, personal communication.